AF622233

PLANT BREEDING
BIOMETRY
BIOTECHNOLOGY

PLANT BREEDING BIOMETRY BIOTECHNOLOGY

Dr Dipak Kumar Kar MSc, PhD
Principal, Asutosh College, Kolkata

Dr Soma Halder MSc, MTech, PhD
Assistant Professor, Asutosh College, Kolkata

New Central Book Agency (P) Ltd
LONDON
HYDERABAD ERNAKULAM BHUBANESWAR
NEW DELHI KOLKATA PUNE GUWAHATI

NCBA

REGD OFFICE
8/1 Chintamoni Das Lane, Kolkata 700 009, India
email: ncbapvtltd@eth.net

OVERSEAS
NCBA (UK) Ltd, 149 Park Avenue North, Northampton, NN3 2HY, UK
email: ncbauk@yahoo.com

EXPORT
NCBA Exports Pvt. Ltd, 212 Shahpur Jat, New Delhi 110 049
email: ncbaexports@ncbapvtltd.com

BRANCHES

New Central Book Agency (P) Ltd
212 Shahpur Jat, **New Delhi** 110 049
email: ncbadel@ncbapvtltd.com

Shop Nos. 3 & 4, Vinayak Towers
681/B Budhwar Peth, Appa Balwant Chowk
Pune 411 002
email: ncbapune@ncbapvtltd.com

Shop No. 15, Bharati Towers, Ground Floor
Forest Park
Bhubaneswar 751 001
email: ncbabub@ncbapvtltd.com

House No. 3-1-315, 1st Floor
Nimboliadda, Kachiguda, **Hyderabad** 500 027
email: ncbahydb@ncbapvtltd.com

GSS Shopping Complex, 1st Floor
Opposite College Post Office, Convent Road
Ernakulam 682 035
0484-4021054 / 53

Radhamayee Complex (Second Floor)
Opp. Axis Bank, Panbazar, Jaswanta Road
Guwahati 781 001
email: ncbaguwahati@ncbapvtltd.com

PLANT BREEDING, BIOMETRY AND BIOTECHNOLOGY • Dr Dipak Kumar Kar, Dr Soma Halder

© Copyright reserved by the Authors

Publication, Distribution, and Promotion Rights exclusively reserved by the Publisher

All rights reserved. No part of the text in general, and the figures, diagrams, page layout, and cover design in particular, may be reproduced or transmitted in any form or by any means—electronic, mechanical, photocopying, recording, or by any information storage and retrieval system—without the prior written permission of the Publisher

First Published: 2006
Reprinted: 2008
Revised and Enlarged Second Edition: 2010
Reprinted: 2013, 2017, 2018, 2019, **2022**

PUBLISHER AND TYPESETTER
New Central Book Agency (P) Ltd
8/1 Chintamoni Das Lane, Kolkata 700 009

PRINTER
New Central Book Agency (P) Ltd
Web-Offset Division, Dhulagarh, Sankrail, Howrah

PROJECT TEAM
Prabir Ghosh and Ashit Ghosh

ISBN: 978 81 7381 633 8

Price ₹ 495.00

CONTENTS

PREFACE TO THE FIRST EDITION

Drawing on our exhaustive experience of being teachers for the last two decades, we felt it necessary to submit a comprehensive textbook series in BOTANY to our undergraduate students covering all the branches of plant science as per their course requirements. Prof. T. B. Jha, Head of the Dept. of Botany, Presidency College, Kolkata, has kindly consented to be the editor of this series—the different volumes are proposed to be written by a group of respective specialised teachers of different colleges and universities of West Bengal. But due to certain limitations the project could not be finished, however, few of us remained steadfast in our endeavour to complete at least some parts of the series. Our exertions found fruition in this book Plant Breeding and Biometry.

PLANT BREEDING, an art as well as science, has played a pivotal role in the Green Revolution of the country by the improvement of existing varieties of crops. This book covers the principles and methods of plant breeding. Elementary principles of Genetics—the foundation-stones of plant breeding—are explained in precise details. Both conventional methods, viz. introduction of acclimatization, selection, hybridization, heterosis and supplementary methods, viz. polyploidy breeding, mutation breeding have been discussed using suitable examples from the Indian context. The recent advances in molecular genetics has resulted in the use of biotechnology as the modern method of breeding in 21st century. Such cutting-edge advances are also under the purview of our book.

BIOMETRY has proved indispensable with the progress in various disciplines of life science. The estimation of probable truth of any biological statement can only be judged by statistical assessments. As the data of bioscience are of a variable nature, the collected data must be processed properly to get concrete conclusion(s) after classification, tabulation, comparison, correlation, and interpretation. The different basic methods of Biostatistics have been exhaustively discussed with objectives, principles, merits and demerits—supplemented by tables, practical sheets, solved problems and a set of illuminating exercises.

This textbook claims little originality in content or presentation and, naturally, heavily drawn from earlier texts on the subject. However, this is a sincere attempt on the part of the authors to present this part of plant science in a comprehensive and precise manner for a meaningful interaction with the erudite teachers and the intelligent taught. The subject matter of the book is concise with uptodate information bearing necessary charts, tables, diagrams, and examples. We have endeavoured to keep the text simple and lucid, arranged in logical and sequential manner.

We express our sincere thanks to Mr Amitabha Sen of New Central Book Agency for his spontaneous acceptance and ardent enthusiasm in publishing this book. It has been a pleasure for us to work with Mr Soumen Paul and other members of the publishing house, specially Mr Prabir Ghosh, all of whom took great care to produce the book.

Constructive criticisms and suggestions from readers—both students and teachers—will be highly appreciated.

10th July 2006
Asutosh College
Kolkata

Dr Dipak Kumar Kar
Dr Soma Halder

Preface to the Second Edition

The second revised and enlarged edition of the book includes a separate section of 'Plant Biotechnology' in addition to 'Plant Breeding' and 'Biometry'.

PLANT BREEDING, an art as well as science, has played a pivotal role in the Green Revolution of the country by the improvement of existing varieties of crops. This book covers the principles and methods of plant breeding. Elementary principles of Genetics—the foundation-stones of plant breeding—are explained in precise details. Both conventional methods, viz. introduction and acclimatization, selection, hybridization, heterosis and supplementary methods, viz. polyploidy breeding, mutation breeding have been discussed using suitable examples from Indian context. The recent advances in molecular genetics have resulted in the use of biotechnology as the modern method of breeding in 21st century. Such cutting-edge advances are also under the purview of our book.

BIOMETRY has proved indispensable with the progress in various disciplines of life science. The estimation of probable truth of any biological statement can only be judged by statistical assessments. As the data of bioscience are of a variable nature, the collected data must be processed properly to get concrete conclusion(s) after classification, tabulation, comparison, correlation, and interpretation. The different basic methods of Biostatistics have been exhaustively discussed with objectives, principles, merits and demerits—supplemented by tables, practical sheets, solved problems and a set of illuminating exercises.

BIOTECHNOLOGY is the controlled use of biological systems for the benefit of human society. Plant Biotechnology has come to the forefront of scientific disciplines simply because of tremendous development in 'Genetic Engineering' on the one hand and 'Plant Tissue Culture' on the other. The book covers the methods by which genes can be chemically resolved, isolated, synthesized and subjected to manipulation through 'Recombinant DNA Technology'. The inherent property of 'Totipotency' in plant cells has opened up the avenue of artificial culturing of plants through '*in vitro* technique' which is also under the discussion of the book. Recombinant DNA technology coupled with advances of *in vitro* technique have resulted in direct transfer of genes from one organism to another to develop the 'Transgenic Plants' or 'GM crops' for their use in agriculture, industry, healthcare and environment. Different techniques and applications of Plant Tissue Culture and Genetic Engineering have been thoroughly depicted in this book alongwith the necessary diagrams, photographs, examples.

This textbook claims little originality in content or presentation and, naturally, heavily drawn from earlier texts on the subject. However, this is a sincere attempt on the part of the authors to present this part of plant science in a comprehensive and precise manner for a meaningful interaction with the erudite teachers and the intelligent taught. The subject matter of the book is concise with uptodate information bearing necessary charts, tables, diagrams, photographs, and examples. We have endeavoured to keep the text simple and lucid, arranged in logical and sequential manner.

We express our sincere thanks to Mr Amitabha Sen and Dr (Mrs) Mita Sen of New Central Book Agency for their spontaneous acceptance and ardent enthusiasm in publishing this book. It has been a pleasure for us to work with Mr Soumen Paul and other members of the publishing house, specially Mr Ashit Kr Ghosh, Mr Prabhat Jas, and Mr Prabir Ghosh, all of whom took great care to produce the book.

Constructive criticisms and suggestions from readers—both students and teachers—will be highly appreciated.

16th January 2010
Asutosh College
Kolkata

Dr Dipak Kumar Kar
Dr Soma Halder

PLANT BREEDING

1 PLANT BREEDING: AN INTRODUCTION

From the beginning of human civilization vis-a-vis agriculture, with the development of scientific knowledge, consciously or unconsciously man has developed a new branch of science and technique, the **Plant Breeding** for the improvement of crop quality.

Smith (1967) has defined **plant breeding as the art and science of improving the genetic pattern of plants in relation to their economic use.**

Frankel (1968) has defined **plant breeding as the genetic adjustment of plants to the social, cultural, economic and technological aspects of the environment.**

Technically plant breeding is an exercise in exploiting and manipulating the genetic system for improvement in relation to crop production. The science of plant breeding deals with the principles and methods required for favourable changes in the genetic constitution of crop plants, which are more suitable in one or more aspects than the existing ones.

1.1 HISTORY OF PLANT BREEDING

The process of plant breeding is assumed to be initiated nearly 7000 years ago with the beginning of human civilization. Bringing a wild species under human management as the source of food which can be referred to as the process of **domestication.** Movement of man from one place to another also helped the movement of cultivated plant species. In this way the **introduction** of new plant species or varieties into new area from other parts of the world became an integral part of plant breeding today. Furthermore man has started the process of **selection** by selecting the best seeds or good grains from the field to be planted in future. The first artificial **hybridization** procedure to get the hybrid was carried out by Knight and described **hybrid vigour.**

After the rediscovery of Mendelism at the beginning of 20th century, the concept of plant breeding underwent tremendous change. The cytogenetical principle behind plant breeding got sound base; different selection methods, back-cross method, and the techniques for inbred line development were established. Polyploid breeding, mutation breeding and ultimately the alien gene transfer method came into use. The non-conventional breeding methods through plant biotechnological approach has become the major thrust of twenty first century.

1.2 ORIGIN AND EVOLUTION OF CROP PLANTS

A. Centres of Origin

N. I. Vavilov has proposed that crop plants evolved from wild species in the areas showing diversity and termed them as primary centres of origin. From these places the crops moved to other areas with the movements of man. But in some areas, certain crop species show considerable diversity of forms although they did not originate there. Such areas are known as secondary centres of origin of these species. Vavilov has suggested eight main centres of origin:

Centre	Main Countries	Crops originated
1. Chinese	Central & Western China	Radish, Apricot, Peach, Litchi, Citrus, Soyabean
2. Hindustan	Parts of India & Burma	Rice, Sugarcane, Cotton, Legumes, Brinjal
3. Central Asiatic	Pakistan, Afghanistan, Punjab, Kashmir, Parts of USSR	Wheat, Pea, Lentil, Apple, Spinach
4. Near Eastern	Middle-East countries	Barley, Wheat, Linseed, Grape
5. Mediterranean	Mediterranean sea	Wheat, Beans, Cauliflower, Cabbage, Sugarbeet
6. Abyssinian	Ethiopia & Eritrea	Coffee, Lady's finger, Seasame
7. Central American	Mexico & Neighbouring countries	Maize, Beans, Chilli, Cotton, Pumpkin, Gourd
8. South American	Peru, Ecuador, Bolivia & Islands	Egyptian Cotton, Tobacco, Sweet potato, Papaya

B. Patterns of Evolution

There are three major lines of evolution pattern for various crops, those can be broadly classified as Mendelian variation, interspecific hybridization and polyploidy.

(A) **Mendelian variation**: Naturally occurring beneficial mutation causes genetic variability and these variability can be retained through hybridization and genetic recombination. Generally the macromutation produces a distinct morphological character or often affects several characters of a plant. Cabbage, Cauliflower, Brussel's Sprouts–all have been originated from a common wild species of *Brassica oleracea* through mutation, they differ from each other with respect to few major genes. Natural selection procedure has acted on variation and several important crops have evolved through Mendelian variation like barley, rice, peas, tomatoes, beans, etc.

(B) **Interspecific hybridization**: Generally the interspecific hybridization produces sterile plants which are unable to produce any seeds. These can be repeatedly back-crossed to one of the parents which help in restoration of genotype of one parent along with few genes from other species. This process is known as introgressive hybridization which leads to transfer of one species to another, thus playing an important role in evolution. For example, modern day maize has been developed by this kind of introgressive hybridization between primitive maize variety and a wild grass *Tripsacum.*

(C) **Polyploidy**: Autopolyploidy (triploid or tetraploid) produces more vigour, larger flowers and fruits, e.g., cultivated banana, apple, watermelon are autotriploid. But allopolyploidy plays major role in crop evolution, it is the process of chromosome doubling of interspecific F_1 hybrid. Many important crop plants have been evolved in this way like *Triticum, Gossypium,* etc. *Triticale* is a man made allohexaploid developed from chromosome doubling of F_1 hybrid between *Secale cereale* and tetraploid wheat.

C. Domestication and its Effects

Domestication is the first step of making the wild weed species o cultivated plants. It is the process of bringing wild species under human managemen according to their needs. In other words domestication of plants is the change of the ide type to adopt them better to man made environment. Under domestication the wild spe ies get changed to cultivated species mainly due to natural selection or human selectio . In nature there is continuous selection by natural forces like temperature, soil, weather, ests, diseases, etc. The genotype which is more suited to a gi en environment leaves behi d others which are less adaptive in nature.

Most of the characteristics of wild ecies have been affected u der domestication which involves three processes like mut tion, hybridization and gen tic recombination under the influence of human selection r natural selection. Some c aracters have got changed, some have lost and many have developed during domestic tion. Some of the important characters which have been aff cted are listed below:

1. Elimination or reduction of sl attering of pods or spikes.
2. Elimination of dormancy per od.

3. Decrease in toxins or other undesirable substances.
4. Increase in size of the grains or fruits.
5. Plant type change like decrease or increase in height, more number of tillers, leaf size, branching pattern, etc.
6. Early maturity.
7. Increase in economic yields.
8. Change in photoperiodic behaviour.
9. Mode of reproduction.
10. Pollination habit.
11. Synchrony in flowering.
12. Loss of defensive adaptation like hairs, thorns, etc.
13. Selection of bisexual variety rather than dioecious.
14. Decrease in variability.
15. Selection of polyploidy.

The crops like rice, wheat, barley, sugarcane, cotton, potato, tobacco, arhar have become cultivated types from the wild species through the process of domestication.

D. Origin of Crop Plants

Origin of Rice: Rice *(Oryza sativa)* had originated probably in 300 B.C. in South and South-East Asia. The cultivated forms might have been arisen from the wild species of rice like *O. perennis* (Fig. 1.1). From India rice moved to China and to Africa and America where during domestication three different forms have been originated : *indica, japonica* and *javanica.* During domestication many morphological, physiological changes have occurred, starting from adaptation to habitats of normal lands and open sun to the shady swamps. Change in leaf shape and size, grain character and other plant type characters differentiated *O. sativa* into three forms.

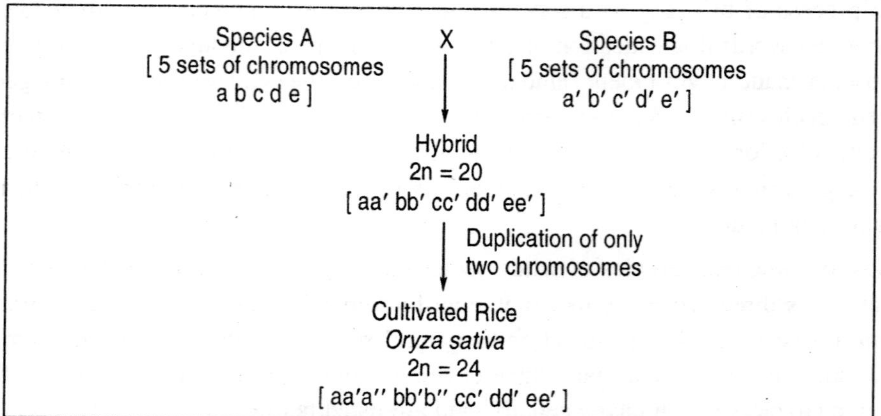

Fig. 1.1: Origin of cultivated Rice, a secondary polyploid with basic chromosome number 5, from wild rice of *Oryza perennis* species complex

Origin of Wheat: *Triticum* might have been originated about 6000 years ago in Afghanisthan and South-Western Himalayas and then moved with man westwards and eastwards. In total, four species *of Triticum* is cultivated in India: *T. durum, T. aestivum, T. dicoccum and T. turgidum* which are allopolyploids (Fig. 1.2). During domestication the plant type has got changed with the major changed character in non-brittle rachis.

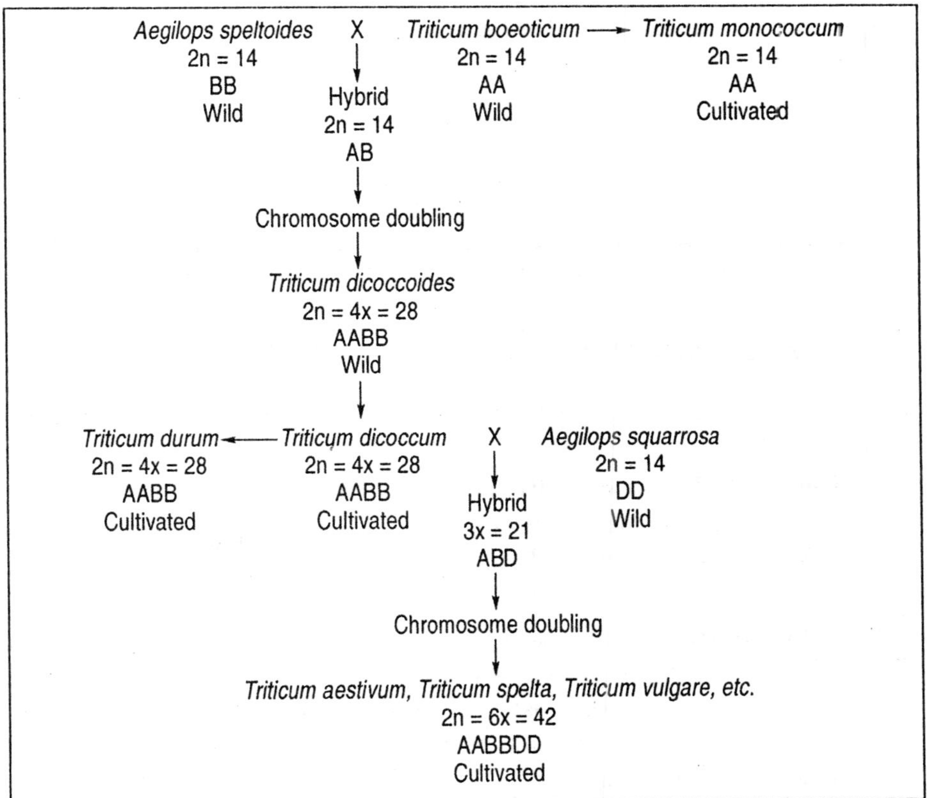

Fig. 1.2: Origin of tetraploid and hexaploid cultivated wheat from their wild ancestors

Origin of Cotton: Out of four species of cotton (*Gossypium*) under cultivation, two are diploid and others two are tetraploid. The diploid cotton have been originated during Indus civilization about 2000 B.C. From India two species moved to Africa and America and there during domestication the tetraploid species have originated. The major character which has been developed during domestication is the corboluted lint and its spinnability. The diploid species are *G. arboreum, G. herbaceum* whereas the tetraploid species are *G. barbadense, G. hirsutum*. New world cotton is an allotetraploid (Fig. 1.3).

Origin of Tea: This crop might have been originated in Central Asia and then differentiated into two types: small leaved China type and broad leaved Assam type. The China type *(Camellia sinensis)* further differentiated and domesticated in South China which has been introduced into India in the early part of 19th century and the Assam type (C. *assamica)* had the secondary centre of origin in North East India.

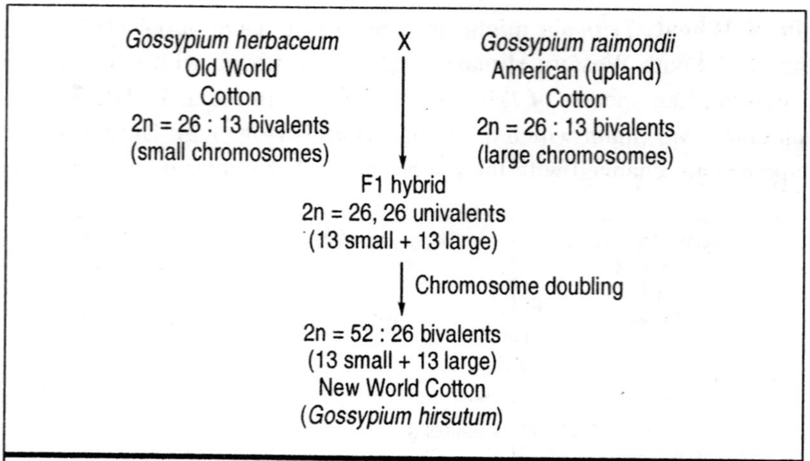

Fig. 1.3: Origin of new world cotton from old world and American cotton

1.3 SCOPE AND OBJECTIVES OF PLANT BREEDING

The different methods of plant breeding for the improvement of crop are directed in the following aspects:

(i) **Higher productive capacity**: This is the first and foremost objective of plant breeding programme, but this higher yield is associated with many other criteria like responsiveness to fertilizer, tillering capacity, resistance to disease or adverse situation, etc.

(ii) **Improved quality**: Besides quantitative characters, the quality of food grains is considered in plant breeding programme keeping an eye to market value, feeding quality and seed quality. High protein content, vitamin content, good amino acid profile, palatability, texture, large and bold grains, good colour– all are qualitative characters.

(iii) **Disease resistance**: In contrast to chemical control of diseases and pests, genetic resistance conferred by resistant crop varieties is stable, safest and cheapest which is highly desirable. The disease resistance can be categorized like insect, fungus, bacteria and virus resistant.

(iv) **Adaptive ability**: The plant breeding programme aims at developing the resistance or tolerance property against adverse environmental situation such as drought, flood, cold, high salinity, etc. All these qualities are influenced and altered by genotype – environment interaction.

(v) **Agronomic properties**: Most of the agronomic properties are often desirable as those are associated with some other characters linked with higher yield, such as dwarfness of cereal crop is associated with lodging resistance; more number of tillers is associated with more yield.

(vi) **Change in maturity duration and synchronous maturity**: The duration for maturity of particular crop can be reduced which helps in multiple cropping and permits new crop rotation. Synchronous maturity is needed mainly in food crop for good harvesting.

(vii) **Physiological properties**: Development of insensitivity against photoperiod and temperature is very much desirable in case of crops like rice, wheat, etc., so that the same variety can be cultivated in different areas and in different seasons.

Dormancy is another criteria which is desirable in case of crop plants because if there is rain at the time of maturity, the seeds get germinated in field before harvesting. But in some other cases it is desirable to remove dormancy.

(viii) **Varieties for new seasons**: The crops cultivated in India can be broadly subdivided into two categories like Rabi and Kharif. But if the crop can be adjusted to grow in new season then the same crop may be cultivated throughout the year.

(ix) **Varieties for new areas**: Newer varieties are developed by plant breeders which are adapted to wider areas and newer agroclimates. Many of the temperate crops by continued selection can be adapted to tropical or sub-tropical conditions.

(x) **Some other desirable properties specific to crop**: In case of leguminous crop non-shattering is one of the desirable characters. Elimination of toxic substances from particular crop sometimes become essential such as khesari seeds have neurotoxin chemical, *Brassica* seeds contain undesirable erucic acid. Removal of these substances increase the nutritional value of crop.

QUESTIONS

1. Define plant breeding. Discuss its scope and objectives.
2. What do you mean by domestication? Discuss the origin of crop plants — rice, wheat, cotton, tea.
3. Write short notes on centre of origin of crop plants and domestication.

2 Reproductive Systems in Crop Plants

2.1	Mode of reproduction — Vegetative or Asexual, Apomixis, Sexual; Significance in generating and fixing genotypic variation
2.2	Breeding systems — Self pollinated crops, Cross pollinated crops, Occasional cross pollinated crops
2.3	Pollination control — Incompatibility and Male sterility — Types and use in plant breeding

In plant breeding programme it is necessary to know about the reproductive system in plant on which the goal of breeding methods and genetic constitution of a plant could be assumed.

2.1 MODE OF REPRODUCTION

The mode of reproduction in crop plants may be broadly grouped into three categories: **Vegetative**, **Apomictic** and **Sexual**.

A. Vegetative or Asexual

In this type of reproduction the vegetative parts of the plants act as propagule in place of seed. This mode is mostly found in all those plants where there is no seed set, long reproduction cycle and heterozygosity exists. The following organs can act as propagule:

1. **Modified stem**: Underground modified stems like rhizome (ginger, turmeric, banana), tuber (potato), bulb (onion, garlic), corn (colocasia, yam), stolon (strawberry), sucker (chrysanthemum, mentha) are used for propagation.
2. **Stem cutting**: In many of the fruit crops the artificially produced clones or stem cuttings are used as propagative material. In sugarcane the nodal portions of stems, in fruit crops like mango, litchi, lemon, grapes – the different methods like layering, grafting, budding are applied to get the stem clone.
3. **Normal or modified root**: Normal roots of woodapple, citrus and many such trees are used as units for propagation. Modified roots such as tuberous root (sweet potato), fasciculated root (dahlia, asparagus) are used as propagule.
4. **Bulbils**: In some plants the flower bud modified into globose bulb which are called bulbils can be used as multiplication unit.

Significance: Asexual or vegetative reproduction leads to perpetuation of the same genotype with great conservation. It is very much advantageous because large number of genetically identical individuals can be obtained irrespective of the degree of heterozygosity of the genotype. At any stage of breeding programme if a breeder gets any desirable clone it can be maintained through vegetative means.

Mutation breeding, i.e., the search for desirable mutants both through natural and artificially induced mutation is very much helpful in case of vegetatively reproducing plants as the sexual reproduction can be avoided.

The method of polyploid breeding is also useful as the induced bud can be used as propagule and polyploidy can be maintained without the intervention of meiotic segregational disturbances.

B. Apomixis

Apomixis is the phenomenon where there is no normal fertilisation of the egg cell, hence no normal development of embryo from the egg cell. However, embryo may develop from an unfertilised egg cell or from a cell other than the egg cell within the embryo sac or from the cell outside the embryo sac (Fig. 2.1). The plants produced by apomixis are called apomictic which are of two types: obligate, producing only apomictic embryos, and facultative, producing both apomictic and normal embryos. This phenomenon where the substitution of sexual process occurs by asexual methods is known as **apomixis** (*apo* = away from, *mixis* = act of mixing), the plants are called **apomictic**.

Types of Apomixis

Non-recurrent Apomixis (haploid seed/plant formation)

Here the seed/plant develops from the unfertilised egg cell or any other unfertilised haploid cell of embryo sac. In this type of apomixis the first event of normal sexual cycle, meiosis takes place, the embryo sac is formed normally, but due to various reasons the fertilisation does not take place. The reasons may be – (i) absence of pollen tube; (ii) inability of the tube to discharge its contents; (iii) no attraction between male and female nuclei; (iv) early degeneration of sperms; (v) maturation of egg and sperm is not synchronised.

(a) **Haploid parthenogenesis**: Here a haploid embryo develops from an unfertilised egg cell of the embryo sac, e.g., in *Orchis maculata, Platenthera chlorantha, Cephalanthera damasonium* – the pollen tube enters but fails to fertilize.

(b) **Haploid apogamy**: It is the development of haploid embryo from any haploid cell of the embryo sac other than the egg cell. In *Lilium martagon, Erythraea centaurium* – two proembryos are formed, one is from fertilized egg (zygotic cell) and another from the synergid cell (haploid cell).

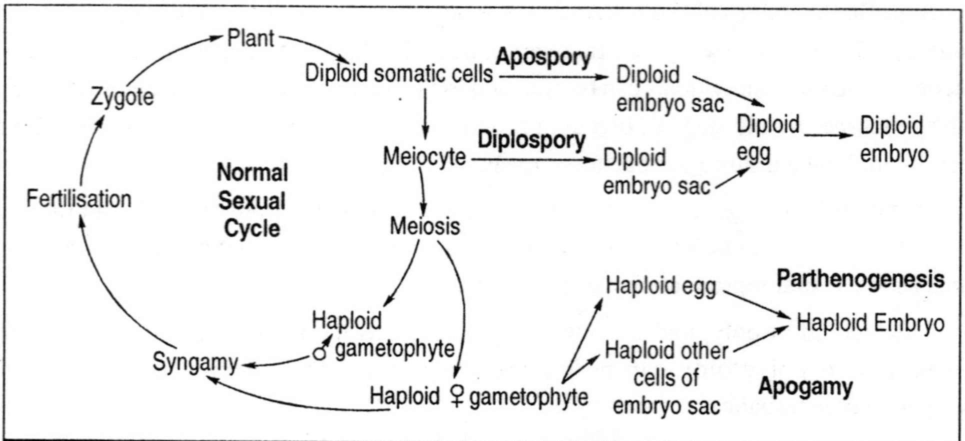

Fig. 2.1: Sexual and apomictic reproduction

Recurrent Apomixis

Here the meiotic event does not take place either in any somatic cell or the meiocyte cell fails to undergo meiosis, thereby produces the embryo sac with diploid cells only. These diploid cells directly give rise to diploid embryos, called **apospory** (Fig. 2.2).

(a) **Generative apospory**: The diploid egg cell develops into embryo without fertilisation, one particular variety of *Parthenium argentatum* exhibits.

(b) **Somatic apospory**: The diploid cell of nucellus or integument develops directly into diploid embryo sac and the diploid embryo is developed from the diploid unfertilised egg cell. This phenomenon is found in *Allium, Meliss, Crepis* etc.

Agamospermy

This term is used to categorise the plants which has the seed habit for propagation but not through normal sexual cycle, either meiosis or syngamy or both are eliminated. This includes either the embryos which may develop from cell of the unreduced female gametophyte or directly from the diploid sporophytic cell of the ovule, such as nucellus or integuments. Diplospory, apospory and adventive embryony, all are included within this category.

Adventive Embryony

In this case the gametophytic generation is completely eliminated. It is very much close to vegetative propagation, but the plants here retained the seed habit, i.e., the diploid embryo is developed from any diploid cell outside the embryo sac but matures into embryo within the embryo sac, zygotic embryo degenerates or competes with apomictic embryo. This kind of embryos are found in *Citrus,* mango, etc.

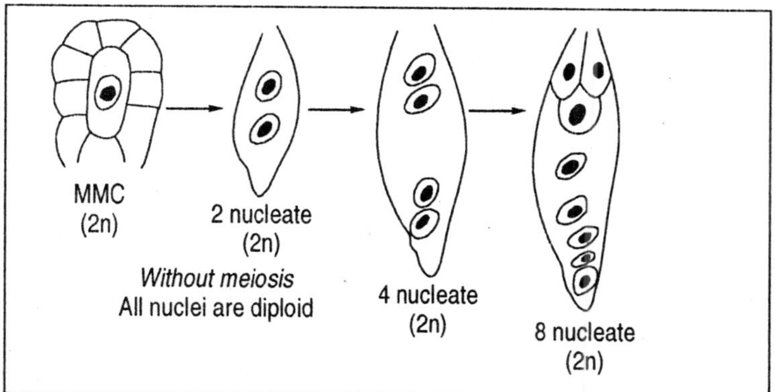

Fig. 2.2a: Aposporous embryo sac in *Eupatorium*

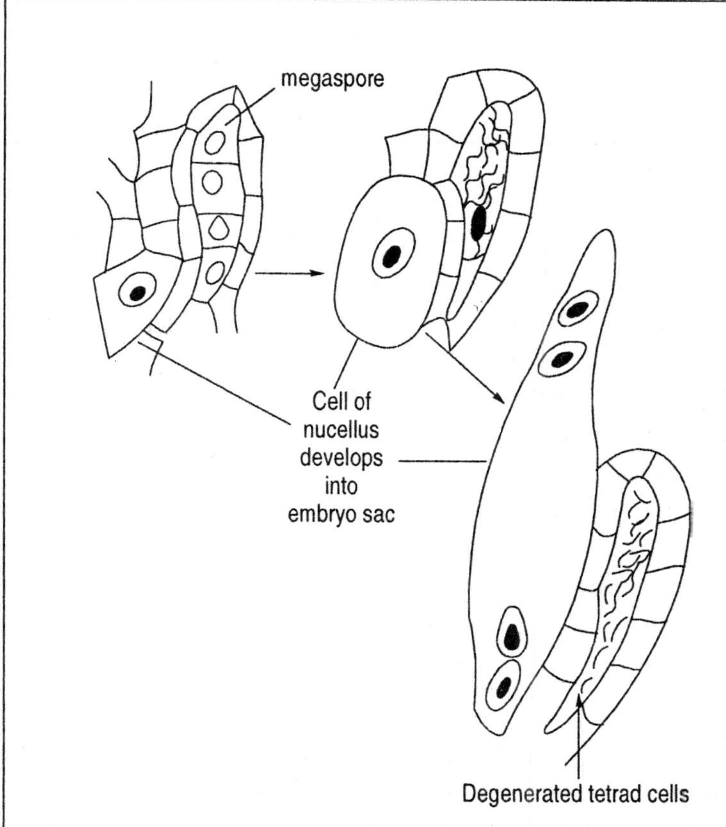

Fig. 2.2b: Aposporous embryo sac in *Hieracium*

Diplospory

Here the MMC differentiates into sexual ovules, but it does not enter into meiosis, it produces the embryo sac directly by mitotic division, all the cells within the embryo sac are diploid, and the embryos are formed.

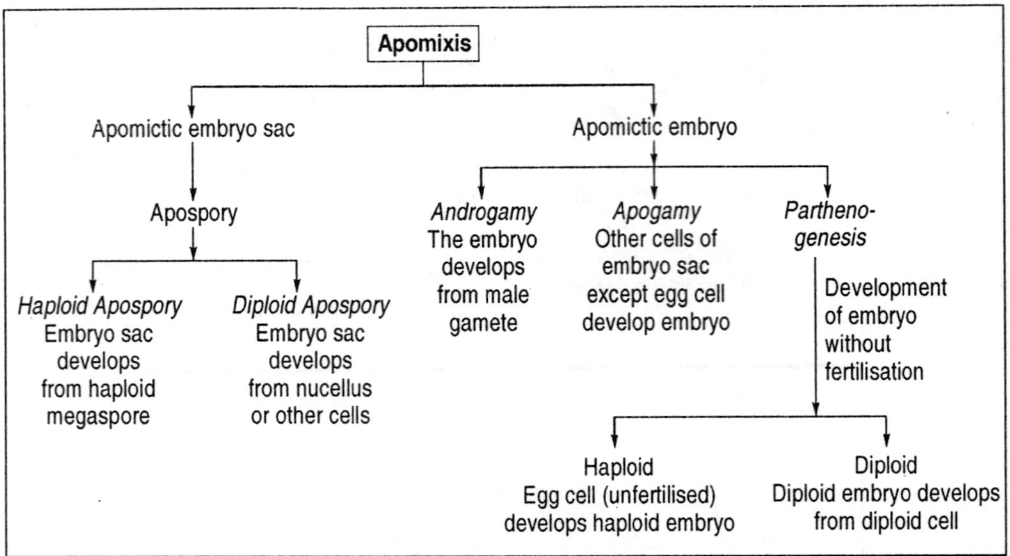

Fig. 2.3: Classification of apomixis

Significance: Apomictic plants tend to conserve the genetic structure and are also capable of maintaining heterozygote advantages. Due to prohibition of fertilisation process, apomixis is the way for exploitation of maternal influence or perpetuation of maternal individuals or maternal properties. The use of apomixis in plant breeding can be summarised :

1. Rapid multiplication of genetically uniform individuals without any risk of segregation.
2. Heterosis or hybrid vigour can be utilised for recurring production of seeds of F_1 hybrids.
3. From generation to generation the maternal effect can be exploited.

C. Sexual

This process of reproduction involves the fusion of male and female gametes and formation of seed, so it is called the process of **amphimixis**. Depending on the nature of pollination the plants can be categorised as self pollinated and cross pollinated. Sexually reproducing crop plants produce a special structure, called flower which bears the essential whorls like androecium and gynoecium. Androecium consists of stamen and gynoecium consists of carpel. The male and female gametes are produced in microspores and megaspores respectively.

Sporogenesis: Production of microspores and megaspores is known as sporogenesis. In anther lobes there are pollen sacs which contain numerous PMC (Pollen mother cell) undergoing meiosis, produces microspores or pollen, the process is called **microsporogenesis.** Inside the ovary the ovules are present in which the MMC (Megaspore mother cell) undergoes meiosis and produces 4 megaspores out of which 3 degenerate and one survives, this process is called **megasporogenesis**.

Gametogenesis: During maturation of pollen, the microspore nucleus divides mitotically to produce generative and vegetative nucleus. The generative nucleus then divides to form male gametes or sperms. The pollen along with pollen tube and sperms is called microgametophyte and the production of sperms is known as **microgametogenesis**.

The nucleus of functional megaspore divides mitotically three consecutive times to produce eight nuclei, which get arranged in the embryo sac. The embryo sac contains generally one egg cell, two synergids, two central nuclei and three antipodals. All these cells are haploid and this process is called as **megagametogenesis**.

Fertilisation: Fusion of one of the two sperms with the egg cell, to produce a diploid zygote, is known as **fertilisation,** and the fusion of the remaining sperm with the secondary nucleus leading to the formation of primary endosperm nucleus known as **double fertilisation** or triple fusion. The zygote divides mitotically to produce diploid embryo. The primary endosperm nucleus produces endosperm by mitotic division, it may be absorbed completely in legumes or may form endospermous seed in cereals.

Significance in Generating and Fixing Genotypic Variation

Among the sexually reproduced crop plants there are two categories, self pollinated and cross pollinated. Both the categories differ in their genetic constituent, as self pollination leads to homozygosity whereas cross pollination tends towards heterozygosity. In self pollination, the genotype AA or aa will remain homozygous, whereas the genotype Aa will segregate into homozygous and heterozygous in 1 : 1 ratio, as a result in each generation the homozygosity will be increased by one half and heterozygosity will be reduced by one half.

Due to cross pollination in the self pollinated crop the heterozygosity develops, or the heterozygosity may develop in a population through natural mutation. Cross pollination is the way for developing the variations in genotype and phenotype in a population. Heterozygosity helps in developing vigor which is known as heterosis or hybrid vigor whereas self pollination continuously will bring in the breeding depression due to attainment of homozygosity.

In plant breeding variation of characters among the population is most wanted. The breeder will select a particular character according to the need for that crop. So the characters which will be selected should be heritable through generations. The total mechanism of heredity is dependent upon the behaviour of chromosomes, which carry the genes. Variations are originated through chromosome rearrangement, genetic recombination, mutation, structural and numerical changes of chromosome.

In heredity the chromosomes are important, as their distribution in germ cells determine the specific heritable character to the progeny. The heritable characters may be regulated by single gene or by multiple genes. The inheritance pattern of a particular character is studied by progeny testing which is a basic procedure in plant breeding. In a breeding programme always the superior characters are selected which may or may not be controlled by dominant gene. Depending on that the homozygosity or heterozygosity in a population is wanted, i.e., if the heritable character is controlled by dominant gene then heterozygosity is attained but if the phenotype is controlled by recessive gene then homozygosity is desirable in a population.

The genes are situated on chromosomes, the recombination phenomenon during meiotic crossing over exchanges the chromosomal segments which controls the genetic distribution. The genes are the determinants for a particular character expression. Depending on their expressability the dominant and recessive nature are determined, the genes occur always in the alternative forms called as alleles. Due to mutation the changes in gene may occur, which also may be heritable or non-heritable. If the change is permanent and if it alters the phenotypic characters expression and if it is heritable through succeeding generations then mutation plays an important role in plant breeding.

In plant breeding when a breeder selects any particular mode of breeding programme then it affects the genetic criteria fixation in that population as he selects a particular desirable character. Normally in a population all the characters may be evenly distributed or the dominant characters are prevailed, but whenever a breeder is choosing for a particular criteria then the genetic constituent of that population gets changed. A particular gene gets advantage to be fixed in that population. Suppose in a breeding programme of barley one breeder is selective for the "white and smooth awn" character, both of them are recessive characters, so in a dihybrid cross programme only one out of 16 plants will bear this character and then selfing of that plant will lead to homozygous population, which will change the total genetic constituent of the population.

Back-crossing is also another phenomenon of changing and fixation of genetic constituent in a population. The parent holding more desirable characters is used as recurrent parent in a back-crossing programme which develops the homozygosity due to repeated back-crossing. Back-crossing programme is more opted when the character controlled by recessive genes are wanted in a population. In dihybrid crossing programme the homozygous recessive is only one out of sixteen, but in a back-crossing programme it is one out of four. So the large number of population is not needed to get the homozygous line.

2.2 BREEDING SYSTEMS

Breeding system in crops is the genetic system which regulates the generation of variability in population and a plant breeder should have an adequate knowledge of breeding systems operative in crop plants to plan for improvement through breeding programme. The significance of breeding systems in plant breeding are:

1. Nature of response of population,
2. Limitations and possibilities of different methods,
3. Regulation of pollination process.

There are three basic or natural systems of pollination which determine the nature of breeding system of a plant:

A. Natural self pollination;
B. Natural cross pollination;
C. Occasional cross pollination.

A. Self Pollinated Crops

Self pollination is the process of transfer of pollen grains from the anther of a flower to the stigma of the same flower. In the self pollinated crop plants the self pollination is not complete, 5% cross pollination may occur which is affected by several factors – variety, environmental condition and location.

Contrivances of self pollination are:

1. **Cleistogamy**: The flowers do not open at all, e.g., wheat, barley, oat, etc.
2. **Chasmogamy**: The flowers open but after pollination as in some varieties of rice, wheat, barley, oat, etc.
3. **Homogamy**: The stamens and carpels of a flower mature at the same time, so there is greater chance for self pollination.
4. Position of anther lobes and stigma are very much close to promote self pollination in crops like tomato and brinjal.
5. The flower opens but the androecium and gynoecium are enclosed within two petals as in legumes.
6. The receptive stigma elongates through the fertile staminal column to get pollinated.

Consequences of self pollination are:

(a) Continuous inbreeding results in progressive loss of genetic variability, i.e., leads to fixation of genotypes.
(b) Intravarietal genetic variation is absent.
(c) Intervarietal genetic variation is large due to random fixation of different alleles in different lines.
(d) Prolonged inbreeding promotes discrete grouping within population.
(e) Inbreeding favours only the homozygotes not the heterozygotes.

B. Cross Pollinated Crops

Transfer of pollen grains from one flower to the stigma of another flower borne in different plants is called cross pollination. The flowers may be bisexual or unisexual, the pollination needs some agencies like wind, water or animals, e.g., insects, birds, snails, etc.

Contrivances for cross pollination are:

1. In **unisexual flower** cross pollination is obligatory.
2. In **bisexual flower** there are some adaptive features for cross pollination:

(a) **Self sterility**: The self sterility or self incompatibility may be due to many factors: (i) incapability of pollen germination, (ii) failure of pollen tube growth, (iii) production of sterile pollen by the plant.

(b) **Dichogamy**: The phenomenon of maturation of male and female sex organs at different times; anthers mature before stigma **(protandry)**; stigma matures before anther lobe **(protogyny)**.

(c) **Herkogamy**: When the structure of male and female sex organs become barrier to self pollination, stigma projects out of the stamen.

(d) **Heteromorphism**: The flowers are of two or three morphological types, may be dimorphic or trimorphic, which usually involves **heterostyly** (styles of different lengths) and **heteroanthy** (different position of anthers).

Consequences of cross pollination are:

(a) Cross pollination always creates and maintains genetic variability and high degree of heterozygosity in the population.

(b) The population becomes fairly dynamic because allogamy prevents genetic stagnation.

(c) Dominant genotypes become prevalent in a population.

(d) Random pollination leads to mixing of outbreeding population which creates the loss of genetic identity.

(e) Genetic correlation among the progenies tend to get reduced as there is mating between phenotypically dissimilar individuals.

(f) Hybrid vigour is lost when there is inbreeding within cross pollinated crops; the inbred lines may be developed and hybrid vigour can be recovered by hybridization between two inbred lines.

C. Occasional Cross Pollinated Crops

Some of the crops like cotton, sorghum are exposed to both cross and self pollination, self pollination is predominant but cross pollination occurs by wind blown pollen grains occasionally. So in these types of crops the advantages of both self and cross pollination can be obtained; but stability achievement in these crops is problematic.

2.3 POLLINATION CONTROL

In most of the breeding schemes, selected plants or strains to be crossed are protected from self pollination as well as cross pollination by means of controlled pollination. There are natural contrivances which help in self pollination and cross pollination. For hybridization programme, two naturally occurring phenomena helping in hybrid production are **self incompatibility** and **male sterility**. Self incompatibility and male sterility prevent self pollination and promote cross pollination; therefore artificial methods are not required for controlling self pollination.

A. Incompatibility

Self incompatibility refers to the situation when in a bisexual flower fertile pollen and ovule are produced but the pollen is unable to fertilise the ovule due to some reasons which may be classified as follows:

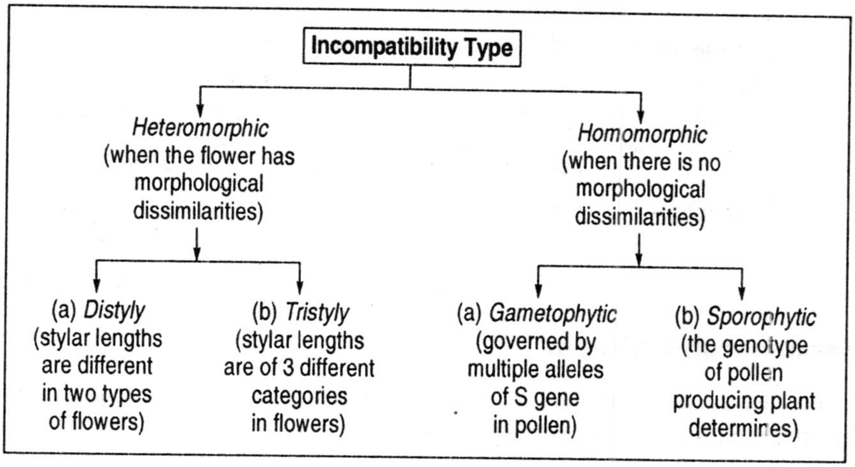

(a) Heteromorphic System

When the species has two or three different kinds of arrangement of floral parts, each type is self incompatible but compatible with others. Here the genes or alleles associated with incompatibility are also linked with length of style and filament.

(i) **Distyly :** In *Primula,* there are two types of flowers:

'Pin type' – long style, short filament, large stigmatic cell, small pollen;

'Thrum type' – short style, long filament, small stigmatic cell, large pollen;

Both are self incompatible but cross compatible. The 'thrum' is governed by Ss and 'pin' by ss. As the reaction of pollen is controlled by the gene of the sporophyte, the pollen of the 'thrum' type will behave like S (Fig. 2.4). The s gametes are incompatible with s type but compatible with S type and the *vice-versa.*

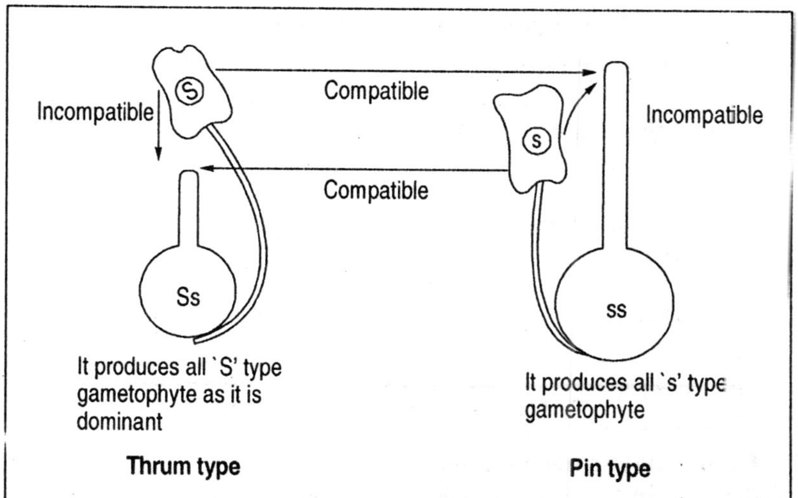

Fig. 2.4: Incompatibility in *Primula*

(ii) **Tristyly**: In *Lythrum,* three types of flowers with different stylar length exist. Here the stylar length is governed by two independent loci M and S. Plants with S have short style irrespective of the nature of other allele. The three different morphological types are self incompatible but cross compatible.

	Long style	Medium style	Short style
Genotype	mmss	Mmss / MMss	MmSs / mmSs/ mmSS / MMSS/ MMSs / MmSS

(b) Homomorphic System

In this case there is no morphological distinction between the self incompatible flowers and the incompatibility is governed by multiple alleles. This system can be of two types:

1. Pollen tube growth is controlled by genotype of sporophyte; and
2. Pollen tube growth is governed by genotype of the pollen.

(i) **Sporophytic incompatibility**: Here the incompatibility is governed by multiple alleles but has the dominant-recessive reaction. Here the pollen tube growth is not controlled by the genotype of pollen but by the genotype of the plant on which it is being produced. Actually the behaviour of gametophyte is controlled by dominant allele of the genotype of sporophyte. For example, multiple alleles for incompatibility are S_1, S_2, S_3, S_4; S_1 is dominant over others, S_2 is dominant over S_3 and S_4, S_3 is dominant over S_4, etc. Irrespective of the gametophytic genotype, the reaction will be of dominant type of the sporophyte.

S_1S_2 will produce all S_1 type gametophyte;
S_3S_4 will produce all S_3 type gametophyte;
S_2S_3 will produce all S_2 type gametophyte;

and all these are incompatible with the same dominant type (Fig. 2.5).

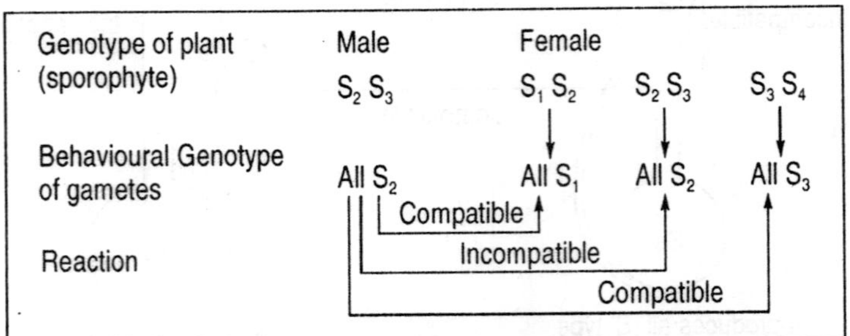

Fig. 2.5: Sporophytic incompatibility

(ii) **Gametophytic incompatibility**: In this system the incompatibility criteria is determined by multiple alleles of S gene and the reaction of the pollen is determined by its own genotype, not by the dominant allele of genotype of the plant on which it is produced. Therefore the pollen tube growth depends on the difference in the genotype of pollen and female parent; here all the plants are heterozygous in relation to this particular 'S' locus.

Genotype		Reaction
male/female	female/male	
S_1S_2	S_1S_2	Incompatible
S_1S_2	S_3S_4	Compatible
S_1S_2	S_2S_3	Partially compatible (50%)

Male parent		S_1S_2		S_1S_3		S_2S_3	
Pollen		S_1	S_2	S_1	S_3	S_2	S_3
Female parent	S_1S_2	—	—	—	S_1S_3 S_2S_3	—	S_1S_3 S_2S_3
		Fully incompatible		Partially compatible		Partially compatible	
	S_1S_3	—	S_1S_2 S_2S_3	—	—	S_1S_2 S_2S_3	—
		Partially compatible		Fully incompatible		Partially compatible	
	S_2S_3	S_1S_2 S_1S_3	—	S_1S_2 S_1S_3	—	—	—
		Partially compatible		Partially compatible		Fully incompatible	

Fig. 2.6: Gametophytic incompatibility in Tobacco

Use of Incompatibility in Breeding

1. When male sterile line is not available, self incompatibility can be used equally to produce the F_1 hybrids in seed crops and grasses. Success has been achieved in case of crucifers and forage grasses. The S homozygotes are produced (S_1S_1 or S_2S_2) and then these are interplanted to produce S_1S_2 hybrids.
2. Double cross hybrids also can be produced with the help of incompatibility. For instance, S_1S_2 and S_3S_4 hybrids can be intercrossed to produce ($S_1S_2 \times S_3S_4$) double cross hybrid.
3. In case of fruit crops or orchard crops self fertility can be induced temporarily or permanently by mutation in the S-allele to S_f through artificial irradiation. This induction will help in uniformity in cropping rather than mixed type of planting required for self-sterile varieties to overcome uneven quality of crops.
4. Self incompatibility sometimes leads to seedless fruit varieties development.

B. Male Sterility

Male sterility is characterised by the deficiency in pollen formation or the nonfunctional pollen production, i.e., there may be some deficiency in the process of microsporogenesis. But in some cases the viable pollen are formed but anthers fail to dehisce. The process of male sterility is totally under genic control which may be nuclear gene or cytoplasmic-gene. Depending on this factor male sterility can be grouped as follows:

(a) Genetic male sterility;

(b) Cytoplasmic male sterility;

(c) Cytoplasmic-genetic male sterility.

(a) Genetic Male Sterility

When the male sterility is governed by a single recessive nuclear gene, *ms* and there is no influence of cytoplasm then it is called as genetic male sterility. In this case the genotypes will be —

Ms Ms — Fertile (R line)
Ms ms — Fertile (B line)
ms ms — Sterile (A line)

A-line or Male sterile line: This term is used to represent a male sterile line which can be used as a female parent.

B-line or Maintainer line: This line is crossed with the male sterile line to produce normal hybrid seeds, F_1 plants fail to produce normal pollen, i.e., remain male sterile. This line should be isogenic line with A line so that by crossing A × B always the A-line can be maintained.

R-line or Restorer line: Restoration means when this line is crossed with the male sterile line then F_1 hybrid seeds will give the plants which will produce the normal viable pollen. But in the F_2 progenies the male sterile plants will reappear due to segregation.

Maintenance of Genetic Male Sterility

Crossing of heterozygous male fertile line with male sterile line	A line (ms ms) ♀	X ↓	B line (Ms ms) ♂	
(Maintained by sibmating, seeds from this line are harvested)	(ms ms) 1 Male sterile ♀	: ↓	(Ms ms) 1 male fertile ♂	(Seeds not harvested)
(Maintained by sibmating, seeds from this line are harvested)	(ms ms) 1 Male sterile ♀	: ↓	(Ms ms) 1 Male fertile ♂	(Seeds not harvested)
		Maintained indefinitely		

Restoration of Fertility in Genetic Male Sterility

Restoration of fertility can be done by crossing this A line with restorer line or R line.

Parents	A line (ms ms)	X	R line (Ms Ms)
F1	All progenies are fertile (Ms ms)		(Selfing)
F2	Ms Ms 1 : Ms ms 2	:	ms ms 1
	male fertile		male sterile

Uses / Importance in Plant Breeding

Genetic male sterility though can be used in plant breeding but the seed from 50% of the plants (heterozygous) cannot be harvested, so this type of behaviour has less value in hybrid seed production. This criteria is utilised for genetical studies and preservation of variability. Genetic male sterility is found in nature in crops like tomato, bean, barley, sorghum, corn, etc.

(b) Cytoplasmic Male Sterility (CMS)

This type of male sterility occurs due to the mutation of mitochondrial gene or some other cytoplasmic factors outside the nuclear genome which make the plant male sterile.

Here the genotypes can be designated as:

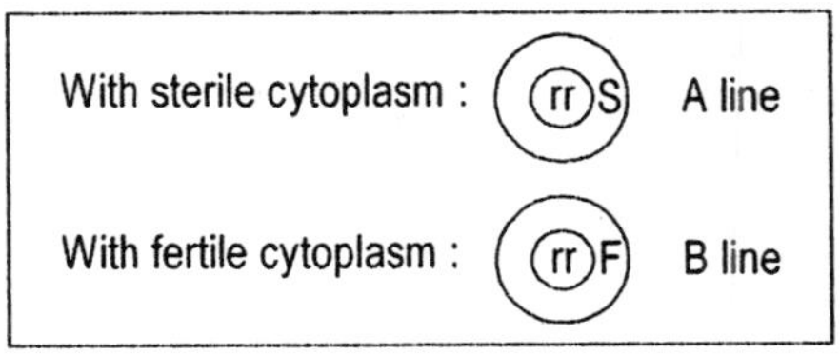

Maintenance of Cytoplasmic Male Sterility

As there are only two types of genotypes and cytoplasm is always inherited from the male sterile (female) plant so the maintenance of sterility is straight-forward here.

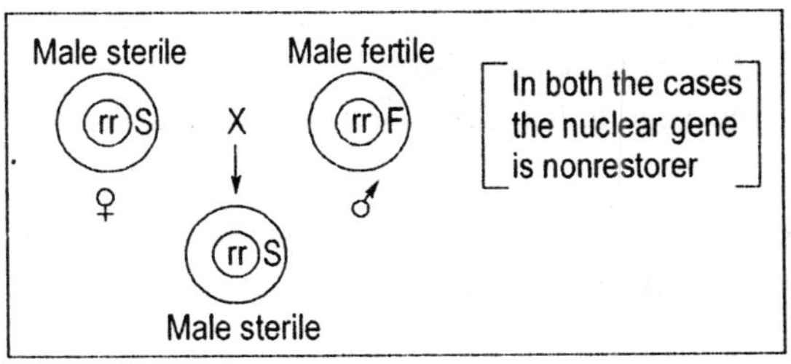

Transfer of Male Sterility to New Genotype

This cytoplasmic male sterility may be transferred easily to a given strain by using that strain as pollinator in successive generations of a backcross programme. After 6-7 generations the genotype of male sterile line would be identical to pollinator line bearing the male sterile cytoplasm (Fig. 2.7). Through more repeated backcrossing total nuclear gene (variety B) will be transferred with the male sterile cytoplasm of variety A.

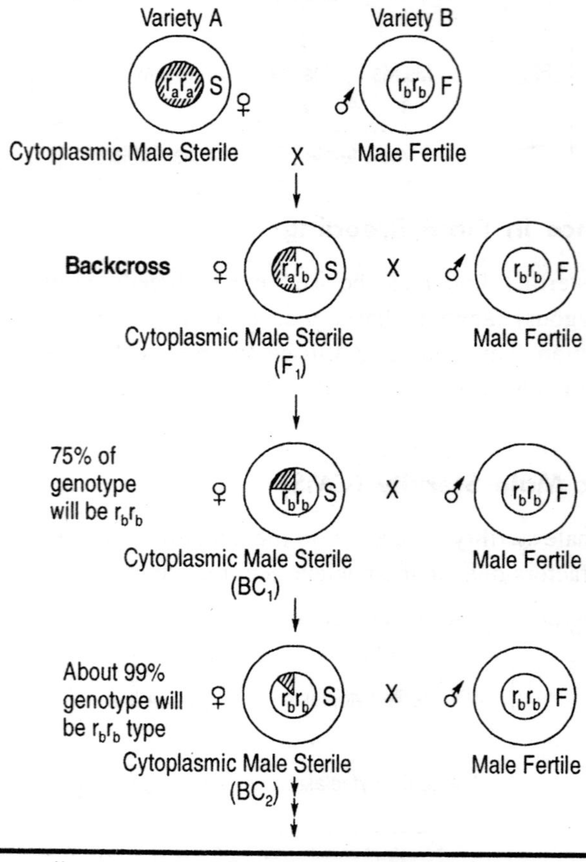

Fig. 2.7: Transfer of cytoplasmic male sterility

Uses / Importance in Plant Breeding

As in cytoplasmic male sterility there is no restorer gene, so restoration of fertility is not possible, therefore this type of male sterility is useful for crops where seeds are not desired product, e.g., the vegetables like onion, broccoli, cauliflower, cabbage or other ornamental plants.

The hybrids developed in this process exhibit maximum advantages of hybrid vigour with respect to longer vegetative duration, larger flower size in ornamental plants and larger bulb size as in onion.

(c) Cytoplasmic-genetic Male Sterility

Here the male sterility character arises from the interaction of nuclear genes and cytoplasmic factor, i.e., when in the cytoplasmic male sterile line fertility can be restored with the help of nuclear gene (restorer gene, R) then this behaviour is called as cytoplasmic-genetic male sterility. In this kind of male sterility the combination of both nuclear genes and cytoplasmic factors determines the fertility or sterility in such plants. Different combinations of these show only one male sterile line:

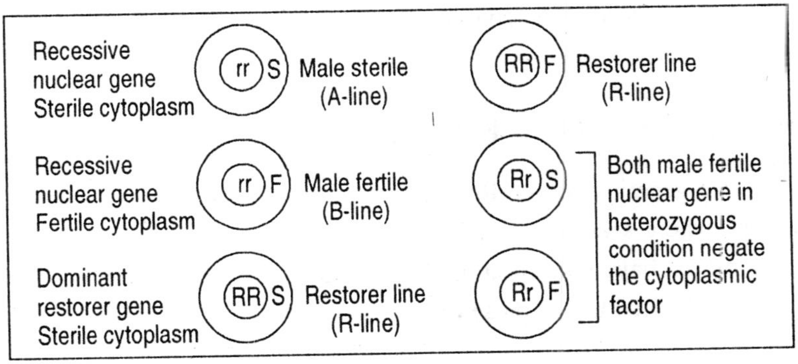

Maintenance of Sterility in Cytoplasmic-genetic Male Sterility

For this purpose, A line and B line should be isogenic. If the nuclear gene is heterozygous in nature then the outcome will segregate into 1 : 1 ratio.

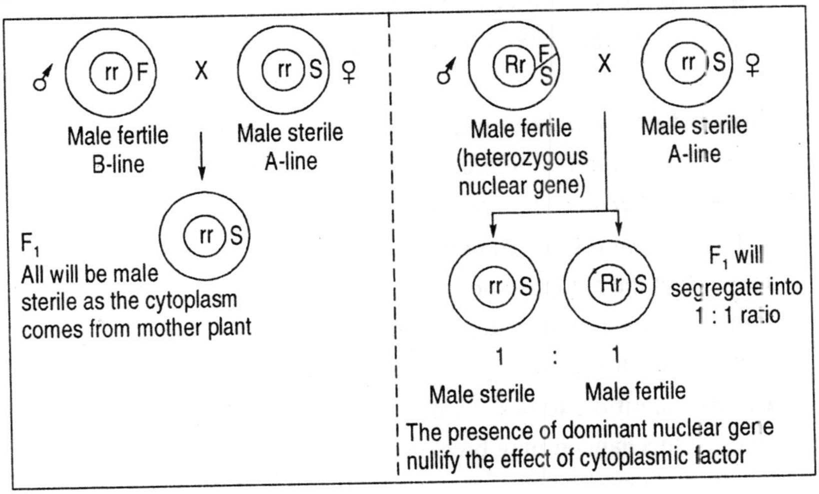

Restoration of Fertility in Cytoplasmic-genetic Male Sterility

Fertility can be restored when a homozygous restorer line, i.e., RR is crossed with male sterile A line, the progenies will become fertile. New male sterile lines can be produced

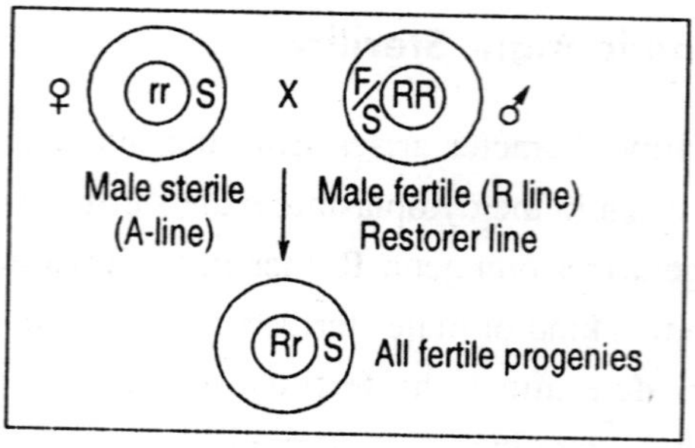

in this kind of male sterility behaviour, but the pollinator should always be recessive homozygous.

Uses / Importance in Plant Breeding

Cytoplasmic-genetic male sterility has much importance in the exploitation of hybrid vigour in crops where seed is desired. By using this cytoplasmic-genetic male sterility hybrid seed in maize, bajra and jowar has been produced. But owing to restriction on cross pollination, it has limited use in these crops.

Limitations of Using Male Sterility in Plant Breeding

1. **Undesirable effects of the cytoplasm**: In many cases it has been found that the male sterile cytoplasm has some undesirable effects, e.g., Cms-T cytoplasm of maize slightly retards growth and more susceptible to *Helminthosporium.*
2. **Unsatisfactory fertility restoration**: In many cases restoration of fertility is not always possible, then those sources cannot be used in production of hybrid seed.
3. **Unsatisfactory pollination**: In case of many self pollinating crops natural cross pollination is a problem. For hybrid seed production if pollination is not 100% then seed production cost increases due to less harvest.
4. **Nuclear gene effect**: Sometimes some of the nuclear genes (modifier gene) may effect the cytoplasmic male sterility, as a result during back-crossing for transfer of male sterile cytoplasm to a new genotype, the male sterility character may get disturbed and fertile pollen grains may be produced.
5. **Environmental factors**: Under certain environmental conditions the male sterility mechanism may breakdown resulting in production of fertile pollen by the male sterile line.
6. **Contribution of male cytoplasm**: During maintenance of male sterility sometimes the small amount of cytoplasm contributed by the pollen may lead to breakdown of male sterility mechanism.

QUESTIONS

1. Discuss the different modes of reproduction of crop plants and mention their significance in generating and fixing variations.
2. State the contrivances and consequences in self pollinated and cross pollinated crops.
3. What is incompatibility? Mention different types and their use in plant breeding.
4. Define male sterility. What are its different types and controlling factors? Discuss their utilization in plant breeding.

3 Breeding Methods: Introduction and Acclimatization

Plant introduction is a process of introducing plants (a genotype or a group of genotypes) from their own environment to a new environment. The process of introduction may involve new varieties of crop or the wild relatives of crop species or totally a new crop species for the area.

Introduction may be classified into two categories:

Primary Introduction: When the introduced variety is well suited to new environment then it is released for commercial cultivation without any alteration of genotype. For example, dwarf wheat varieties like 'Sonora-64', 'Lerma rojo' and dwarf rice varieties like 'Taichung Native 1', 'IR-8' are the examples of primary introduction.

Secondary Introduction: When the introduced variety is subjected to selection or used in hybridization programme with local varieties to get the improved varieties with some new characters introduced called secondary introduction. For example, the varieties like 'Kalyan Sona' and 'Sonalika' of wheat have been selected from material introduced from CIMMYT, Mexico.

The process of plant introduction is the successful compliance of two important aspects, viz., **domestication** and **acclimatisation**. Domestication is the process of bringing of a wild species under cultivation by making them changed in behaviour suitable for new environment.

Acclimatization is the ability of a crop to become adapted to a new climatic and edaphic conditions. The process of acclimatization leads to increase the frequency of most adapted genotypes which depends on three factors:

(a) breeding system,
(b) genetic variation,
(c) duration of the crop.

The success of acclimatization depends on two factors like:

(i) place effect,
(ii) the selection of new genotypes.

3.1 PURPOSE

(a) **Use in agriculture, forestry and industry**: New varieties of plants or crops are introduced from various places for use as food, fibre, wood or medicinal purposes as well as the breeding material for hybridisation work.

(b) **For aesthetic interest**: Various ornamental plants are introduced for beautification.

(c) **For germplasm conservation**: The spread of high yielding varieties causes a danger to old varieties to get lost from an area. But germplasm collection and conservation help to maintain lines, clones, mutants, cultivars, etc. from as many sources as possible.

(d) **For studying origin and distribution**: The distribution of crop plants and their various forms in different parts of the world gives an idea of their origin and evolution.

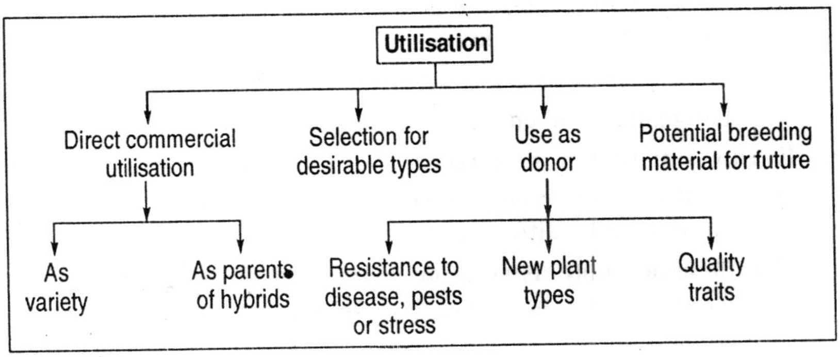

3.2 PROCEDURE

Plant introduction procedure consists of following **steps**:

(a) Plant procurement or procurement of Germplasm,

(b) Packaging and despatch,

(c) Entry and plant quarantine,

(d) Cataloguing,

(e) Evaluation,

(f) Multiplication and distribution.

(a) **Ways of procurement of plants or germplasm**: Requisition for introduction of new crop plant or new varieties should be submitted to NBPGR (National Bureau of Plant Genetic Resources) within the country or to IBPGR (International Bureau of Plant Genetic Resources). The material may be obtained on an exchange basis from friendly countries either directly or through F.A.O. or the material can also be purchased or obtained as free gift from individuals or organizations.

(b) **Packaging and despatch**: The part of the plant for propagation of that species is known as propagule. The propagule may be seeds, tubers, runners, suckers, stolons, bulbs, root cuttings, buds or seedlings depending upon the crop species. Depending on the type of propagule those are cleaned from other weed-seeds and contaminants and treated with fungicides, packed carefully and despatched so that it can reach the destination in viable condition.

(c) **Entry and plant quarantine**: On receipt of the material the *entry inspection* is done by the country for other contaminants and the presence of insect, diseases, nematodes are checked. The materials are treated with insecticides, fungicides or nematicides and then released to the user. The general objective of all 'quarantine and regulatory' measures is to prevent pests and diseases from entering into the country as well as to check spreading further.

(d) **Cataloguing**: After quarantine the introduced specimen is given a number regarding species, variety, place of origin and the data are recorded. The plant materials are classified into three groups:

1. Exotic collection – prefix 'EC'
2. Indigenous collection – designated as 'IC'
3. Indigenous wild collection – marked as 'IW'.

(e) **Evaluation:** To assess the potential of new introductions, their performance at different substations are evaluated as well as resistance to diseases and pests are evaluated under different environments.

(f) **Multiplication and distribution**: Promising introduced materials are propagated and then released as varieties after necessary trials.

3.3 AGENCIES IN INDIA

The central plant introduction agency in India is **National Bureau of Plant Genetic Resources (NBPGR)**, which has its headquarters at New Delhi, but has substations for testing the plant materials. NBPGR has the gene bank for long term storage and future use, it helps to assess the introduced plant material, coordinates the work of other agencies and imparts training in plant collection, introduction and maintenance in India.

The substations under NBPGR are :

Simla: It represents the temperate zone of approx. 2300m above sea level, place for germplasm collection in northern hills, station for acclimatization of material introduced from temperate zone and high altitude.

Jodhpur: Exclusively meant for exploring and acclimatizing plant material for arid zone, this is under Central Arid Zone Research Institute.

Amravati: This substation carries out acclimatization and multiplication of introduced material for central zone of India.

Kanya Kumari: It represents the tropical zone situated at South India.

Shillong: This centre has been created for collection of germplasm from North-East India.

Other agencies engaged for this purpose are —

Forest Research Institute, Dehradun: The plant introduction organization set up at the institute looks after introduction, acclimatization, conservation and testing of forest trees.

Botanical Survey of India: This body is meant for introduction of medicinal plants and also plants for botanical importance.

Other sources: Different Central Research Institutes as well as Agricultural Universities play the role for collection, introduction and maintenance of germplasm of crop plants.

3.4 MERITS AND DEMERITS

A. Merits

(a) It provides entirely new crop plants to a place.

(b) Superior varieties may be originated directly or after selection or hybridisation.

(c) Germplasm collection, maintenance and protection of genetic variability are possible through the ways of introduction and exploration.

(d) It is the most quick and economical method of crop improvement when introduced material can be used directly.

(e) Introduction of some varieties to newer areas may protect them from some diseases.

B. Demerits

The demerits or disadvantages associated with plant introduction are introduction of weeds, diseases and pests.

(a) The **weeds** like *Argemone, Eichhornia, Lantana* have been introduced from other countries with the introduction of crop plants.

(b) The **fungal diseases** like late blight of potato, flat smut of wheat, coffee rust, bunchy top of banana – all have been introduced in India alongwith plant materials.

(c) Many **insect pests** like potato tuber moth, woolly aphis of apple, fluted scale of *Citrus* were introduced in India along with plant introduction.

However, in most of the cases, the introduction of weeds, diseases and insect pests occurred during a period when quarantine was almost non-existent.

3.5 ACHIEVEMENTS

The process of introduction from the pre-historic times helped the whole world to achieve newer crop species to new places as well as the new varieties of crop plants. The examples of achievements are listed below:

(A) **New crop species**: The crops like potato, maize, groundnut, chillies, coffee, rubber, guava, grape, pineapple, many oranmentals like Gulmohor, Phlox, Salvia, Aster–all have been introduced in India from outside.

(B) **New crop varieties**:

(a) **Direct multiplication and released as new varieties**: Dwarf wheat varieties like 'Sonora 64', 'Lerma Rojo'and dwarf rice varieties like 'TN-1', 'IR-8' are introduced in India and used directly. There are many more examples of direct release of varieties in other crop plants like oat, soyabean, tomato, cauliflower, onion, etc.

(b) **Selection of desirable varieties**: Many varieties have come up after selection from introduced varieties, e.g., 'Kalyan Sona', 'Sonalika' were selected after introduction of Mexican varieties.

(c) **Introduced varieties as donor in Hybridization programme**: Sometimes the introduced varieties, superior than the existing varieties, may be used as donor for disease resistance or pest resistance or stress resistance for plant type and quality characters in hybridization. In case of rice the dwarfing gene from TN-1 or IR-8 have been used for developing other dwarf rice varieties in rice. All the sugarcane varieties have been derived from introduced noble cane variety of *Saccharum officinarum*.

(d) **Mutation Breeding**: If the introduced material is agronomically superior but lacking a few characters, may be treated with mutagen to rectify the defect. For example, 'Sonora-64' introduced with red colour grains not accepted by the farmers, the new mutagenic product 'Sharbati Sonora' with amber colour grains was produced by M.S. Swaminathan by Gamma-ray treatment.

3.6 GERMPLASM MAINTENANCE

The sum total of hereditary material or genes present in a species is known as the germplasm of that species. Germplasm collection is the collection of a large number of genotypes of a crop species and its wild relatives, it is the richest source of variability. Crop improvement depends on the availability of variability to be utilized in breeding programme. With modernisation of agriculture there is gradual loss of wild and 'desi' varieties which leads to gradual loss of variability in the cultivated forms and in their wild relatives, referred to as **genetic erosion**.

Establishment of **IBPGR (International Bureau of Plant Genetic Resources)** was directed for coordinated germplasm conservation activities throughout the world. Germplasm collections contain all the land varieties, various wild forms, primitive races, exotic collections and highly evolved varieties. As most of the seeds lose viability quickly, it is necessary to grow the seeds in every few years. Cold storage of seeds helps in easy maintenance. In **NBPGR (National Bureau of Plant Genetic Resources)**, India has this kind of cold storage facilities known as **National Germplasm Repository**.

The germplasm resources can be **preserved** in the following ways:

1. **Long term storage**: At 20°C and 5% relative humidity, the seeds can be stored for long time.
2. **Live plants**: Perennial plants, trees or bulbous plants can be maintained in growing condition.
3. **Tissue culture**: For some problematic plants, tissue culture has been suggested as an aid for maintenance.

The whole process for germplasm utilization and maintenance can be shown by following diagram.

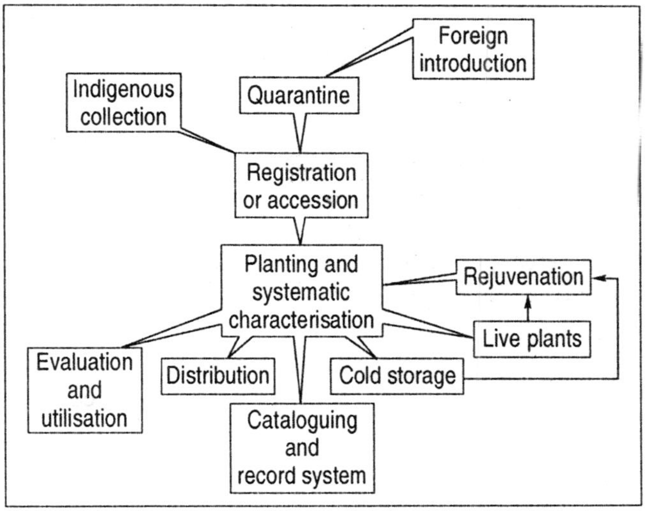

For germplasm conservation or maintenance various devices have been developed like gene bank, seed bank or pollen and spore bank.

Gene bank: It is the concept where all available genotypes of a genus are collected and kept according to its own specific requirement. This is the storage facility for *ex situ* conservation of a whole plant, seeds, pollens, eggs, ovules and a variety of tissues. It preserves permanently and potentially viable germplasms of species in a suitable *in vitro* environment.

Gene sanctuaries: This may be defined as an area of diversity protected from interference from man. It conserves the germplasm *in situ*, in the environment where it naturally grows. This concept is not only for conservation but also for proceeding of evolution in its natural course, it allows appearance of new gene combinations and new alleles which were not originally present. *Musa, Citrus, Saccharum, Mangifera*, etc., can be preserved in this way.

Seed bank: In case of sexually propagated plant species seeds represent as the storage form of genetic variability. Such storage of full complement of genetic combination of a plant species through seeds are referred to as seed bank. Seeds are cleaned and dehydrated upto 5% moisture content and stored at –20°C with 5% relative humidity. Removal of excess water from seed is to prevent rupturing cell membranes by ice crystals formed inside seeds. Seeds can be stored for long time in this way, though periodic germination is needed to get fresh viable seeds, but the interval of obtaining fresh seeds differs from species to species.

Pollen and spore bank: Pollens from flowering plants and spores from non flowering plants are stored cryogenically in very small containers. Freeze-drying technique is now-a-days applied for their storage where drying and freezing stages combine together within a single quick-freeze process. Many repositories have started successfully pollen or spore bank.

QUESTIONS

1. What do you mean by introduction and acclimatization? State the purpose. Mention the agencies involved in this process.
2. Discuss the procedure for introducing a new variety. Write down its merits and demerits. Mention the achievements.
3. State the method of germplasm maintenance.
4. Write short notes on : Seed bank, Gene bank, Pollen bank, NBPGR.

4 BREEDING METHODS: SELECTION

The procedure selection involves the retention of superior phenotypic plants from mixed population. This can be done in different ways in different crops.

4.1 MASS SELECTION

When a large number of plants of similar phenotype are selected and their seeds are mixed together to constitute a new variety is called **mass selection**. The population obtained from the selected plants would be more uniform than the original population.

Purpose

In case of *self pollinated crops,* mass selection procedure has following objectives:

(a) **Improvement of local varieties**: In local varieties sometimes there are mixtures of several types which may differ in flowering maturity time, disease resistance, plant height, etc. Elimination of poor quality plants help to get uniform performance of the variety.

(b) **Purification of existing pureline varieties**: Maintenance of purity of the existing pure line varieties is done through this method. The pure line varieties sometimes tend to become variable with time due to mechanical mixtures, natural hybridisation and mutation. Through regular mass selection the purity of the pure line varieties is maintained.

(c) **Production of new varieties from heterogeneous local land races**: By increasing the frequency of superior genotype, the population character can be changed. This change is a function of heritability and the number of genes conditioning the trait under selection.

In case of *cross pollinated crops,* inbreeding must be avoided as it leads to loss in vigour and yield. But in mass selection several plants are selected and their seeds are mixed together to raise the next generation, so inbreeding is avoided or kept minimum.

Procedure

The process involves selection or retention of a large number of individual plants on the basis of superior phenotype, compositing the harvested seeds without progeny-testing, the mixture of seeds is sown *in mass* to raise the next generation. This process continues till the desired level of uniformity is achieved.

The procedure for mass selection may be outlined in brief as follows:

First year: A large number (200-2000) of plants are selected on their performance and phenotypic characters like vigour, plant type, disease resistance or other desirable characteristics. The seeds are composited together.

Second year: The composite seeds are planted in a **preliminary yield trial** along with standard varieties as checks as well as the variety from which the selection is made to determine whether there is any improvement or not.

Third to Sixth year: The composite seeds from previous year collection are evaluated in coordinated yield trial in several locations within the same agroclimatic zone. **Initial evaluation trial** for one year then **uniformity regional trial** for two or more years are done.

Seventh year: If the result is suitable for cultivation and performance is well then certification of seeds is necessary and released by central or state variety release committee.

Schematic representation of mass selection in self-pollinated crops:

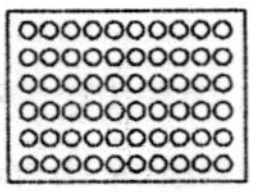

FIRST YEAR

(a) From a variable population, 200-2000 plants are selected with similar and desirable traits.

(b) Harvested seeds are mixed together.

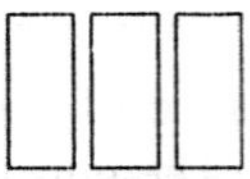

SECOND YEAR

(c) The mixed seeds (composite) are planted along with standard checks and preliminary yield trial is done.

(d) Phenotype of the selected population is critically evaluated.

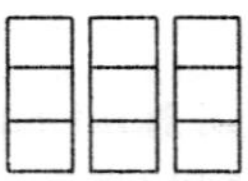

THIRD TO SIXTH YEAR

(e) Promising selections are evaluated in co-ordinated yield trial in several locations.

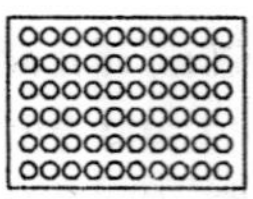

SEVENTH YEAR

(f) If the performance is satisfactory in each location then may be released as new variety.

(g) Seeds are multiplied for distribution.

The procedure of mass selection whenever is coupled with progeny testing then that helps in maintaining the purity of pure line varieties. For this procedure, in the second year the individual plant progenies are planted individually and poor performers are rejected. The remaining progenies are mixed together to constitute the variety. The process may be repeated every few years to keep the variety pure as often as it is found necessary.

The **modified procedure** of mass selection is represented below schematically:

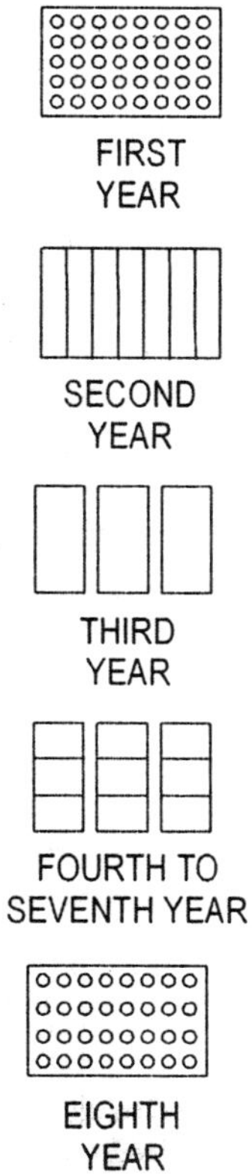

(a) Similar and superior phenotypes are selected (200-2000 plants).

(b) The seeds are harvested from individual plant separately.

(c) The harvested seeds from individual plant are sown in rows.

(d) The inferior or segregating progenies are rejected.

(e) The seeds from the remaining selected plants are harvested together.

(f) The composite seeds are used for preliminary yield trials along with standard checks.

(g) Multilocation co-ordinated yield trials are performed.

(h) Superior phenotypes are released as new varieties.

(i) Seeds are multiplied and distributed.

MERITS AND DEMERITS

Merits

1. Since a large number plants are selected, the variety is more stable in performance over different environments as it is more adapted than a single pure line. So the varieties developed through mass selection are more widely accepted than pure line.
2. Extensive and prolonged field trials are not necessary. This reduces the time and cost needed for developing a new variety. This method is also less labour consuming.
3. Mass selection retains considerable genetic variability, so another mass selection after few years improves the variety.
4. This method can be applied to cross pollinated crops.

Demerits

1. The varieties developed through mass selection show variation and are not uniform as pure line varieties, and strict selection for uniformity would lead to inbreeding depression.
2. The improvement of variety through mass selection is poorer than through pure line selection.
3. When the mass selection is done without progeny test, the homozygosity of the selected plants is not assured. In the self pollinating species as there are chances for cross pollination to some extent, some plants must be heterozygous. In this method the genetic superiority either may be due to environment or due to genotype.
4. Pure line selection is more used than mass selection, only improvement of local old varieties are done through this method.
5. In seed certification programme, it is more difficult to identify the mass selected variety than the pure line.
6. This method utilises only the variability already exists in the population, improvement is done only through selection. So the limitation is that it cannot generate new genetic variability.

Achievements

Mass selection is effective when the population has the following characters:

(a) High genetic variability for different traits.
(b) The characters should be highly heritable in nature.
(c) The crop is grown under low population density.
(d) If only one particular character is chosen.

At present use of mass selection is limited to purification of pure line varieties of self pollinated crops, because the superiority of the pure lines would be lost quickly if their purity is not maintained through mass selection.

Although no proper records are available, but the following varieties in different crops have been developed by mass selection in India:

Maize — Jaunpur local, Tinpakhia, Basri
Pearl Millet — Bajra-207, Bajri-28-15, Bichpuri local, Pusa moti
Mustard — Gurgaon brown sarson
Sorghum — RS1, T22
Cotton — Dodahatti local, Dharwar American, Combodian cotton
Peanut — TMV-1, TMV-2

Except these there are large number of examples of mass selection in maize, sugarbeet, alfalfa, redclover available in other countries.

4.2 PURE LINE SELECTION

In this case, a large number of plants are selected from a population of self pollinated crop, harvested individually and their individual progenies are evaluated and the best progeny is released as a pure line variety. So the **pure line** is a progeny of a single homozygous self pollinated plant.

Purpose

In self pollinated crops pureline selection method has several applications:

(a) **Improvement of local varieties**: It is more favourite method of improvement of local varieties which have considerable genetic variability.

(b) **Pure line selection for introduced varieties**: Introduced varieties are often subjected to pure line selection to develop suitable varieties for that place.

(c) **Improvement of old pure line varieties**: Pure lines become variable with time, i.e., offtype plants appear in pure line. From such genetically variable pure lines often again pureline selection method is applied to make it more stable.

(d) **Selections for new characteristics in a pure line**: Sometimes new favourable characters which were not considered previously can be selected from the existing variety through pure line selection method.

(e) **Pure line selection in hybridization programme**: Pure line selection method is most applicable in hybridization programme, i.e., selection from segregating generations of crosses. The pure lines in such cases may be selected using pedigree, bulk or backcross methods.

Procedure

The general procedure of pure line selection involves three **steps**:

1. Selection of individual plants is done from a local variety. The superior and similar observations are selected assuming those are homozygous.
2. Visual evaluation of individual plant progenies is performed and the superior progenies are harvested separately.
3. Replicated yield trials are carried out for critical evaluation of the progenies. The most promising strain is identified and released as variety.

The procedure for pure line selection may be outlined in brief as follows:

First Year: A large number of plants (200-3000) are selected from 'desi' or local variety or any other mixed population. The seeds are harvested from individual plant separately. The selection is made on easily observable characters like flowering, maturity times, disease resistance, presence of awns, plant height, etc.

Second Year: Progenies from individual selected plants are grown separately with proper spacing. The progenies are observed visually and poor, weak or defective progenies are rejected, selection is made for very much desirable characters. Disease epiphytotics may be created to test the progenies for disease resistance.

If necessary this process may be repeated for one or more years.

Third Year: This step consists of replicated yield trials for critical evaluation of the progenies. The selected progenies are planted in replicated trial using the best variety as check. If enough seeds are available then preliminary yield trial may be conducted.

Fourth Year: Replicated yield trials are conducted by the breeder using the best variety as check. The promising strains are observed and harvested for coordinated yield trial for further evaluation.

Fifth to Eighth Year: The promising strains are evaluated at several locations alongwith other strains. The most promising strains are identified and released as new varieties.

Ninth Year: The best progeny seeds are multiplied and released as a new variety through seed certification.

Schematic representation of pureline selection in self-pollinated crops:

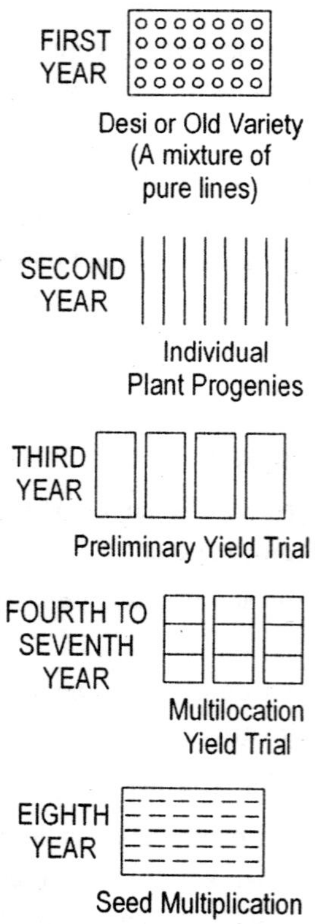

(a) 200-3000 plants are selected on the basis of their superior and similar phenotype.

(b) Individual plant progenies are grown separately.

(c) Undesirable progenies are rejected.

(d) Remaining progenies are planted for preliminary yield trial.

(e) Low yield or inferior progenies are rejected.

(f) Replicated yield trials conducted at several locations.

(g) Inferior progenies are rejected.

(h) Disease resistance and quality tests are done.

(i) Best progeny released as a new variety.

(j) Seed multiplication and distribution are performed.

The total time required for the whole programme may vary depending upon several factors. If enough seeds are available in the second year then preliminary yield trial is done in the third year. Depending on the performance in the coordinated multilocation yield trial the time period can be lessened.

MERITS AND DEMERITS

Merits

1. Pure line selection can achieve maximum possible improvement over the original variety. This is because the variety is the best pure line present in the population.
2. The pure line variety is extremely uniform since all the plants in the variety have the same genotype. Such uniformity is liked both by the farmers and the consumers.
3. Due to extreme uniformity, the variety is easily identified in seed certification programme.

Demerits

1. As the pure line variety is genetically more homogeneous, the adaptability to various environment is less.
2. This procedure requires more time, space, labour and more expensive yield trials.
3. The upper limit of improvement is set by the genetic variation present in the original population.
4. Once a pure line is developed, the variability is frozen and further selection becomes less effective.

ACHIEVEMENTS

Pure line selection is the most extensively used breeding method in early days of crop improvement in India because at that time many genetically variable 'desi' varieties were available. A large number of improved varieties were developed by this method in many self pollinated crops:

Wheat	—	NP_4, NP_6, NP_{12}, Pb-8, C13, K46
Rice	—	M-351, Vidisa 60-1, Patni 6, Aispuri, BP 53
Cotton	—	Coimbatore 2, Gadag 1, MCU1
Barley	—	C-251, C-50, K-12
Tobacco	—	NP28, NP63, NP70

There are many more examples of achievements through pureline selection in India as well as in abroad. Apart from the above examples there are many more pure lines available in case of pulses like *Cicer arietinum, Vigna radiata, Vigna mungo, Arachis hypogea,* oil yielding *Brassica* (both 'rai' and 'toria'), jute and in many vegetables also. For long time many pure line varieties are dominating in Indian agriculture.

COMPARISON BETWEEN PURE LINE SELECTION AND MASS SELECTION

Mass selection and pure line selection differ in the procedure, product as well as the quality of improvement gained.

Pure line selection	Mass selection
A. Procedure	
1. The plants are selected for their desirability of characters that may not be of similar phenotype.	1. The selected plants should be of similar phenotype as the seeds are mixed together.
2. The selected plants are subjected to progeny test.	2. In normal mass selection progeny test is not performed.
3. The procedure is more effective as careful progeny test and yield trials are conducted.	3. As a large number of plants are selected, extensive yield trials are not necessary; less demanding process.
4. Generally 9-10 years are required to develop a new variety.	4. Generally 5-7 years are required to develop a new variety.
5. This method is used in self pollinated crops only.	5. This method is used in both self and cross pollinated crops.
B. Product	
6. The new variety is a pure line.	6. The new variety is a mixture of pure lines.
7. The new variety is highly uniform, genetic variation is very less.	7. The variety has genetic variation, so in general appearance it is less uniform.
8. Pure line variety is expected to have a narrower adaptation and lower stability in performance as there is less genetic variation.	8. The variety has wider adaptation and greater stability than pure line varieties.
9. The variety is easily identified in seed certification programmes.	9. The variety is relatively more difficult to identify in seed certification programme.
C. Quality of improvement	
10. The improvement of quality of the variety is generally the best which is present within the existing variety of the original population, i.e., this method brings about the best improvement over the original variety.	10. The variety is inferior to the best pure line because most of the lines included within it will be inferior to the best pure line.

4.3 RECURRENT SELECTION

This selection method is a variation of progeny selection, the main difference is that the selected progenies are allowed for all possible intercrosses through open pollination to provide all kind of recombination. So the method can be defined as reselection generation after generation with interbreeding of selects to provide for genetic recombination.

There are four main **types** of recurrent selection:

(a) **Simple recurrent selection**: In this method a number of plants with desirable phenotypes are selected and self pollinated. In next generation the progenies from the selected plants are grown separately, and are allowed to intercross in all possible combinations. Equal amount of seeds from each plant are composited to produce the next generation. This completes the **original selection cycle**. For recurrent selection several desirable plants are selected from the composite population and are self pollinated. Progeny rows are grown and all possible intercrosses are made by hand. Equal amount of seeds from all the intercrosses are composited to produce next generation. This is the **first recurrent selection cycle**. This population may be subjected to further more recurrent selection cycle in the same way.

Recurrent selection is effective in increasing the frequency of desirable genes in the population. It is most suited for characters with high heritability. Simple recurrent selection is considerably more efficient than selection with self pollination.

(b) **Recurrent selection for general combining ability**: This method is based on the combining ability of the lines or plants. In this system a number of superior plants are selected from the source population, the selected plants are called S_0 plants, which are selfed and crossed to heterozygous tester of broad genetic base. This selfed seed is kept in cold storage. The crossed seeds of S_0 plants with tester is used to evaluate the combining ability of various S_0 plants. Now the S_0 plants with good performance are grown in third year from the selfed seeds kept in storage. Those are intercrossed in all combinations and composite of intercrossed seed is then used to establish a new population for further selection. This cycle may be repeated.

(c) **Recurrent selection for specific combining ability**: This method is same as recurrent selection for general combining ability except that the tester is an inbred line, i.e., the tester here is with narrow base. So here the differences in the performance of S_0 plants in crosses are due to specific combining ability alone. Great care should be taken in inbred line selection which will be used as tester, if it is of inferior quality then the whole scheme will fail.

(d) **Reciprocal recurrent selection**: This is the method to improve simultaneously two populations from two diverse sources, A and B, one population serves as a tester for the other. Both the populations are selfed as well as outcrossed with the other, the superior progenies are selected.

4.4 CLONAL SELECTION

A **clone** may be defined as a group of plants derived from a single plant by vegetative propagation. The clone can be characterised by the following characters:

(a) Clone is homogeneous,

(b) Individuals of a clone are either homozygous or heterozygous,

(c) Clones are stable in nature,

(d) Variability can be induced through mutation,

(e) Clone is propagated vegetatively.

Normal stem, runner, sucker, stolon, tuber, rhizome, bulb and root or root cuttings can be used as clones.

The **procedure** of clonal selection is the selection of desirable clones from the mixed population of vegatatively propagated plants. The technique lies in selecting and propagating the best clone based on its performance. The steps may be summarised as follows:

1. The collection of all possible clonal variability.
2. Critical evaluation of each clone and each member of a clone for yield and quality following test for disease and pest infestation.
3. The disease free and high yielding clones are selected and evaluated and multiplied as variety.

Advantages of clonal selection are:

1. **Easy maintenance**: In this method there is no chance for outcrossing or loss of seed viability.
2. **Very quick**: As there is no time requirement for seed development, single clone can be identified very easily and can be multiplied straight way to give new variety.
3. **Permanent hybrid**: Heterotic clones can be exploited for any length of time without the need to produce hybrid seed as in seed crops.

QUESTIONS

1. Define mass selection. Write down the purpose & procedure of the process. Point out its achievements.
2. What is pure line? Discuss the method of pure line selection. In which type of crop this method is preferred. Mention the significant achievements.
3. Compare mass selection and pure line selection. Point out the merits and demerits of both the process.
4. What is recurrent selection? Write down about different types of recurrent selection.
5. What type of selection practice is done in vegetatively reproducing plant? Write down the procedure and merits of this process. Define clone.

5 Breeding Methods: Hybridization

After domestication and selection, hybridization is the most potential breeding method for improvement of crop. The previous two methods utilise the existing genetic variability within the particular species or variety. Hybridization aims at to create new genetic variation of characters. It is the method of crossing of two pure line plants of two dissimilar genotypes, which will produce the F_1 hybrids and then the subsequent generations will be segregating generations.

5.1 PURPOSE

The objectives of hybridization can be categorised under 3 major criteria:

1. **Development of productive varieties**: One of the objectives is to create new variability which is further used for development of improved varieties with respect to existing varieties. This is achieved through the following methods of breeding or selection of the segregating progeny of F_1 hybrid.
   (a) Pedigree method
   (b) Mass-pedigree method
   (c) Bulk progeny
   (d) Bulk population breeding.
2. **Removal of bottle-neck genes**: The appearance of new problems (pathogen, insect) to yield can be protected through the creation of new recombinants which will help to come out from this kind of situation.
3. **Development of heterotic F_1 hybrids**: In order to exploit hybrid vigour, hybridization between two chosen inbred parents can be resorted to produce F_1 hybrids. In case of self pollinated crops two methods of breeding are used for attaining this objective.

(a) **Combination breeding**: The main objective of this method is to transfer one or more characters into a single variety from other varieties. In this breeding the genetic divergence between the two parents is not the major consideration, here among the two parents one must have the character to be transferred and another parent is the popular variety.

(b) **Transgressive breeding**: This method aims at improving yield or contributing characters through transgressive segregation. It is the production of plants in the F_2 generation which will be superior to both the parents for one or more characters. Assembly of all productive genes in transgressive progenies requires the following genetic situations:

(i) The character must be polygenic;

(ii) Parents should be completely homozygous;

(iii) Parents should be complementary to each other;

(iv) There should not be any linkage between the chosen characters.

5.2 GENERAL TECHNIQUE

Before the beginning of any hybridization programme the breeder should decide the objective of the programme. The whole hybridization programme involves the following procedures (Fig. 5.1):

(a) Choice of parents
(b) Crossing schedule
(c) Emasculation
(d) Bagging
(e) Tagging
(f) Pollination
(g) Harvesting and storage of F_1 seed.

Choice of Parents

Parents which are chosen for hybridization provide the requisite variability for isolating desirable segregants in subsequent generations. The following criteria are essential for choosing parents for hybridization.

(i) **Agronomic base**: Any well established local cultivar used as seed parent or female parent should have agronomic base for that area.

(ii) **Complementary pollen parent**: The male parent should be complementary to the female parent, i.e., it must possess the complementary attributes in their intense form.

(iii) **Homozygosity of parent**: The parent chosen for hybridization must be homozygous for the character, i.e., pureline. Though there may be cryptic variability.

(iv) **Multiple parentage**: When the desired attributes are not present in the above two parents, more than two parents can be chosen for crossing.

(v) **Combining ability**: Experimental evidences suggest that cross-combinations involving high and low combiners (H × L) or both high combiners (H × H) often provide better opportunity for superior recombinants.

(vi) **Genetic divergence**: Choice of parents based on genetic divergence will be helpful to get the divergent parents in segregating generations.

In conclusion, the breeder should have a well defined, clear cut objective, based on the present and expected future needs, in developing a new variety. Breeder should select the parents accordingly and use them in a suitable hybridisation programme.

Crossing Schedule

Crossing schedule entails development of hybrids by emasculation and pollination which helps in increasing the spectrum of variation among the hybrids, this heterogeneity is narrowed down by developing homozygous pure line.

Type of crosses: The parents involved in hybridization should belong to the same species; they may be of two strains, varieties or races of the same species. When only two parents are crossed to produce the F_1 then it is called **simple cross**; the F_1 hybrid may be selfed or back crossed with homozygous parent.

More than two parents are also crossed to produce the F_1 hybrid, which is then used to produce F_2 or is used in a back cross. Such a cross is also known as **convergent cross** because this crossing programme aims at converging, i.e., bringing together genes from several parents to a single hybrid. This is the **complex cross** where multiple convergence is attempted.

Three parents (A, B, C)

A × B

↓

F_1 (A × B) × C

↓

(A × B) × C Complex hybrid

Four parents (A, B, C, D)

A × B C × D

↓ ↓

F_1 (A × B) × (C × D)

↓

(A × B) × (C × D) Complex hybrid

Number of crosses: How many (single, double or multiple) crosses should be developed would depend upon the objectives defined for the breeding programme, floral biology of the crop and the quality of parents chosen. Based on the recent concept of genetics, the correct choice of parents can cut short the number of crosses to develop the wanted hybrid.

Emasculation

In case of hybridization programme for self pollinated crops, it is very much essential to emasculate the plant to avoid the self pollination. Removal of male sex organs or killing that part of the flower without any damage or disturbances to female reproductive organ is known as **emasculation**. Before going to hybridization programme the efficiency of emasculation technique may be tested by bagging the emasculated flowers without pollination. The amount of seed thus set would indicate the frequency of self fertilisation occurring during emasculation.

There are various **techniques** of emasculation:

(i) **Hand emasculation**: A general procedure for hand emasculation is as follows:

(a) Emasculation is done before the anthers mature and the stigma has become receptive to minimise self pollination.

(b) The corolla of selected flowers is opened with the help of fine tip forceps and the anthers are removed.

(c) In case of epipetalous stamen the corolla lobes are removed carefully keeping the gynoecium not injured at all.

(ii) **Suction method**: This method is useful in case of small flowers where hand emasculation is not possible. The petals are generally removed with forceps exposing the anthers and stigma. A thin rubber tube or glass tube attached to a suction hose is used to suck the pollen grains from its surface. The suction may be produced by an aspirator or by a small suction pump. Care must be taken that suction should be enough to suck the stamen and pollen grains but not the gynoecium. But this method is not very efficient, as 15% of self pollination takes place. Washing the stigma with water may also reduce self pollination.

(iii) **Hot water emasculation**: Pollen grains are more sensitive to hot water than the female reproductive organs. So treatment with hot water at particular temperature and for fixed time period is helpful for killing the pollen grains without damaging the female organ. Treatment with water at 42-48°C for 10 minutes is effective in jowar, treatment at 40-44°C for 10 minutes is effective in rice; in both the cases the whole spike is immersed in thermoflask containing hot water.

(iv) **Alcohol treatment**: It is not very popular method; a particular concentration of alcohol is used for a fixed time period to kill the pollen grains. But a little bit more exposure, i.e., few seconds more than the recommended time period will reduce the female receptivity, i.e., seed set, as female organs would also be killed by this treatment.

(v) **Cold treatment**: Like hot water treatment, cold treatment can also kill pollen grains without damaging the gynoecium. Keeping rice plant at 0-6°C kills the pollen grains, and also wheat plants at 0-2°C for 15-24 hrs. kills the pollen grains. But cold treatment is less effective than hot treatment.

(vi) **Genetic emasculation**: Genetic or cytoplasmic male sterility may be used to eliminate the necessity of emasculation, i.e., male sterile plants are naturally emasculated. For self incompatible species emasculation is not necessary, but in certain genotypes male sterility can be induced by gene manipulation in cytoplasmic or nuclear genome.

Bagging

Immediately after emasculation, the flower or the inflorescence is enclosed in suitable bag to prevent random cross pollination. The bags may be made up of paper, butter paper or parchment paper, and tied at the base of the inflorescence with thread, pin or wire. As the moisture and temperature become higher within the bag so fungus may develop, which may be prevented by removing the bag after 2-3 days after pollination.

Tagging

The emasculated flower or inflorescence is tagged after bagging. The tags are made up of light weight tin-plate and are written in carbon pencil. The tag should bear the information:

(a) Date of emasculation;

(b) Date of pollination;

(c) Names of female and male parents, consecutively in the cross.

Pollination

Mature, fertile and viable pollen from donor plant (Male parent) should be placed on the receptive stigma to bring about fertilisation. During pollination pollen viability is a major factor and also is the time duration for receptivity of stigma. Both timing should match during pollination. The detail procedure for pollination vary in case of different crops.

Harvesting and Storing F_1 Seeds

The seeds are harvested from crossed heads or pods. The seeds should be dried and properly stored to protect them from storage pests. Proper care should be taken to avoid contamination of the hybrid seed with other seeds. The seeds from each cross should be kept separately and preferably, the seeds should be kept alongwith the original tags.

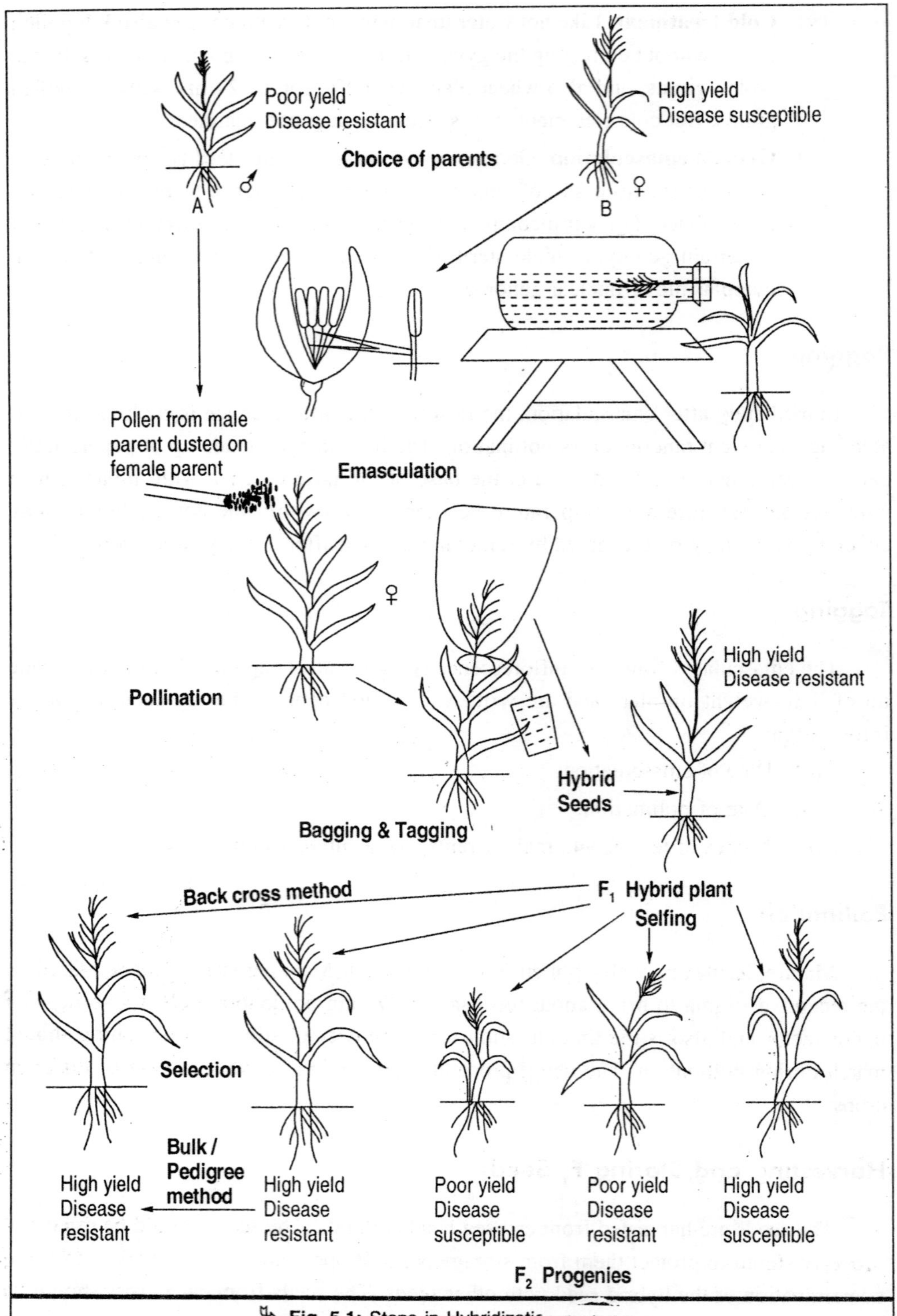

Fig. 5.1: Steps in Hybridization technique

Raising F_1 Generation

In a hybridization programme it is expected that the F_1 generation will be in heterozygous condition. The size of the F_1 generation is a crucial matter, large F_1 population is difficult to handle, but the large F_1 population has the greater opportunity of rare recombination to occur. Usually F_1 is allowed to self pollinate to get the F_2 but in back cross programme it is crossed to one of the parents.

Selfing

The technique of selfing varies from one crop to the other depending upon their mode of reproduction. In self pollinated crops selfing is not very troublesome as it is the natural mode of reproduction. But in cross pollinated crops the flowers are bagged to prevent cross pollination. If the inflorescence bears bisexual flowers then bagging is very much helpful for self pollination also.

Consequences of Hybridization

Segregation and recombination of genes would produce several new genotypes in F_2, in addition to parental types. The number of recombinant genotypes increases rapidly with the increase in number of segregating genes. The Fig. 5.2 indicates the reduction in heterozygocity by 50% in each successive generation on selfing.

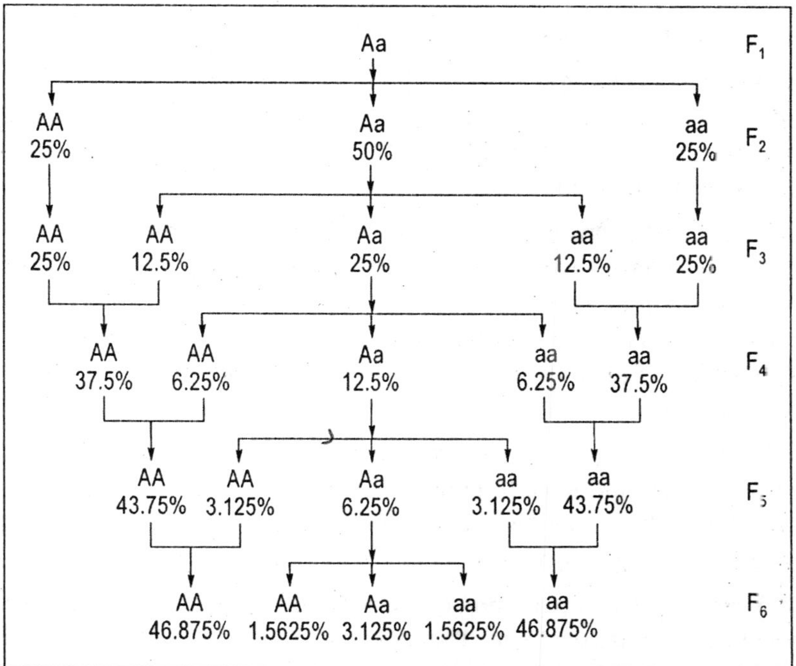

Fig. 5.2: Genetic consequences of selfing of F_1 hybrid

5.3 METHODS IN SELF POLLINATED CROP

For selection of a new variety from a segregating population after hybridization the methods followed in self pollinated crop are of three types:

(A) Pedigree method, (B) Bulk method, (C) Back cross method.

A. Pedigree Method

Individual plant progeny is selected from F_2 and subsequent generations, and their progenies are tested. During this process the record of parents as well as offsprings is kept, for which it is known as **pedigree method**. The pedigree is defined as the description of the ancestors of an individual and it is generally helpful in finding out the amount of relatedness among two individuals, i.e., whether they are related by common parent in their descent ancestry or not.

Procedure

First Year: The **hybridization** is done among two selected parents, after emasculation one become female parent and another male parent. After seed set and maturation, the F_1 seeds are harvested separately from each plant individually. On the basis of choice of parents, the type of cross will be of two types — it will be simple cross or complex cross.

Second Year: **F_1 generation** seeds are space planted and selfing is allowed, each F_1 will produce more F_2 seeds. From 15-30 selected F_1 plants, the F_2 seeds are collected to get a reasonable size of F_2 population and variation.

Third Year: In **F_2 generation,** 2000-10000 plants are space planted, 100-500 plants are selected and their seeds are harvested separately. If the parent plants are closely related varieties then the number of selected F_3 plants would be smaller whereas in case of distantly related varieties the number of F_3 progenies will be of relatively larger numbers.

Fourth Year: In **F_3 generation** also, the individual plant progenies are space planted. Each progeny should have about 30 or more plants. Individual plants with desirable characteristics are selected, disease and lodging susceptible progenies to be eliminated, and also the progenies with undesirable characters are rejected even from the selected plants. During this selection if the number of superior progenies are very small then the whole cross programme may be rejected.

Fifth Year: The selection procedure is same as previous year, only if two or more progenies coming from the same F_3 progeny are similar and comparable, then only one may be saved and others may be rejected. The emphasis is given on the selection of desirable plants from superior progenies.

Sixth Year: Individual plant progenies of **F_5 generation** are planted according to recommended commercial seed rate. Three or more rows for each progeny will help in comparison among progenies. Many progenies may have become reasonably homozygous genotype and may be harvested in bulk. If the progenies show variation then the individual plants are selected. The number of selected progenies should be reasonable so that preliminary yield trial with 25-100 progenies can be done.

Seventh Year: Individual plant progenies of **F_6 generation** are planted in multi-row plots and evaluated visually. Progenies harvested in bulk since they become homozygous. The segregating progenies may be discarded and the preliminary yield trial may be done for the progenies which are reasonably homozygous and have enough seeds.

Eighth Year: **Preliminary yield trial** with three or more replications are conducted to identify few superior lines. The progenies are evaluated for plant height, lodging, disease resistance, flowering time, maturity time, etc. Quality test is done to serve as an additional basis for selection.

Ninth to Tenth or Thirteenth Year: The superior lines are tested in **replicated yield trials** at several locations. The above mentioned criteria are evaluated for these lines. The line which is superior than the best commercial variety may be released as new variety.

Eleventh or Fourteenth Year: The selected strain should get **multiplied** to release as a new variety. Breeder has the responsibility to supply the seeds to the state seeds corporation for production and marketing of the seeds.

Schematic representation:

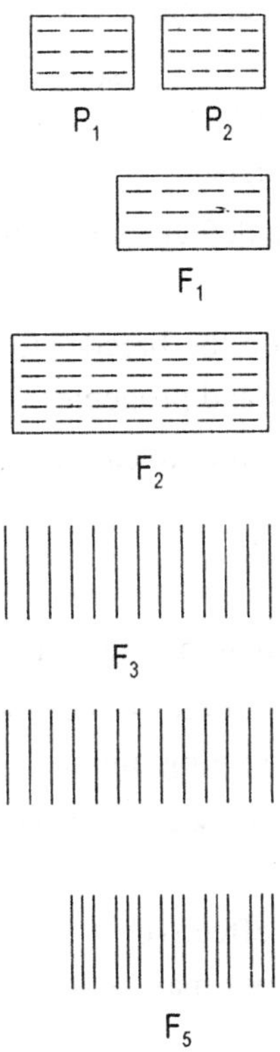

First Year: Selected plants of parental variety are hybridized to get the F_1 hybrid seeds.

Second Year: The hybrid seeds (10-30) are planted and the seeds of F_2 harvested in bulk.

Third Year: 2000-1000 seeds are space planted. 100-500 superior plants are selected and seeds are harvested from those as pedigree.

Fourth Year: Individual plant progenies are planted in row and space planted.

Superior plants are selected.

Fifth Year: The same procedure as in fourth year.

Sixth Year: The individual selected plant progenies are in multi-row plots.

Superior plants are selected from those progenies.

Seventh Year

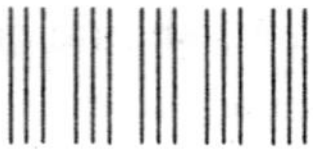

F_6

The same procedure as in previous year.

Enough seeds to be collected from superior plants.

Eighth Year

F_7

The preliminary yield trial and quality test has to be done.

Ninth to Thirteenth Year

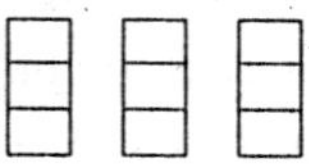

$F_8 - F_{12}$

Co-ordinated yield trials to be done.

Disease and quality tests have to be done.

Fourteenth Year

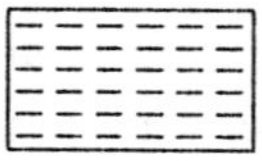

F_{13}

Multiplication of seeds for distribution and release as a new variety.

Merits

1. This method is most useful as transgressive segregation for yield and other quantitative characters may be recovered in addition to improvement of specific characters.
2. This method is well suited for improvement of characters which can be easily identified and simply inherited.
3. Through the maintenance of pedigree record the breeder may be able to obtain the information about inheritance of characters.
4. Plants or progenies with weaker and visible defects are eliminated at an early stage in the breeding programme.
5. This method gives maximum importance on the breeder to use his/her skill and judgement about the selection of plants and progenies.
6. This method takes less time than bulk method to release a new variety.

Demerits

1. The success of the method is mainly dependent on the skill of the breeder.
2. To keep the individual pedigree record is laborious and time consuming, it may be the limiting factor for large breeding programme.
3. Selection of large number of progenies in every generation is also laborious and time consuming.
4. In F_2 and F_3, the selection for yield is not effective. If sufficient number of progenies are not retained, valuable genotypes may be lost in early segregating generations.

Achievements

Pedigree method is useful in selection of new superior recombinant types from a hybridization programme. This method is suitable for improving specific characteristics, such as disease resistance, plant height, maturity time, etc. as well as yield and quality characters.

Many improved varieties have been developed through pedigree method in many crops like wheat, rice, barley, pulses, oil seeds, cotton, tobacco, jowar, vegetables, etc.

Wheat

K 65 (tall variety) ← C 591 × NP 773

K 68 (good quality grain) ← NP 773 × K 13

WL 711 (dwarf, high yield) ← (S308 × Chris) × Kalyan Sona

Malviya 12 (good grain) ← NP 876 × Cno 66

Rice

'Jaya' and 'Padma' (short duration, finer grain) ← Taichung Native 1 × T 141

Cotton

'Laxmi' (fibre quality, early maturing, resistant to leaf blight) ← Gadag 1 × CC 2 (Combodia Coimbatore 2)

Tomato

Pusa early dwarf (more yield) ← Meeruti × Red cloud

B. Bulk Method

After hybridization programme when the F_2 and subsequent generations are harvested in mass or as bulk to raise the next generation, it is called **bulk method.** At the end of bulking period, individual plants are selected and evaluated in the same manner of pedigree method.

This method is applicable to any self pollinating crop like cereals, legumes and oil seeds. Three different **advantages** may be achieved from this method: (1) isolation of homozygous lines, (2) waiting for selection by environmental disaster, (3) the long period of bulking may be helpful for natural selection to change the composition of population.

Procedure

First Year: **Hybridization** is done between the selected parents with the desirable attributes. The cross may be simple or complex depending upon the number of selected parents.

Second Year: The **F_1 generation** seeds are space planted and seeds from these plants are harvested in bulk.

Third Year to Seventh Year: Starting from **F_2 generation** onwards the seeds are planted at commercial seed rate and spacing. The seeds of subsequent generations are harvested in bulk. Many environmental factors, disease outbreak, etc., may select out the particular genotype from this bulk population. The population size in each generation should be large always, 30,000 - 50,000 plants in each generation.

Eighth Year: **F_7 generation** seeds are space planted and superior phenotypes are selected and harvested separately. The selection is made on disease reaction, grain character, etc.

Ninth Year: Individual plant progenies of **F_8 generation** are grown in single or multi-row plots. Weak or inferior progenies are rejected, only 100-300 plant progenies are selected which are superior as well as showing no segregation.

Tenth Year: Preliminary yield trial is conducted with **F_9 generation** with standard commercial variety as check. The undesirable progenies are rejected.

Eleventh To Fifteenth Year: **Replicated yield trial** at several locations is carried out using standard commercial variety. The lines are evaluated for some important characteristics in addition to yield, disease resistance, etc.; superior lines are released as new variety.

Sixteenth Year: The seeds are **multiplied** and **distributed** to the farmer.

Schematic representation:

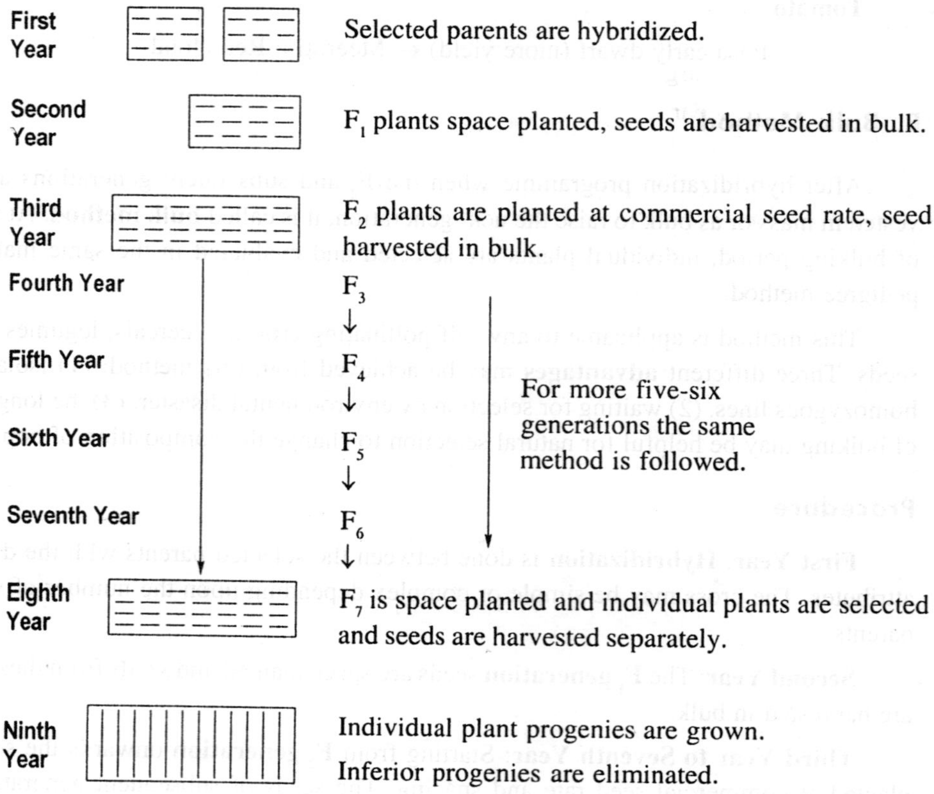

Tenth Year

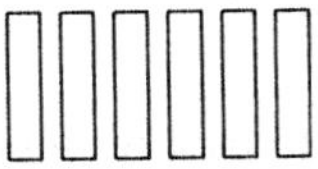

Preliminary yield trials using standard varieties as check.

Eleventh to Fifteenth Year

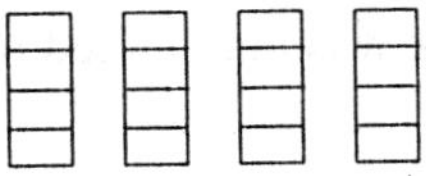

Multilocation yield trials for the selected progenies.

Sixteenth Year

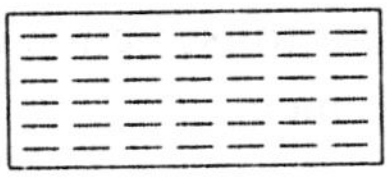

Multiplication of seed for distribution.

Merits

1. This method is simple, convenient and less expensive.
2. The natural disease epiphytotics, winter killing, etc., eliminate the undesirable types and increase the frequency of desirable types, which is more helpful for isolation of desirable types.
3. As natural selection acts here, so the progenies selected from long term bulk are far superior than those selected from F_2 of short term bulk.
4. Less attention and labour is needed for this method.
5. Since large populations are grown, the transgressive segregants are more likely to appear and there is greater chance of isolation of transgressive segregants in this method.
6. Artificial selection may be practised to increase the frequency of desirable types.
7. Survivability of any gene or a particular genotype can be studied in this method.

Demerits

1. This method takes a long time to release a new variety, as the method of natural selection becomes operative only after F_8 or F_{10}.
2. The short term bulk method is useful for isolation of homozygous lines but the effect of natural selection has little effect on the genetic composition of the population.
3. Breeder has little opportunity to show his skill in this method, practically selection is mainly based on natural selection.
4. Inheritance of characters is not individually maintained as in pedigree method.
5. At the end of bulking period a large number of progenies are being selected.

Achievements

This method has been used to a limited extent due to lack of popularity. In U.S.A the barley varieties like Arivat, Beecher, Glacier and Gem have been originated from a bulk population derived from cross Atlas × Vaughn, the bulk was maintained in this case for 7 generations.

As this method requires more time to get the effect of natural selection for developing a new variety and the lack of using any skill of breeder for selection of superior types, this method is less used by the breeder for crop improvement.

Comparison between Bulk and Pedigree Method

Pedigree Method	Bulk Method
1. The desirable plants are selected in F_2 and individual plant progenies are grown in subsequent generations.	1. No selection is made in F_2 and the subsequent generations are maintained as bulk.
2. Artificial selection, artificial disease epidemics, etc., are the integral part of this method.	2. Natural selection is operative here which determines the composition of population at the end of bulking period.
3. Pedigree record of each progeny is maintained in such a way that even the finally selected progeny can be traced back to the particular F_2 plant.	3. No pedigree records are maintained.
4. It requires close attention of a breeder from F_2 generation onward as in each generation selection and pedigree records are maintained.	4. It is simple, convenient and overall requires less attention of the breeder during the period of bulking.
5. More laborious, time consuming and expensive.	5. Less laborious and less expensive.
6. The undesirable segregants are rejected in each generation specially in F_2.	6. As there is no selection in F_2 onwards, so there is no rejection of segregants also.
7. The size of the population is usually smaller than in the case of bulk method.	7. Generally large populations are grown. Natural selection is expected to play the role to select the transgressive segregants.
8. This method is more scientific, popular and widely used.	8. This method is not so popular and less used by the breeder.

C. Back-cross Method

Back-cross is the event of crossing of F_1 with either of the parents, but here in the back-cross method of breeding the hybrid is to be crossed with the superior parent whose genotype is to be transferred to the local variety. For example, in a particular area there is a well adapted high yielding local variety but susceptible to a particular disease, this variety will be considered as **recipient parent**. The disease resistant variety will be the **donor parent**. The recipient parent will be used in this method of breeding repeatedly to get all the genes to be transferred which is also known as **recurrent parent** and the donor parent is known as **non-recurrent parent**.

Genetic consequences of repeated back-crossing

Repeated back-crossing results in decreasing the frequency of heterozygosity by 50% in each subsequent generation and rapid increase in homozygote, upto 98% by sixth back cross progeny, will be achieved. The genotype of the back-cross progeny becomes increasingly similar to that of the recurrent parent.

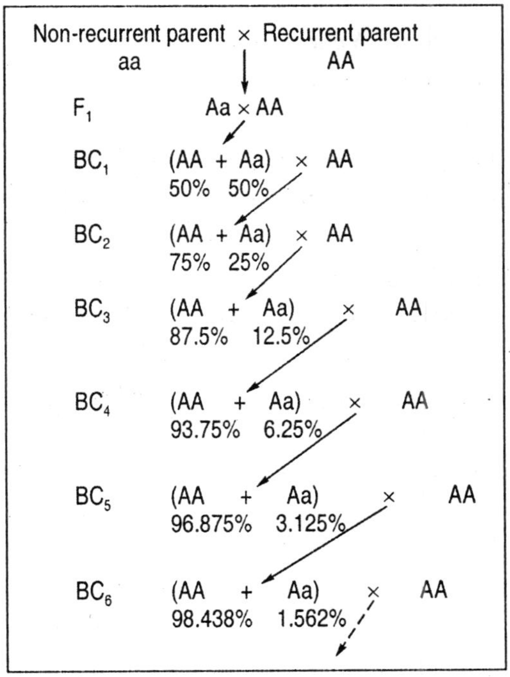

Procedure

The plan of back-cross method differs on the gene which is to be transferred, whether it is dominant or recessive.

(a) Transfer of dominant gene

Selection of parents : Variety A — Recurrent parent (well adapted, high yielding); Variety B — Non-recurrent parent, dominant gene controlled (stem rust resistant).

Hybridization (First Year): The recurrent parent (variety A) should be used as female parent and the non-recurrent parent (variety B) as male parent.

F_1 generation (Second Year): F_1 plants are back-crossed to variety A, in this generation all the plants are heterozygous for rust resistant character.

BC_1-generation (Third Year): 50% of the BC_1 plants which are resistant are selected and back-crossed to the recurrent parent, i.e., variety A.

BC_2-BC_5 generation (Fourth-Seventh Year): In each back-cross generation, segregation would occur for rust resistance. Rust resistant plants are selected and back-crossed to recurrent parent, variety A.

BC_6-generation (Eighth Year): The plants of this generation will have 98.4% genes from variety A. Rust resistant plants are selected and selfed. Their seeds are harvested separately.

BC_6-F_2 generation (Ninth Year): Individual plant progenies of selfed plants are grown separately. The rust resistant plants are selected and the seeds are harvested separately.

BC_6-F_3 generation (Tenth Year): Individual plant progenies are grown. The plants homozygous for rust resistance and similar to the plant type variety A are harvested in bulk.

Yield test (Eleventh Year): The new variety is tested in replicated yield trial along with the variety A as a check, as the new variety will be same in performance with variety A only with rust resistant character.

Twelfth Year: Seed multiplication is done and distributed to the farmers.

Schematic representation:

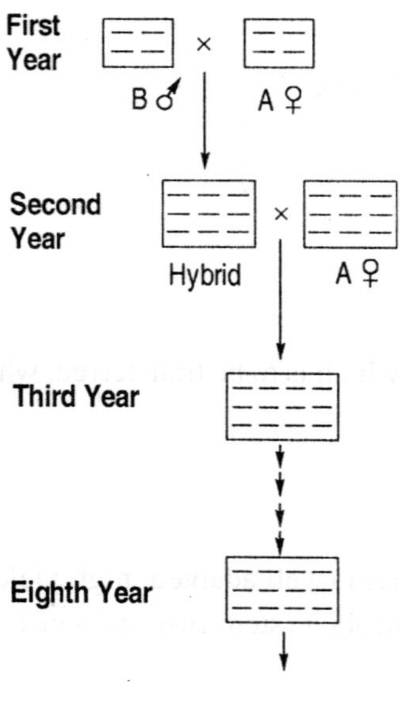

Hybridisation between variety A and B.

A is the recurrent parent, B is non-recurrent.

The hybrid plant is back-crossed with recurrent parent A.

The plants are selected which are rust resistance as well as similar to recurrent parent A.

BC_2 – BC_5 — Same procedure is followed for every year.

BC_6 — Rust resistant plants are self pollinated. The seeds are harvested separately.

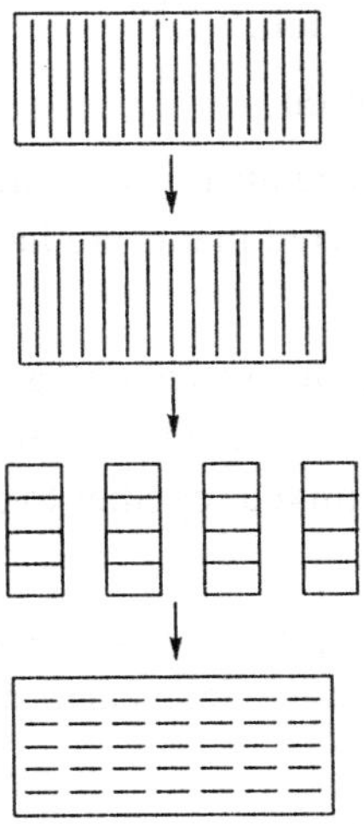

Ninth Year: Individual plant progenies are grown.

The rust resistant plants similar to plant type A are selected.

Tenth Year: Individual plant progenies are grown.

Homozygous progenies similar to recurrent parent A are harvested and bulked.

Eleventh Year: Replicated yield trial with recurrent parent as one of the check.

Twelfth Year: Seed multiplication and distribution.

(b) Transfer of recessive gene

Procedure for transferring a **recessive gene** will be different from that of dominant gene, as the recessive gene will be expressed only in homozygous condition. So the selection for that recessive gene requires the F_2 generation, i.e., selfing is needed after every two back-crosses, and testing for the presence of that character has to be done in that generation.

Selection of parents: The parents are selected in this programme as the previous one, only difference is that the rust resistance is a recessive gene controlled character.

Hybridization (First Year): The recurrent parent is used as female parent and hybridized with the rust resistant donor plant.

F_1 generation (Second Year): The F_1 plants won't be rust resistant, these will be back-crossed to the recurrent parent.

BC_1 generation (Third Year): Here also no rust resistant plants will appear, so no need to do any rust susceptibility test. All the plants are selfed.

BC_1-F_2 generation (Fourth Year): The plants are tested for rust resistance character. These plants are selected and back-crossed with recurrent parent.

BC_2 generation (Fifth Year): There is no rust resistance test, the plants are selected only on resemblances with recurrent parent A, and again back-crossed with recurrent parent.

BC_3 generation (Sixth Year): Again the plants are self-pollinated, but without any selection for resistance character.

BC_3-F_2 generation (Seventh Year): The selection for rust resistance is done in this generation. The rust resistant plants having resemblance with variety A are selected and back-crossed.

BC_4 generation (Eighth Year): Same as BC_2 generation.

BC_5 generation (Ninth Year): Same as BC_3 generation.

BC_5-F_2 generation (Tenth Year): Plants are selected against rust resistance and also the characters of variety A. The selfed seeds from the selected plants are harvested separately.

BC_5-F_3 generation (Eleventh Year): Individual plant progenies are grown and subjected to rust resistance test. Rigid selection is done for resistance to stem rust and the characteristic of variety A. Seeds from selected plants are mixed together to constitute the new variety.

Yield test (Twelfth Year): The variety is tested in replicated yield trial along with the variety A as check. The new variety should be same in yield performance except bearing the new character introduced.

Thirteenth Year: Seed multiplication and distribution to the farmers are carried out.

Schematic representation:

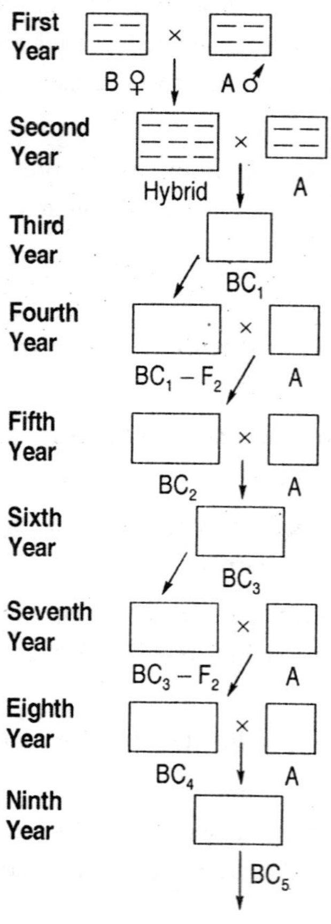

Hybridization between variety A and B.
A is recurrent parent, B is with recessive rust resistance gene.

F_1 is back-crossed with recurrent parent.
No rust resistance test.

No rust resistance test.
The plants are all self pollinated.

The rust resistant plants are selected and back-crossed with recurrent parent.

No rust resistance test.
The plants are back-crossed with recurrent parent.

As in BC_1.

As in $BC_1 - F_2$.

As in BC_2.

No rust resistance test.
The plants are self pollinated to raise $BC_5 - F_2$ generation.

Tenth Year

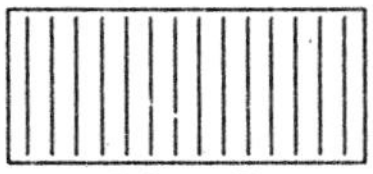

$BC_5 - F_2$

The plants are subjected to rust resistance test and selection is made on variety A.
The selfed seeds are harvested separately from the selected plants.

Eleventh Year

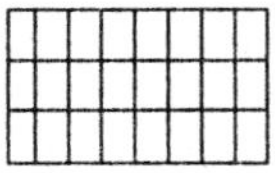

$BC_5 - F_3$

Individual plant progenies grown.
Progenies similar to variety A selected and composited.

Twelfth Year

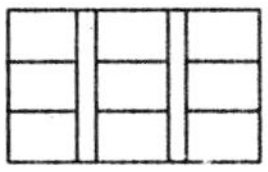

Replicated yield trial alongwith variety A as check.

Thirteenth Year

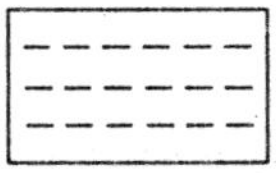

Seed multiplication and distribution to farmers.

Comparison between the transfer of dominant and recessive gene during back-cross method

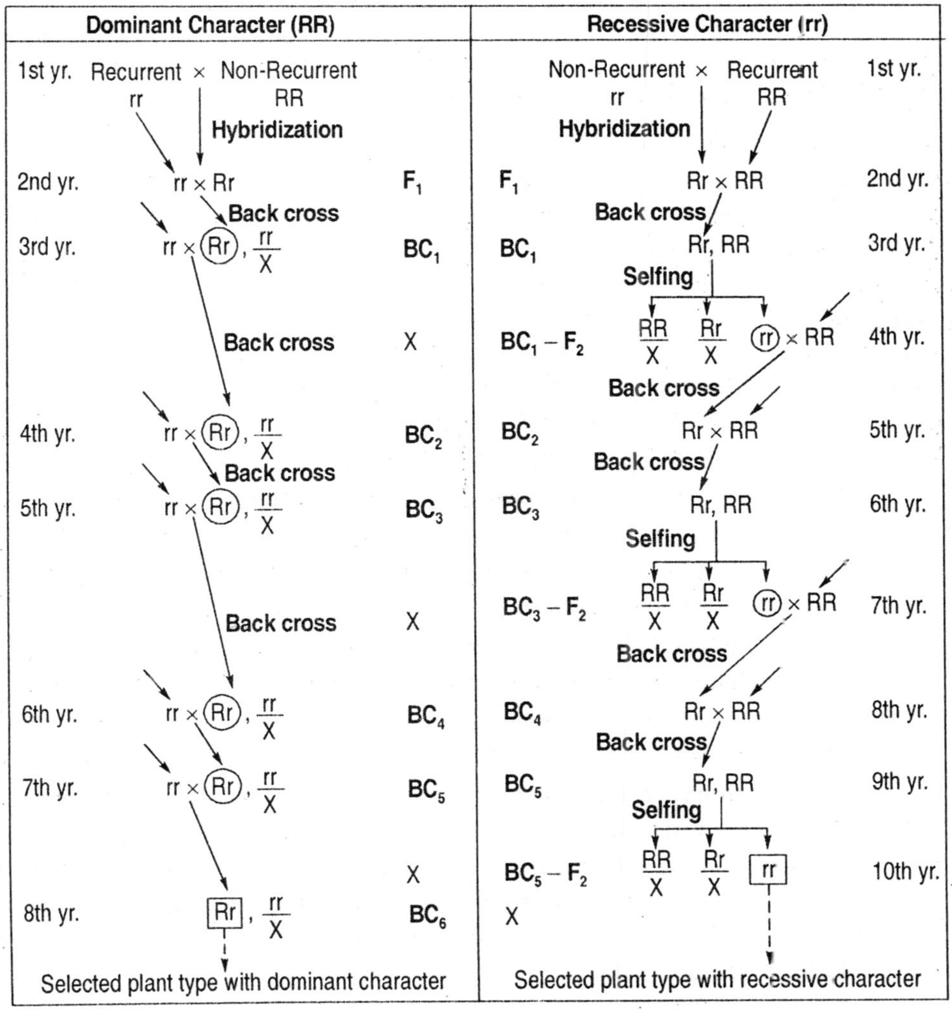

(c) Transfer of quantitative character

Quantitative characters are governed by polygenes such as grain size, plant height, maturity time, etc. As during transfer intensity of characters will be lost, so the non-recurrent parent must have the characters more intense from than it is desired in the new variety. For transferring this type of character, each back-cross generation should be selfed and F_2 generation is grown which should have large population. Then the rigid selection is done for the character to be transferred. The system of alternate back-crossing and selfing goes on till the desired results are obtained. At the end of back-cross programme, F_2 is handled like pedigree method. The progenies which are more likely to recurrent parent bearing the desirable trait are selected and mixed to make up the new variety.

(d) Transfer of two or more characters in a single parent

For simultaneous transfer of many characters in a single back-crossing programme, the characters from different non-recurrent parents are brought together by multiple crossing and then back-crossing to recurrent parent is performed. Larger back-cross progenies would be needed than in case of transfer of a single character, and also it may take long time as in each step all the desired characters may not get expressed.

The multiple characters can be transferred in a stepwise method. The recurrent parent is first improved for one character then it is used as recurrent parent for introduction of another new character. Thus in stepwise manner more than one characters can be introduced into a single recurrent parent.

Applications

Generally the back-cross method is used for transferring disease resistance character to a good and well adapted variety. Other quantitative characters can also be transferred through back-cross method in both cross- and self-pollinated crop.

1. **Intervarietal transfer of simply inherited characters**: The characters like disease resistance, seed colour, plant height, etc., which are controlled by one or two major genes are most suited for transfer through back-cross method from one variety to another of the same species. The successful transfer depends on the minimum linkage between desirable and undesirable trait.
2. **Intervarietal transfer of quantitative characters**: Grain characters like seed size and shape, earliness, plant height–all can be transferred from one variety to another with the criteria of high heritability.
3. **Interspecific transfer of simply inherited characters**: Mainly the character like disease resistance can be transferred from related species to cultivated species. This specially requires the relatedness between the species where the chromosomes can pair during meiosis. The transfer will be unsuccessful if the genetic environment of recurrent parent is not suitable for functioning of the gene of desirable character from donor parent.

4. **Transfer of cytoplasm**: In case of transfer of male sterile character from one parent to another requires back-cross method. The variety or species from which the cytoplasm is to be transferred is used as female parent. The recurrent parent should be the male parent. After 6-8 back-crosses, the cytoplasm will be of the donor parent with the genotype of recurrent parent.
5. **Transgressive segregation**: Modification of back-cross method will produce transgressive segregants. Few (1 to 3) back-crosses with F_1 allow much heterozygosity to appear, or two or more recurrent parents may be used in back-cross programme to accumulate genes from them into the back-cross progeny. Such kind of modification would produce new variety which won't be like the recurrent parent.
6. **Production of isogenic lines**: Back-cross method is useful for production of isogenic lines, i.e., the lines of a crop which are identical in their genotype except for one gene. These isogenic lines are useful for studying the effect of individual genes.

Merits

1. This method does not change the genotype of the popular established variety, only it helps a single desirable character to be transferred in the existing variety.
2. As the recurrent parent is an established variety, so it is not necessary to test the yield performance which ultimately saves five years time, as well as expenses.
3. In case of short duration plant, 2-3 generations can be raised within a year as the selection is based on inheritance of a particular character, not on performance. So this method drastically reduces the time period required to develop a new variety.
4. Back-cross method requires smaller populations than the pedigree method.
5. Defects of an established variety can be removed using this method, may be only by introducing a single character.
6. Interspecific gene transfer only can be done through this method.
7. Transgressive segregation can be obtained in case of quantitative characters by modified back-cross method.
8. This method is very much useful for cytoplasmic gene transfer to the recurrent parent, i.e., new variety will bear the cytoplasm of donor parent and genotype of recurrent parent.

Demerits

1. The new variety is not superior in performance than the existing variety except the introduction of single character.
2. During transfer of such kind of single desirable gene, sometimes some undesirable genes may get transferred.

3. For introduction of more than one gene controlled character, multiple crossing programme is necessary which is often difficult, time taking and costly.
4. Once a recurrent parent taken in a back-cross method may get replaced by another superior variety of high yielding ability.

Achievements

Back-cross method is very much useful for transferring of simply heritable character like disease resistance to well adapted popular local variety.

In case of **wheat,** 'Kalyan Sona' is the popular established variety to which the leaf rust resistance character has been transferred from diverse sources like Robin, KI, Bluebird, Tobari, etc. using back-cross method.

In case of **Bajra,** Tift 23A, a male sterile line which was susceptible to downy mildew has made resistant through back-cross method.

For interspecific transfer of genes, back-cross method is widely used for crop improvement. Cultivated **sugarcane** (*Saccharum officinarum*) is susceptible to pests and disease, crossed with *S. spontanium* which is resistant. This brings resistance but with undesirable characters like more fibre, low sugar, thin stem, etc. By back-crossing with noble cane these undesired characters are removed.

In **cotton,** the hybridization between *Gossypium hirsutum* and *G. arboreum* yielded highly sterile F_1 hybrid, few tetraploid seeds were obtained. These plants were then back-crossed with *G. hirsutum* and two varieties have been selected from the back-cross progeny which are now being widely cultivated in Gujrat.

Except the characters of disease resistance other characters can also be transferred by the back-cross method. Such as in **cotton,** the ginning quality of fibre has been improved by back-cross method from the local variety. By using BD8 (wilt resistant, high spinning value but low ginning out turns) as recurrent parent and Goghari A26 (high ginning out turn) as non-recurrent parent, new variety **Vijay** was developed.

Comparison between Pedigree method and Back-cross method

Pedigree Method	Back-cross Method
1. The aim of this method to produce the new variety which is totally different from the parents bearing superior characters.	1. In this method the new variety is totally similar to recurrent parent except it bears a new character transferred from donor parent.
2. The method is used generally to improve the yield and other characters of the variety.	2. This method is used to improve the specific defect of a well adapted popular variety.
3. This method is useful to improve both qualitative and quantitative character.	3. This method is suitable for introducing a specific highly heritable character.

Contd.

Pedigree Method	Back-cross Method
4. Gene transfer from related species is not suitable through this method.	4. It is the most suitable method for gene transfer from related species to produce addition or substitution lines.
5. Hybridization is limited only to produce the F_1 generation.	5. Hybridization with the recurrent parent is done in every back-cross generation.
6. F_1 and subsequent generations are allowed to self pollinate.	6. F_1 and subsequent generations are generally allowed to back-cross with recurrent parent.
7. F_2 and subsequent generations are much larger than those of the back-cross method.	7. The progeny of back-cross generations are smaller than pedigree method.
8. The method of breeding does not get modified whether the character is dominant or recessive gene controlled.	8. This method is different for transferring dominant and recessive gene controlled character.
9. The new variety is subjected to extensive testing and yield trial before releasing as new variety.	9. As the back-cross method always aims at improving some particular character of recurrent parent, so extensive yield test is not required before releasing as new variety.

5.4 METHODS FOR CROSS POLLINATED CROPS

The hybridization method, in case of cross pollinated crops, aims at inbreeding at low level to avoid its undesirable effect and exploit heterosis which is the basis for breeding hybrid and synthetic varieties. Hybrid varieties are the first generations (F_1) from crosses between two purelines, inbreds, open pollinated varieties, clones or other populations that are genetically dissimilar.

The **operations** involved in production of hybrid varieties are as follows.

(a) *Development of inbred lines:* The inbred lines are developed from a genetically variable population by continued inbreeding. Self-pollination is desirable, if possible, which leads to homozygosity very rapidly. The inbred lines can also be developed from haploid plants originated through parthenogenetic development.

(b) *Evaluation of inbreds:* The inbred lines are evaluated on the basis of phenotypic performance; effective yielding ability is tested in replicated yield trial.

(c) *Production of hybrid seed:* Production of hybrid seeds requires two major criteria in the breeding programme: (1) easy emasculation method of female parent, (2) effective pollen dispersal from the male parent to ensure satisfactory seed set in the female parent.

There are **various methods** (programme type) for production of hybrid seeds:

1. **Production of single cross hybrids**: Crossing between two inbreds or varieties is called single cross. Such crosses express maximum amount of heterosis, but if the inbreds are poor then very little amount of seeds are produced. So, it is not widely used for commercial production.

Total number of single crosses by a given number of inbreds (n) can be calculated by the formula n(n–1) and excluding reciprocals by n(n–1)/2. For example, with 4 inbred lines, 12 single crosses and 6 straight crosses can be made.

For commercial production of single crosses, a male sterile line is used as female, the male parent is the restorer line, the seed set in the female parent is the hybrid seed, but the seeds of male parent are selfed seed.

2. **Production of double cross hybrids**: Crossing between two single cross hybrids is called double cross, which also involves four different inbred lines. For example, if A, B, C, D represent 4 inbred lines, one of the possible double crosses will be the crossing between two single crosses, A × B and C × D.

The number of possible double cross hybrids can be calculated by the formula

$\frac{n(n-1)(n-2)(n-3)}{8}$, where n is the number of inbred lines. So with 4 inbred

lines, $\frac{4\times3\times2\times1}{8}=3$ double cross hybrids can be produced.

Two different schemes can be taken for production of double cross hybrids:

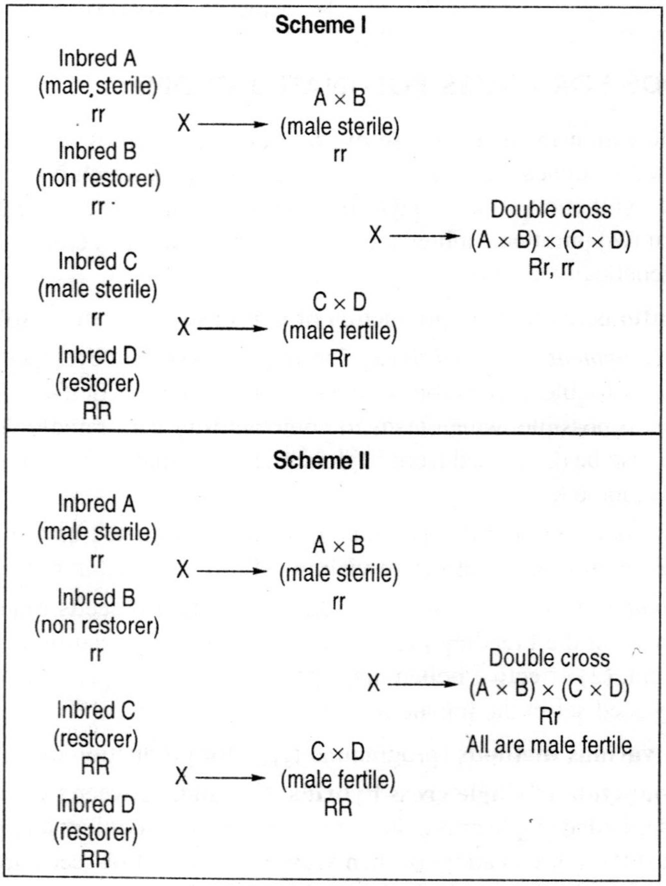

3. **Production of top-cross seed**: This scheme is also known as **inbred-variety cross**, (Inbred × Variety), i.e., inbred line is crossed with an open pollinated variety. Also the cross between a single cross and with another inbred is known as **double top cross** or **3 way cross**. This method is used not only for developing a hybrid but to test the combining ability of the inbred. The variety, or the single cross product is used as male parent and the inbred line to be tested is used as female parent (emasculated), the seeds from the inbred line is harvested.

4. **Production of synthetic cross**: A synthetic variety is produced by crossing in all combinations a number of lines that combine well with each other, this synthetic variety is maintained by open pollination in isolation.

   It is produced by open pollination between several inbred lines together in isolated place. Seeds of these inbreds are mixed in equal proportion and sown in isolated plots, natural cross pollination takes place and the seeds are harvested together.

   The term **poly cross** is used in case of forage crops where crossing is difficult and also it differs from synthetic cross that the inbred lines are not pretested.

   Another term **composite variety** is used as synonym for synthetic variety but by definition it is the production of mixture of several phenotypically similar outstanding lines by open pollination.

QUESTIONS

1. Define hybridization. Point out its types. Write down the purpose of hybridization.
2. Describe the general technique and the steps involved in hybridization programme.
3. What is emasculation? Write briefly the different methods of emasculation. What do you mean by natural emasculation?
4. Define pedigree method of breeding. Write the procedure in brief.
5. Define bulk method of breeding. Write the procedure in brief.
6. Give a comparative account of pedigree method and bulk method of breeding. Point out the merits and demerits of both the methods.
7. What is back cross method of breeding? Mention the genetical consequences of the method. Write down the procedure and point out its merits and demerits.
8. Discuss the hybridization in self-pollinated crops.
9. Discuss the hybridization in cross-pollinated crops.

6 HETEROSIS AND HYBRID SEED PRODUCTION

The progress of crop improvement through plant breeding depends on better understanding and proper exploitation of **heterosis,** *the gain in vigour on crossing of two inbreds,* i.e., the superiority of F_1 hybrid over both the parents in terms of yield and other characters. Heterosis may manifest in either of the two directions — positive or negative.

(a) *Positive heterosis* – increase in grain yield, tiller number, protein content;

(b) *Negative heterosis* – early maturity, dwarf height.

The term **'hybrid vigour'** is being used as synonym of **'heterosis',** though the hybrid vigour describes only superiority of the hybrids over the parents, i.e., positive heterosis, but the negative heterosis is also desirable in some cases of agricultural practices.

6.1 HYBRID VIGOUR VS. INBREEDING DEPRESSION

Inbreeding depression, i.e., decrease in vigour due to inbreeding and hybrid vigour, i.e., increase in vigour due to crossing between two inbred lines are the manifestations of the same phenomenon. Inbreeding increases the degree of homozygosity in the progeny reducing the degree of heterozygosity, whereas hybridization develops heterozygosity.

Comparison in effect between Inbreeding Depression and Hybrid Vigour

1. **Increase in homozygosity vs. development of heterozygosity**: Due to inbreeding each line becomes increasingly homozygous, as a consequence the variation within a line decreases rapidly. After 7-8 generations of selfing, the lines become almost uniform (99% homozygosity) which are called **inbred lines**.

   Hybridization always favours heterozygosity; the species which reproduce by cross-fertilisation are highly heterozygous. Due to heterozygosity the effects of many recessive allele are not expressed in heterosis, only the dominant effects or the multiple effects are expressed.

2. **Appearance vs. no/less expression of some lethal and sublethal alleles**: Inbreeding may result in appearance of many harmful characters due to accumulation of harmful recessive alleles after selfing, e.g., chlorophyll deficiency (albina, chlorina), rootless seedlings, defective floral parts, etc.

   This type of effects are not found in case of heterosis as most of the lethal characters are expressed in homozygous condition. Heterosis or hybrid vigour prevails heterozygosity, so appearance of such characters does not happen.

3. **Reduction vs. increase in vigour, yield and reproductive ability**: Due to inbreeding there is a general reduction in vigour of the population, plants become shorter and weaker. The hybrids are generally more vigorous, healthier and increased in size.

   The reproductive ability also decreases in the population rapidly due to inbreeding, many lines reproduce so poorly that these cannot be maintained. The hybrids exhibiting heterosis show an increase in fertility or reproductive ability.

   Inbreeding generally leads to loss in yield, the inbred lines yield much less than the open pollinated varieties from which they are derived. Heterosis is generally expressed as an increase in yield of the hybrid. Commercially this phenomenon is of great importance as an objective of plant breeding.

4. **More susceptible to disease vs. increase in disease resistance property**: Due to inbreeding as homozygosity increases, there may be rapid loss of vigour as well as disease resistance property. Whereas due to heterosis the hybrids are known to exhibit a greater resistance to insects or diseases than the parents.

5. **Less adaptable to changed environment vs. greater adaptability in hybrids**: The inbred lines are homozygous, so they are less adaptable to changed environment as the modification of characters are not possible according to need. Whereas the hybrids are generally more adaptable to environmental changes than the inbreds. Variation due to heterozygosity offers the hybrids more adapted to environmental variations.

6.2 MANIFESTATION OF HETEROSIS

Performance or expression of any character or trait is influenced by many genetic factors — some are positive (stimulating) and others are negative (decreasing). Expressivity of the genes or the degree of manifestation of a character is the result of genetic balance in the action of differently directed factors.

The various manifestations of heterosis may be summarised as follows:

1. **Increased yield**: Increase in yield which may be measured in terms of grain, fruit, seed, leaf, tuber or the whole plant is one of the most important manifestations of heterosis.

2. **Increase in size and general vigour**: Heterosis results in more vigorous growth which ultimately leads to healthier and faster growing plants with increase in size than the parents.
3. **Better quality**: In many cases heterosis yields better quality which may be accompanied with higher yield.
4. **Greater adaptability**: Hybrids are generally more adapted to environmental changes than the inbred lines due to heterozygosity.
5. **More disease resistant**: Heterosis sometimes results into development of more disease resistant character in the hybrids.
6. **Increased reproductive ability**: Hybrids exhibit heterosis by expressing high fertility rate or reproductive ability, which is ultimately expressed in yield character.
7. **Increase in growth rate**: In many cases the hybrids show faster growth rate than the parents, but that does not always produce larger plant size than the parents.
8. **Early flowering and maturity**: In many cases the hybrids may show earlyness in flowering and maturity than the parents, for some crops these are the desirable characters for crop improvement.

All these manifestations of heterosis can be traced at all levels of hybrid plant organisation.

Molecular level: Heterosis is manifested in increased rate of DNA reduplication, transcription and translation influencing the formation of genetic information, enzymatic activity, other regulatory mechanisms and also hybrid protein molecule formation.

Functional level: Heterosis is expressed as an effective regulation in metabolic processes and morphogenesis in hybrid organism.

Cellular level: Due to change in electrokinetic properties of hybrid cell nuclei, the heterosis is manifested by increased mitosis.

Organism level: Heterosis is expressed as increased growth and differentiation of vegetative organs, synthesis and accumulation of nutritional substances and utilisation of metabolic process for yield formation.

6.3 GENETIC BASIS OF HETEROSIS

There are two main theories to explain the genetic cause of heterosis.

(A) Dominance Hypothesis

This hypothesis was proposed by Davenport and further expanded by others. This hypothesis suggests that at each locus dominant allele has the favourable character, whereas the recessive allele has the unfavourable character. When they are combined together; i.e., in heterozygous condition in the hybrids, the favourable characters get expressed whereas the unfavourable characters are masked. So the heterosis results from the masking of harmful effects of recessive alleles by their dominant alleles.

Dominance hypothesis has assumptions:

(a) Dominant genes are beneficial and recessive genes are deleterious.

(b) The loci show addition effects, non allelic interactions are absent.

(c) No recombination barrier between the genes.

With the help of following example heterosis can be explained:

In a cross between Inbred A (AAbbCCdd) with Inbred B (AAbbCCdd), there will be no heterosis in F_1 hybrid, there is no masking of recessive gene in hybrid. But in another cross, Inbred A (AAbbCCdd) is crossed with Inbred D (aaBBccDD), where the F_1 hybrid is (AaBbCcDd) with all the genes having dominant allele. As a result the harmful effects of a, b, c, d are hidden by the dominant alleles A, B, C and D. Thus some parents produce heterotic progeny while others do not. Generally parents of diverse or different origin are more likely to produce heterotic progeny than those of similar origin.

Inbred A x Inbred B	Inbred A x Inbred D
AAbbCCdd ↓ AAbbCCdd	AAbbCCdd ↓ aaBBccDD
Hybrid	Hybrid
AAbbCCdd	AaBbCcDd
No heterosis	Heterosis

Objection

1. **Failure in isolation of inbreds as vigorous as hybrids**: According to dominance hypothesis it should be possible to get the inbred line with all the dominant genes. Such inbreds should be as vigorous as the F_1 hybrids, but such inbreds have not been isolated.

2. **Symmetrical distribution in F_2**: According to dominance hypothesis, the quantitative characters should not show symmetrical distribution as because dominant and recessive alleles should segregate in the proportion of 3 : 1, but generally the F_2 shows symmetrical distribution.

Above two objections can be explained by linked genes. Many of the quantitative characters are governed by linked genes together, so to get the inbred line with all dominant genes require several precisely placed crossovers. In another explanation it can be showed that if the number of genes governing the quantitative characters are large, symmetrical distribution would be obtained even without linkage.

(B) Overdominance Hypothesis

This hypothesis was independently proposed by East and Shull. This is sometimes known as single gene heterosis, superdominance, cumulative action of divergent alleles and stimulation of divergent alleles. According to this hypothesis, heterozygotes are superior to both the homozygotes. So the heterozygote Aa would be superior to both the homozygotes AA and aa. Consequently, heterozygosity is essential for the cause of heterosis. In case of maize, the gene *ma* affects maturity. The heterozygote *Ma/ma* is more vigorous with late maturity than the homozygotes *Ma/Ma* or *ma/ma*.

Another proposal by East was that there are several alleles, e.g., a_1, a_2, a_3, a_4 ...., etc. with increasingly different functions. Heterozygotes between more divergent alleles would be more heterotic than those involving less divergent genes, e.g., a_1a_4 is more heterotic than a_1a_2, a_2a_3, a_3a_4, etc. In these cases due to presence of divergent alleles the hybrids have the capacity to perform different functions which is not possible by any of the heterozygotes.

Objection

1. There are many examples where the superiority is due to the epistatic affect of several nonallelic genes, not due to overdominance (which is the interaction between allelic genes).
2. There is another objection against overdominance hypothesis that there are many examples where the homozygotes are superior to the heterozygotes.

Comparison between Dominance and Overdominance Hypothesis

Both the hypotheses have the similar expectation like the degree of heterosis would depend upon the genotypes of the two parents. The greater the genetic diversity of the parents, higher the heterosis.

The two hypothesis differ in the view that according to overdominance hypothesis the heterozygotes are superior than homozygotes, but according to dominance hypothesis the heterozygotes are as good as dominant homozygotes.

There are number of genes which show the superiority in heterozygotes. It is generally accepted that heterosis, to a large extent, is due to dominant gene action, but epistasis and overdominance are also involved.

6.4 PHYSIOLOGICAL BASIS OF HETEROSIS

Hybrid vigour, the product of heterotic mechanism, is essentially a physiological manifestation. This better physiological efficiency of hybrids is derived chiefly from:

1. Better initial growth.
2. Greater uptake followed by better utilisation of nutrients by hybrids.

The initial growth activities include the different physiological processes during germination:

(a) efficient water absorption,

(b) better activity of enzymes,

(c) rapid mobilization and utilization of stored food matter,

(d) transformation and building up of active protoplasmic synthesis.

To explain all these processes different hypotheses have been put forwarded:

1. **Initial capital and physiological stimulus**: Large embryo and seed size in hybrids provide initial advantage to the hybrid during germination and early growth of seedlings. This hypothesis is debatable due to two reasons : the greater seedling vigour always not associated with maturity and also hybrid seeds are need not to be always with large size to attain hybrid vigour.
2. **Balanced metabolism and heterosis at molecular level**: The hybrids are endowed with a more balanced metabolism than their inbred parents. Many of the enzymes of heterotic plants exhibit greater efficiency over those of their better parents. The hybrids show better and rapid unfolding of balanced metabolic processes.
3. **Mitochondrial complementation and heterosis**: ATP-ase activity of the mixture of mitochondria from different inbred lines of maize sometimes exceed that of the mitochondria of individual lines. This heterotic effect is called as mitochondrial complementation. The mitochondria of heterotic hybrids absorb more O_2 and have high P/O index, i.e., phosphorylation/oxidation ratio than those of inbred lines and non-heterotic hybrids. This suggests that high level of oxidising phosphorylation and synthesis of high energy ATP bonds create favourable conditions for biosynthetic processes and important requisite for heterotic development.
4. **Greater ability for uptake and utilisation of nutrients**: Heterosis in post germination seedling growth is associated with better absorption and assimilation of several specific substances essential to the fundamental growth processes of the organism; such as nutritional factors, water absorption and other factors. Efficient uptake and assimilation of nutrients by heterotic hybrid seedlings confer the following advantages —
    1. Larger number of leaf primordia.
    2. High carboxylase and photophosphorylation activity.
    3. Greater leaf area and larger number of leaves.
    4. More branches per panicle and more grains per branch.
    5. High grain weight, etc.

6.5 APPLICATIONS OF HETEROSIS IN PLANT BREEDING

Heterosis is observed in almost every crop species studied, the application of this phenomenon for its commercial exploitation depends on the expression of the degree of heterosis. This phenomenon is commercially used to produce hybrid or synthetic varieties, which needs the maximisation of its expression and also fixation.

Maximisation of hybrid vigour (HF_1) can be achieved by increasing either 'd' (directional dominance) or 'y' (initial differences in gene frequency between parents), i.e., choosing genetically divergent parents, $HF_1 = \Sigma dy^2$.

Fixation of hybrid vigour is needed for its commercial application which can be done by vegetative propagation, or by stable apomictic reproduction, or by transferring heterozygosity to polyploid or fixation by obtaining structural heterozygotes.

Fixation of heterosis in crops like potato, sweet potato, sugarcane, sugarbeet and many ornamental plants can be achieved by vegetative propagation as there seeds are not essential. Incorporation of genes conditioning vegetative apospory or diploid parthenogenesis in hybrid seed crops may lead to a permanent heterozygote advantage. Heterozygosity can be maintained or saved from being lost due to segregation by converting diploid heterozygotes into tetraploid or hexaploid. Incorporation of *gaudens-velans* combination helps to survive only the heterozygotes not the homozygous combination — such mechanism may be introduced in crop plants, thereby hybrid vigour can be fixed with great success.

Heterosis or hybrid vigour have been commercially utilised in both cross pollinated and in some self pollinated species. In most of the cases the utilisation of this heterosis phenomenon is not successful because of difficulty in production of large quantities of hybrid seeds. This is particularly difficult in self pollinated species.

Few examples where the heterosis has been utilised for improvement of the crop plant are :

Crop species : Asexually propagated species and also cross pollinated species like maize, jowar, bajra, sunflower, legume, cotton, etc.

Vegetable crops : Tomato, brinjal, onions, cucurbits, etc.

Fruits : In almost all the fruit trees.

6.6 HYBRID SEED PRODUCTION

The process of exploitation of hybrid vigour is the hybrid seed production. The production of hybrid is possible in both the groups of plants, i.e., self-pollinated and cross pollinated crops which involves two parents, male and female. For this purpose a suitable male sterile line is employed as seed parent where the artificial pollination can be successfully substituted by natural pollination.

The development of hybrid mainly consists of following steps:

(a) choice and development of seed parent,

(b) choice and development of pollen parent,

(c) production and improvement of F_1 hybrid seeds.

A. Choice and Development of Seed Parent (A-line)

The seed parent or the female parent of a commercial hybrid should be a male sterile line. But in case of non-availability of suitable male sterile line, the fertile line or a self incompatible line can be used where manual labour will be needed.

In case of food grain crops where the number of seeds produced in each pollination is restricted, there the cytoplasmic-genetic male sterile line should be used as seed parent. But in case of vegetable crops where seed is not the economic product and also only a single pollination will produce many seeds, there fertile pure line can be used as seed parent. In case of monoclinous plants like maize, the male flowers can be removed easily as these are borne on apices of the plants.

After identification of a male sterile line (A-line), it should be maintained by an isogenic B-line. Strains can be identified to have the B-line reaction by crossing them to the A-line. Then a potential B-line is converted to A-line by a process of repeated back-crossing till the A and B lines become similar (isogenic).

Transfer of male sterility is required when disease susceptibility or unwanted agronomic traits are associated with male sterile character. This can be done in two ways:

1. **Single phase repeated back-crossing**: This method follows the repeated back-crossing of existing male sterile line (non-recurrent parent) possessing the desirable characteristics. In case of transferring cytoplasmic male sterility, it is convenient through six generations of back-crossing. But in case of developing cytoplasmic-genetic male sterility, it should possess required agronomic characters and maintenance of male sterility is needed through B-line or maintainer line.
2. **Two phase limited back-crossing**: In this method instead of using A-line as non-recurrent parent, the B-line is back-crossed with a parent of desirable donors and the new line formed is termed as I_{BC}. Then these developed lines are again crossed with A-line to seek out the male sterile line. These lines are then back-crossed repeatedly with I_{BC} lines which will help to develop the male sterile line with desirable attributes.

The desirable attributes which should be considered for a male sterile line or A-line development are:

1. **Plant height**: Male sterile line should not be too tall or too dwarf. Since plant height of the hybrid is the function of both the parents, so plant height of A-line should be lower than that of R-line (male fertile parent).
2. **Duration and span of flowering**: This is an important attribute for commercial hybrid seed production. Successful hybrid seed production depends on the synchrony between the flowering of seed parent (A-line) and male fertile parent (R-line).
3. **Tillering ability**: Profuse and synchronous tillering on a male sterile line is advantageous both for harvesting male sterile seeds and also hybrid seeds.
4. **Productivity potential**: As the main objective is more hybrid seed production, thus male sterile lines should possess a high productivity potential.

5. **Stable male sterility**: An ideal male sterile line should maintain its sterility under all environmental conditions.
6. **Free from diseases**: The male sterile line and also the R-line should be resistant to probable diseases particularly seed borne diseases.
7. **Combining ability**: The male sterile line under use must have a high general combining ability, so that it can combine well with R-lines to produce a large number of hybrids.

B. Choice and Development of Restorer or Male Parent (R-line)

The second component which is required for hybrid production is the male parent or restorer line which are essentially inbred lines, like A-lines. Their development is done through pure line selection method where forced selfing is practised. The procedure is as follows:

(a) From a heterozygous gene pool, the S_0 is selected on the basis of desirable attributes such as vigour, productivity, disease resistance, etc.

(b) After self-pollination, the S_1 plants are obtained and growing them rowwise, after repeated selfing S_2 seeds are obtained.

(c) This process is continued upto S_5 or S_6 generation.

A number of inbred lines are developed for a successful hybrid programme as to improvise the genetic potential of the source material which can impart hybrid vigour. The inbreds are evaluated for their agronomic traits and on the basis of general combining ability. The unwanted inbred lines are discarded in the beginning of inbreeding process and also the prospective R lines should be chosen for the development of F_1 hybrids on the basis of combining ability in S_5 or S_6 generation.

The choice of parents for restorer line specifically depends on the presence of few attributes like —

(a) Ability to produce abundant pollen grains (profuse pollen production),

(b) Maximum genetic diversity from the A-line choosen,

(c) High specific combining ability of the cross between A-line and the R-line.

C. Maintenance and Multiplication of Parental Seeds

In case of cross-pollinated crops, the hybrid programme is based on 3 lines : A-line (ms), B-line (Maintainer) and R-line (Restorer) — all homozygous inbreds are maintained by the breeder. Any kind of change in the population should be discarded by the breeder which may occur due to random pollination and mutation.

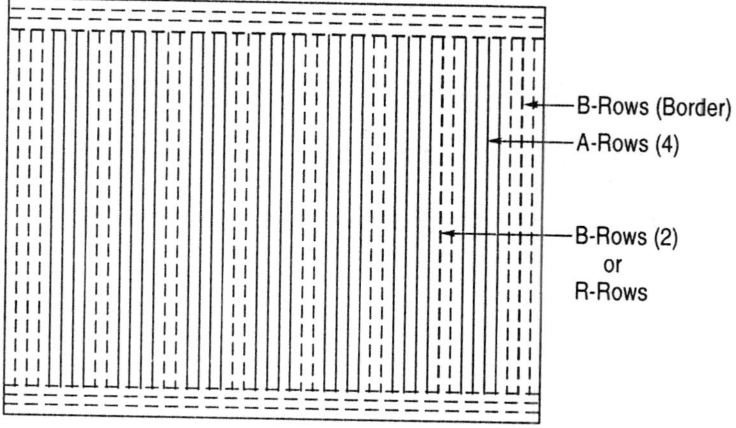

Fig. 7.1: Maintenance block of A, B and R lines and hybrid seed production (in latter case B-lines are replaced by R-rows)

For proper maintenance, the A-lines and B-lines are grown in field in ideal agronomic condition and in rows in 4 : 2 ratio. In the border rows also, the B-lines are grown which supply sufficient pollen for pollinating A-line. On maturity the seeds of B-line are harvested first and kept separately. Then the seeds of A-line are harvested. Thus in the same field both A-lines and B-lines are maintained and multiplied in the same block.

For maintenance of R-line, the plot should be completely isolated from others. The natural open pollination characteristic of cross-pollinated crop will be helpful for its maintenance. There is no need to maintain the R-line separately except for the first use, as R-line is automatically maintained in the hybrid production block.

D. Production and Improvement of F_1 hybrids

For F_1 hybrid seed production, A-line and R-line are grown together in 4 : 2 ratio in hybrid production block (Fig. 7.1). The arrangement is same except the B-lines have been substituted by R-lines. At maturity the seeds of A-rows are harvested separately and carefully as these hybrid and also the R-line seeds are harvested for next use.

For improvement of F_1 hybrids, the parental lines should be improved. If the incidence of diseases (for which the inbred lines are not resistant) occurs, breakdown of male sterility in A-line and genetic weakness (like combining ability, nutritional quality) appear, then these drawbacks can be removed by changing the parental lines.

In many crops like pearl millet, sorghum and maize, hybrid seed production programme has been pursued and many of the hybrid varieties have been released.

QUESTIONS

1. What do you mean by hybrid vigour and inbreeding depression? Discuss their manifestations.
2. Define heterosis. Write down the causes of heterosis. How is this phenomenon utilised in plant breeding?
3. Define hybrid seed. Discuss the method of hybrid seed production.
4. Discuss the dominance and overdominance hypothesis to explain hybrid vigour.

7 BREEDING METHODS: SUPPLEMENTARY PROCEDURE

In general breeding for crop improvement follows the conventional method of introduction, selection or hybridization when the desired type of variability exists in nature. But to introduce some new criteria or to get more effect of existing criteria, the method of **mutation breeding** and **polyploid breeding** are used. Application of **Biotechnology**, modern non-conventional breeding method, is the need of present century.

7.1 MUTATION BREEDING

Mutation is a sudden heritable change in the genotype of an organism. The process by which the change does occur is called **mutation** and the individual in which it is observed is called **mutant**. Mutation occurs in natural population but at a very low rate, which is called **spontaneous mutation**, but the mutation can be induced by chemical or physical **mutagen** is called **induced mutation**. Utilisation of characters which are the result of induced mutation in crop improvement programme is known as **mutation breeding.**

Objectives

As in mutation process the ratio of beneficial and harmful character is very small, so it is very difficult to get the desirable trait within a small population Any mutation breeding programme should have well defined and clear-cut objectives on which the *material type* and also the *type of mutagen* to be used will depend. Furthermore it should be kept in mind that whether the character is governed by *oligogene* or *polygene* – the handling of treated material will be different.

Selection of Material

The mutation should be attempted in the variety which is the best one in that species, as getting a desirable mutant in a poor variety is not useful for breeding programme.

Though sometimes it may happen that for selection of particular character poor variety is selected, such as, for searching a 'dwarf' mutant some tall varieties should be considered which may not be best one.

Many plant parts, such as, seeds, pollen grains, vegetative propagules (buds or cuttings) may be used for mutagenesis. Plant parts to be used for mutagen treatment depend on the mode of reproduction of plant and the type of mutagen to be used for mutagenesis.

Selection of Mutagen

The choice of mutagen depends on the material type used, its effectiveness and the proper dose should be considered during treatment. The mutagen may be **chemical,** such as, base analogues (5-Bromo-uracil), antibiotics (azasorine), alkylating agents (EMS, MMS, DES) and also miscellaneous compounds (dyes, hydroxylamine), etc. or may be **physical,** such as, ionising (X-ray, β-ray, γ-ray) or non-ionising (UV) radiations.

An optimum dose is the one which produces the *maximum frequency of mutation* and causes the *minimum killing*. Overdose will cause death or detrimental to most whereas the underdose will produce few mutations.

Procedure of Mutation Breeding

A. In self pollinated crops

First treated generation (M_1): The common effect of mutations are death of seedling, growth inhibition, morphological or developmental abnormalities, and the heritable changes in qualitative or quantitative characters. Among these only the heritable changes either in cytoplasm or chromosomes or genes which are observed in M_1 generation are selected as **mutants** and these are promoted to M_2 generation.

M_2 generation: In this generation a large population is essential and maximum number of mutants are recovered due to recombination and segregation. The process of recovery would depend on the efficiency of screening procedure in M_2 and also in the later generations. Mutagen treatment of seeds and vegetative propagules produce **chimeras,** where some parts of plant get changed or mutated but other parts remain unchanged. Chimera may be periclinal or sectorial, the inner chimera only will be transmitted to next generation but not the outer one.

When seeds are produced in mutated M_1 plants, the mutated tissue has to pass through two sieves of selection pressure. Mutated tissue should compete with normal tissue during formation of vegetative stage and reproductive organs. This selection is called **diplontic selection**. If the M_1 flower is chimeric, mutant or normal pollen will be formed and the stigma will receive two types of pollen and both will compete for fertilising the ovule — this is called as **haplontic selection**.

After competing in both the selection pressure, i.e., diplontic and haplontic selection, the mutant type would appear in M_2 generation.

M_3 and later generation: Selected plant progenies of M_2 generation are grown and evaluated critically in M_3 generation.

Progenies of selected M_3 plants are grown as M_4 generation for further purification, evaluation and multiplication.

Most mutants get purified by M_5 generation and those are then evaluated in field trials along with parental variety as standard check variety.

B. In cross-pollinated crops

In this type of crops genetic variability occurs more naturally, so there is minimum need for mutation breeding. Also as these are open pollinated, the detection of mutants is more difficult.

C. Asexually propagated crops

Mutation breeding is more useful in this type of crops as other conventional breeding methods cannot be applied due to lack of their sexual cycle. Somatic mutant can easily be created in this type of plants, propagated and directly used.

Since such plants are more heterozygous, any mutation from dominant to recessive may be detected and used. Any phenotypic effect of chromosomal arrangements may also be utilised due to vegetative mode of reproduction.

After mutagen treatment, formation of chimera is regular occurrence. This can be avoided by irradiating the youngest possible stage of bud, i.e., less differentiated primordia. Various propagation methods like budding, grafting and prunning are used for selection of mutated parts.

Merits

1. ***De novo* (New) creation of genetic variability**: When the desired character does not exist, mutations are used to get the new character.
2. **Breaking undesirable linkage**: When two undesirable characters are linked together, mutagen treatment may be used to break the linkage.
3. **Production of haploids**: Haploids may be produced using X-rayed pollen for pollination; unfertilised egg develops into haploid which are variously used.
4. **Increase or decrease of chiasma frequency**: Mutation may change the chiasma frequency which is directly related with recombination and segregation of characters.

5. **Production of transitory sexuality in apomicts**: Apomictic plants breed like asexually propagated plants and breeding is difficult. Transitory sexuality can be induced so the better apomictic type can be selected after crossing with various sexual types.
6. **Reduction of incompatibility in wide crosses**: Though actual mechanism is not known but irradiation induces better pollen tube growth in few genera which enables interspecific crosses with the irradiated pollen.
7. **Variation in F_1 hybrids**: F_1 hybrids from intervarietal crosses may be treated with mutagens in order to increase genetic variability by inducing mutation and to facilitate recombination of linked genes.
8. **Production of distant hybrid with translocation**: Irradiation of interspecific (distant) hybrid causes chromosomal segment translocation carrying some desirable genes which may help in transfer of character from one species to another.

Demerits

1. **Unpredictability**: The frequency of desirable mutants is very low, 0.1% of total mutations. So it is required to grow a large population of M_2 generation and it is laborious to screen out a few economically wanted mutated progenies.
2. **Difficulty in detection**: Mutation breeding is difficult if the detection procedure is elaborate. The disease resistance or quality characters cannot be detected easily which needs elaborate tests. Thus mutation breeding is not easily applicable to improve this type of character in crop. Moreover most of the mutations are recessive in nature, it is very difficult to detect it in polyploid species.
3. **Undesirable expression**: Desirable mutations are sometimes associated with undesirable side effects or chromosomal aberrations. To remove these kinds of deleterious effects, back-crossing is done which requires more time and expenses.
4. **Unrepeatability**: Since the process of mutation is not well understood, so there is no control on the outcome or result of mutagen treatment. Unless directed mutagenesis is applied, it is difficult to obtain the same or desired mutants every time.

Achievements

Point mutation as well as chromosomal mutation both have been used to produce many improved varieties which represent the improvement in oligogenic as well as polygenic characteristics.

'Sonora 64', a good variety of **wheat** had red colour grain, γ-radiation produced 'Sharbati Sonora' with white colour grain. Chromosome segments of *Aegilops, Secale, Agropyron* carrying rust resistance genes have been translocated to wheat by using X-rays.

High yielding **groundnut** variety NC_4 X modified into NC_4 by micromutation which looks like the parent.

Point mutations have produced superior **sugarcane** varieties TS-1 and TS-2 which are the mutants of CO 419. In sugarcane, CO 8152 is a γ-ray induced mutant from CO-527 which gives 40% higher yield, but CO 8152 has two chromosomes less than CO-527.

Duplication in a segment of chromosome 6 of **barley** carrying gene 'orange lemma' associated with α-amylase activity improves the matting quality of barley.

'Jagannath' **rice** variety is a γ-ray induced semidwarf mutant from tall cultivar T-141, which has the improved resistance to lodging, high yield, responsive to fertiliser.

JRO-3690, a high yielding **jute** variety has been developed by hybridisation between two low yield mutants of jute.

7.2 POLYPLOIDY BREEDING

Irregularities occurring during mitotic or meiotic division may lead to changes in chromosome number which might have some role in improvement of crop plants.

The type of changes in chromosome number can be broadly subdivided into two categories:

1. **Aneuploidy**: Addition or loss of one or few chromosomes.
2. **Euploidy**: Addition or loss of total genome set.

Aneuploids are either hypoploid (**Monosomy** or **Nullisomy**) or hyperploid (**Trisomy** or **Tetrasomy**). Euploids may be haploid, diploid, triploid, tetraploid, etc. Polyploids are euploids above the level of diploid which are either **autopolyploids** or **allopolyploids**.

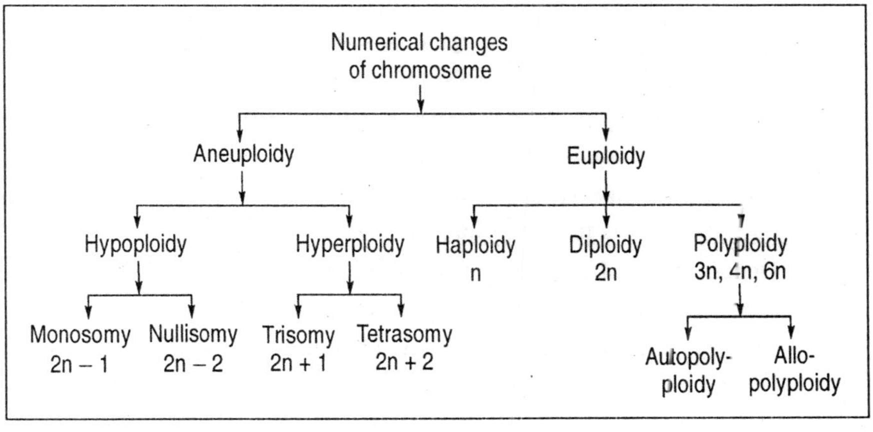

Effects of Polyploidy

Primary effect: The increase in cell size in growing tissue is the primary effect of polyploidy, though the magnitude of cell elongation and increase in cell number depend on the type of plant parts. It is reflected in the plant parts where there is determinate growth habit (sepals, petals, anthers, fruits, etc.).

Secondary effects: **Changes in growth rate**: Polyploidy results in slower growth rate associated with reduction in branching, etc.

Effects on fertility and genetic behaviour: Reduction in pollen viability and seed fertility occurs due to irregular chromosome distribution during meiosis.

Changes in cell composition: As polyploidy is associated with cell volume expansion, so it is associated with high water content and lower somatic tension.

Effects on segregation: Multivalent formation causes differential behaviour in meiosis which effects the segregation pattern of genes.

Mating barrier: Polyploidy creates a barrier on crossing between a polyploid and the diploid progenitor.

Induction of Polyploidy

Polyploidy is induced artificially with the help of agents like:

1. **Colchicine drug**: It is a non-toxic water soluble alkaloid obtained from *Colchicum autumnale* which is applied by different methods:
   (a) *Tube method* — Treatment of shoot apex and cotyledons of rooted seedlings through a tube.
   (b) *Dropper method* — A small drop of colchicine solution is applied on the shoot apex situated between cotyledons.
   (c) *Immersion method* — The seeds are allowed to germinate in immersed condition in colchicine solution.
   (d) *Wilk method* — The young seedlings are allowed to soak colchicine solution through roots.
2. **Physical agents**: Heat or cold treatment, X-ray or γ-ray irradiation may produce polyploids in low frequencies.
3. **Decapitation**: In some plants decapitation induces polyploid cell development at cut ends and the adventitious bud formation.
4. ***In vitro* culture**: Ploidy level may change during *in vitro* cultural condition. During protoplast culture it is a normal feature to get the auto-fused product. In anther culture also, the polyploidy can be induced.

Breeding Procedure

The procedure for polyploid breeding is mainly based on following aspects:

1. Artificial induction and collection of naturally occurring polyploids.
2. Detection of different kinds of polyploids.
3. Hybridization and selection of polyploids.

Haploid/Monoploid Breeding

Haploidy is very much important in plant breeding as it is the quickest procedure of attaining homozygosity, so it helps to get the expression of recessive genes in double haploids. Haploids are helpful to get the expression of induced mutations and can be used to determine the homology between genomes.

Haploidy can be induced by parthenogenesis and through anther or pollen culture technique. Through wide hybridization between species and through chromosome elimination technique haploidy can also be attained.

The haploidy breeding is concentrated on three items:

(a) High frequency of haploids produced.

(b) Easy detection of haploids.

(c) Diploidisation of desirable haploids.

Autopolyploid Breeding

The production of triploids (3n) and tetraploids (4n) are most important among autopolyploids. Triploids are obtained through crossing between tetraploid and diploid ($4n \times 2n \rightarrow 3n$). Owing to their self-sterility seed yield is low, but in case of seedless fruits (watermelon) or where the vegetative parts are commercially used, the triploids can be exploited successfully. The triploids or tetraploids can easily be detected by their greater vigour for which they are economically viable.

Amphidiploid or Allopolyploid Breeding

When the hybrids are developed by inter-crossing between two distantly related taxa then to restore the normal segregation and chromosome balance doubling is needed. Amphidiploid breeding offers a short-cut method of transfer of alien gene/chromosome into cultivars by overcoming the natural crossing barrier. One of the most important example is *Triticale* (*Triticum aestivum* × *Secale cereal*).

Aneuploid Breeding

Though nullisomics or trisomics have no direct value in plant breeding but they are extensively used to bring the alien variation into the cultivated varieties Tertiary trisomics have been used successfully in production of hybrid barley seeds.

Applications of Autopolyploidy in Crop Improvement

Monoploids: These are weaker than diploids and have no direct use in agriculture. But they are of great interest for crop improvement to develop diploid homozygous line, for isolation and detection of mutant line.

Triploids: Triploids are obtained after crossing between 4n and 2n plants, as they are highly sterile to produce seedless fruit like watermelon, thus have been used successfully. Another example is sugarbeet where triploidy has contributed more sugar content than diploids.

Tetraploids: Autotetraploids have been produced in large number of crop species, these are very much useful in breeding for improved quality, overcoming self incompatibility and helpful in making distant crosses.

Autotetraploid maize has 43% more carotenoid pigments and vitamin A than diploid, but the performance of tetraploids may not always be superior than diploids.

Sometimes the distant hybridization programme fails due to self incompatibility in case of diploids. But autotetraploids can overcome this problem as experimented in *Brassica oleracea,* or in *Solanum tuberosum.*

The most successful autotetraploids used in agriculture are tetraploid red clover, ryegrass, berseem, etc. There are also attempts for production of larger grains with high protein content and higher yield in case of Jowar and Barley.

Application of Allopolyploidy in Crop Improvement

In nature many crop species have been evolved due to successful production of allopolyploids. So the allopolyploids have immense importance in plant breeding as this can be used for three broad objectives:

(a) **Utilisation as bridging species**: Amphidiploids can serve as a bridge in transfer of characters from one species to a related species, may be from wild species to cultivated species. In most of the cases, the F_1 hybrid between two widely related species may become sterile which on diploidization produces the amphidiploid fertile plants.

For example, F_1 hybrid of *Nicotiana tabacum* and *N. sylvestris* is sterile, diploidization of which produces allohexaploid *N. digluta* which is fertile. This new bridging species is back-crossed with *N. tabacum* to produce 'Tabacum' like plants which is TMV resistant. The objective was to transfer TMV resistant character from *N. sylvestris* to *N. tabacum.*

(b) **Creation of new species**: *Triticale,* is the most successful synthetic allopolyploid which is produced by crossing wheat and rye. This has the combination of characters like yielding ability and grain qualities of wheat and the hardiness of rye. The successful commercial variety of *Triticale* has the yield as much as the best varieties of common wheat.

Raphanobrassica is another promising allopolyploid obtained from the crossing between *Brassica oleracea* and *Raphanus sativus*. There is possibility of improvement of *Raphanobrassica* by further hybridization and selection at the polyploid level.

'*Varalakshmi*', a hybrid variety of cotton which is the interspecific hybrid between *Gossypium hirsutum* (American cotton) and *G. barbadense* (Egyptian cotton), made by diploidisation of the F_1 hybrid.

(c) **Widening the genetic base of existing allopolyploids**: The genetic base of some natural allopolyploids may be narrow, and it may be useful to introduce variability by producing new allopolyploids.

Brassica napus is a natural allopolyploid which has very narrow range of genetic variability. The synthetic allopolyploid *B. napus* can be obtained by crossing *B. campestris* (n = 10, AA) and *B. oleracea* (n = 9, CC) to produce the amphidiploid, *B. napus* (n = 19, AACC).

Application of Aneuploidy in Crop Improvement

1. Aneuploids are useful in studying the effects of loss or gain of an entire chromosome on the phenotypic character expression of any individual.
2. By using secondary or tertiary trisomics, the location of a particular linkage group can be detected on a particular chromosome.
3. Homologous chromosomes can be identified among the different genomes of a particular crop, i.e., nullisomic condition of a particular chromosome can be compensated by tetrasomic condition of another chromosome of other genome set.
4. Aneuploids are useful in production of substitution line which are helpful to study the effect of individual chromosome.
5. Aneuploids can also be helpful to identify the translocated chromosomes in an individual.

7.3 BIOTECHNOLOGY IN PLANT BREEDING

In the conventional plant breeding programme, the development of a new variety or hybrid takes about five to twelve years, starting from inbred production and then hybridization and selection of F_1 hybrids. To overcome the sexual barrier (pre-fertilisation and post-fertilisation), there is the need of modern non-conventional breeding methods. One of the approaches is the use of **'Biotechnology'** through different cell and tissue culture techniques and genetic engineering methods. Somatic hybrid production by protoplast culture-fusion technique, use of different molecular biological techniques and alien gene incorporation into the genetic background of cultivated species thus become obvious.

A. Tissue culture techniques and their applications in plant breeding

The culturing of plant cells or tissues in synthetic medium and their development into mature plants has immense potential for plant improvement. There are few major avenues which were opened by plant tissue culture can be listed below:

Micropropagation

The technique of micropropagation or regeneration of plantlets from any tissue has been successfully achieved in case of wheat, rice, sugarcane, maize, barley and many other crop plants. But this technique is specially useful for propagation of medicinal plants which grow slowly and cannot be bred in the conventional methods. Moreover the vegetatively propagating plants, such as banana which multiply by rhizome and one plant can yield about 10 plants per year, through micropropagation as many as 2,00,000 plantlets can be obtained. Equally this technique is applicable to tree plants like teak, eucalyptus, etc.

The meristem culture helps to get the disease free plant and also the vegetatively propagated crop plants can be maintained in disease free condition for long time. Clonal propagation method used for some heterozygous plants, especially the ornamentals, helps a lot in breeding programme. Maintenance and multiplication of self-incompatible inbred line (male sterile line) is possible by tissue culture methods very easily. Mutagens can be applied to single cell and the effect can be detected easily, isolated and utilised fully for new variety production through tissue culture.

Embryo culture

Distant hybridization programme sometimes yields non-viable embryo, then the embryo culture method and embryo rescue help to obtain the viable hybrids.

Embryo rescue or embryo culture is with the objective to rescue the embryo which aborts at an early stage of development, i.e., no mature seed can be obtained. The hybrid embryos are excised and put on a synthetic medium so that they can develop seedlings. Most extensive use of this technique has been observed in raising the interspecific and intergeneric crosses within the tribe Triticeae of Poaceae. Distant hybridization programme sometimes eliminates the chromosomes of one parent, thus the culture of hybrid embryo allows to develop the haploid plant. For example, the intergeneric cross between maize (*Zea mays*) and wheat (*Triticum aestivum*) has resulted in production of monoploid wheat plant.

Protoplast culture and fusion

Protoplast culture technique itself has an immense potential for crop improvement programme, as the alien gene introduction or incorporation is more easier in this way and transgenic or genetically modified crops can be regenerated.

Protoplast fusion, i.e., somatic hybrid production shows a new path to overcome the sexual barrier between distantly related wild and crop plants. This will help to transfer some useful characters like disease resistance, salt tolerance, drought tolerance, etc.

Somatic hybrids between rice (*Oryza sativa*) and barnyard grass (*Echinochloa oryzicola*) has been obtained. The most extensive programme has been in progress in the family Brassicaceae where the different traits like drought tolerance from *Eruca*, pathogen resistance from *Sinapis*, cytoplasmic male sterility or CMS from *Diplotaxis* have been transferred in cultivated *Brassica.*

Cybrids or cytoplasmic hybrids are obtained through following methods : (i) fusion of normal protoplast with enucleated protoplast from other parent, (ii) fusion of normal protoplast from one parent and protoplast with non-viable nuclei from other parent. In *Brassica,* the success has been achieved using this technique, i.e., CMS line with 'Ogura' cytoplasm, herbicide (atrazine) resistant trait have been transferred to cultivated variety.

Haploid culture

Haploid plants from anther and pollen culture and diploidisation of these haploids help a lot to get the homozygous inbred lines which are to be used in breeding programme.

The haploid plantlet production is aimed through anther culture or pollen culture, where the embryoids developed from this culture (haploid) may be treated with colchicine to get diploid homozygous plants which may be used in breeding programme. In China more than 100 rice varieties developed using the technique to give an increased yield. This type of haploid production technique has been successfully used for breeding of barley, maize, sugarcane, oilseed rape and some other crops.

The greatest usefulness of anther culture lies in the rapid production of haploid plants which are of great value in plant breeding and genetics. An unlimited number of haploid plants can be produced within short time, success has been achieved in barley, rice, wheat, potato, tomato, etc. Mutants of any nature can be detected easily as allelic interactions are non-existent. Anther culture avoids natural loss of inbred lines due to excessive inbreeding depression. The non-viable gene-combinations causing sterility are promptly exposed, so selection is automatic.

Somaclonal variations

From the observation of Larkin and Scowcroft (1981) it is obvious that natural variability in tissues, i.e., somaclonal variation can be utilised at selection level.

Somaclonal variation can be generated through tissue culture technique and the selected clone can be produced in mass scale. There are various reports in many crops where the different somaclones have been reported.

Plants	Characters considered for improvement
1. *Oryza sativa*	Plant height, maturity time, seed fertility, grain number and weight
2. *Triticum aestivum*	Plant and ear morphology, awn character, amylose content, grain weight and yield
3. *Hordeum vulgare*	Plant height and tillering
4. *Saccharum officinarum*	Diseases (eyespot, fijivirus, downy mildew) resistant, plant height and yield
5. *Solanum tuberosum*	Tuber shape, maturity date, plant morphology, photoperiod sensitivity, vigour, plant height, etc.
6. *Lycopersicon esculentum*	Branching habit, fruit colour, pedicel length, male fertility and growth

B. Gene transfer techniques in plant breeding

In plant breeding, techniques involving gene transfer through sexual and vegetative propagation are well established. The aim being to introduce genetic diversity into plant population and to select superior plants carrying the desired traits and to introduce some new characters into the cultivar, with the rapid improvement of genetic engineering techniques based on the knowledge of gene structure and function, plant breeding method has been changed.

The directed desirable gene transfer from one organism to another and the subsequent stable integration and expression of foreign gene into the genome is referred as **genetic transformation**. The gene is called **transgene** and the changed plants carrying the stably integrated desirable gene are **transgenics**.

In vitro gene transfer technique allows transferring desirable genes across taxonomic boundaries into plant from other plants, animals, microbes or any artificial, synthetic or chimeric gene also. The techniques presently rely on natural plant **vectors** as well as vectorless systems, which include directed physical and chemical methods for delivering foreign DNA into plant cells. One of the most important thing of these techniques is the potentiality of the recipient cell to express the introduced gene. Only a small fraction of cell get transformed i.e., DNA gets stably integrated into the chromosome of the cell. The DNA introduced into majority of the cells is lost with cell division. Stable transformation occurs when DNA is integrated into the plant nuclear or plastid genomes, expression occurs in regenerated plant and is inherited in subsequent generations (Fig. 7.1).

Different transfer techniques

(a) **Agroinfection or *Agrobacterium* mediated gene transfer method** is widely used with engineered Ti plasmid (modified T-DNA) in case of most dicot plants as well as monocot.

(b) **Direct or physical transfer method** is commonly used in case of cereals as these are naturally reluctant to *Agrobacterium* infection. There are many types of delivery systems like (i) **Biolystic** or **Particle bombardment**, (ii) **Electroporation**, (iii) **Microinjection**, (iv) **Pollen transformation**, (v) **Liposome mediated transfer**, (vi) **Silicon carbide fibre (SCF) mediated transfer**, (vii) **PEG mediated transfer**.

[For details refer Chapter 24]

All these different techniques have been applied in different plant materials using different kinds of genes for many desirable traits of agronomic values e.g., the genes for **stress tolerance** i.e., drought, salt and temperature stress (physical or abiotic stress); genes for **disease resistance** such as resistance against any pathogen by producing toxin, PR protein, plantibodies, or RNA mediated resistance gene; herbicide resistance, insect resistance; genes for development of **male sterile line** and also the restorer line; genes for better **nutritional quality** as in many cereal crops (rice, wheat, maize), oil seed crops

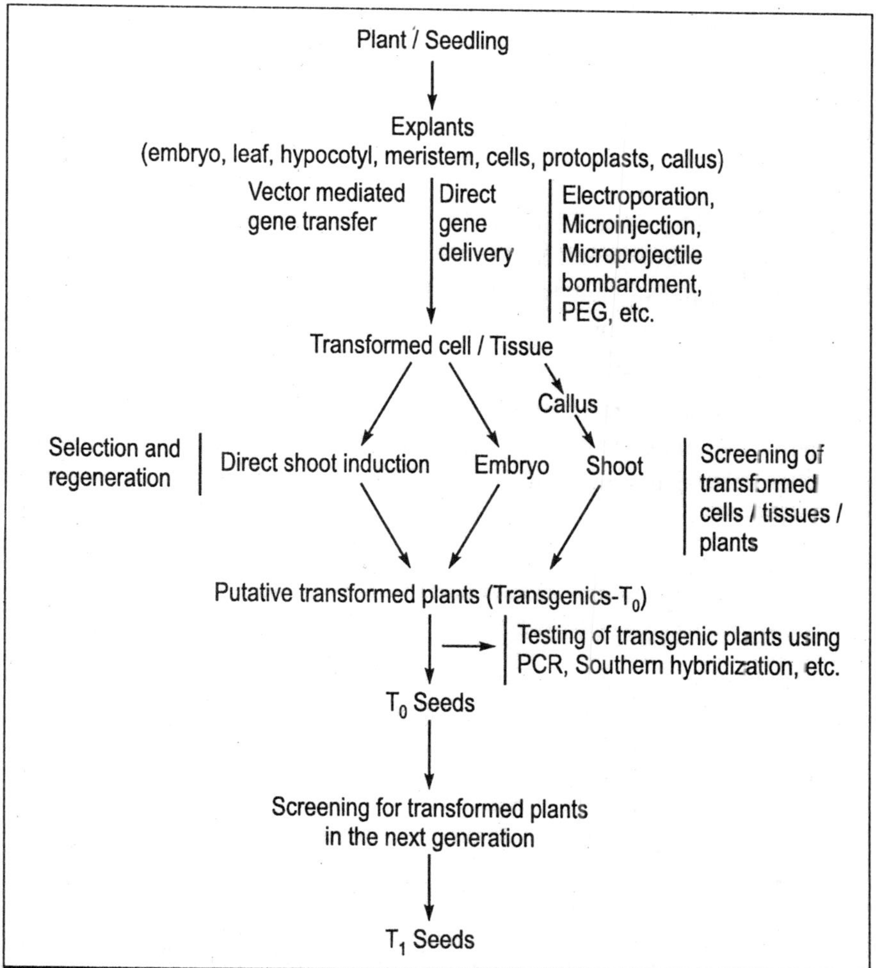

Fig. 7.1: Gene transfer techniques in plant breeding

(*Brassica*), pulses and vegetables (potato, tomato); etc. Many transgenic plants bearing the desirable traits have already been released as variety (refer Chapter 24 for details). Few of the major achievements has been listed below:

Gene	Attribute	Crop plant
(a) *Bt* (toxin from *Bacillus thuringiensis)*	Insect resistance	Rice, tomato, potato, cotton, etc.
(b) *cpti* (cow-pea trypsin inhibitor)	Insect resistance	Rice
(c) *cp* (coat protein)	Virus resistance	Tomato, potato, Alfalfa
(d) *ncp* (nucleocapsid protein)	Virus resistance	Tomato, tobacco, lettuce
(e) *acp* (acyl-carrier protein)	Altered fatty acid composition	*Brassica*
(f) *bar* (Barnase-Barstar)	Male sterility and fertility restoration	Maize

C. Molecular breeding technique (Use of DNA markers in plant breeding)

Molecular breeding using DNA markers often provide a wide array of applications in the field of plant improvement. Molecular markers are used for the analysis of genetic variation in germplasm available for plant improvement. Molecular marker aided breeding strategy involves the potentiality of molecular markers in plant breeding, particularly helps in marker assisted selection procedure which speeds up the whole breeding process.

Molecular markers are DNA sequences whose inheritance pattern can be established. Some of the unique features of molecular markers are: (i) they exhibit polymorphism, (ii) they show codominant inheritance which helps in distinguishing homozygous from heterozygous, (iii) they are easy for detection, (iv) they are distributed frequently throughout the genome.

Different molecular markers in use are:

(a) RFLP (Restriction Fragment Length Polymorphism)
(b) RAPD (Randomly Amplified Polymorphic DNA)
(c) AFLP (Amplified Fragment Length Polymorphism)
(d) VNTR (Variable Number Tandem Repeats)
(e) STS (Sequence Tagged Sites)
(f) SCAR (Sequence Characterised Amplified Region)
(g) SNP (Single Nucleotide Polymorphism)

[For details see Chapter 22 (22.7)]

Molecular marker development and its implementation in breeding programme has made the whole breeding exercise less time consuming and offers selection of desirable combination of traits. This approach is done by establishing linkage between molecular marker and traits to be selected. In this process the whole breeding procedure can be conducted in laboratory not waiting for the phenotypic expression in field, e.g., resistance property to plant pathogen can be evaluated in the absence of disease.

Marker assisted selection (MAS)

(a) **Mapping of plant genome**: Among several crop plants, rice has been the most wanted target plant. RFLP markers from closely related species are good markers for constructing gene map.

(b) **Linkage of molecular marker to desired trait**: Identification of genes responsible for useful trait may be established by a linkage analysis with markers on a genetic map of plant genome. Polymorphic markers are generally used to

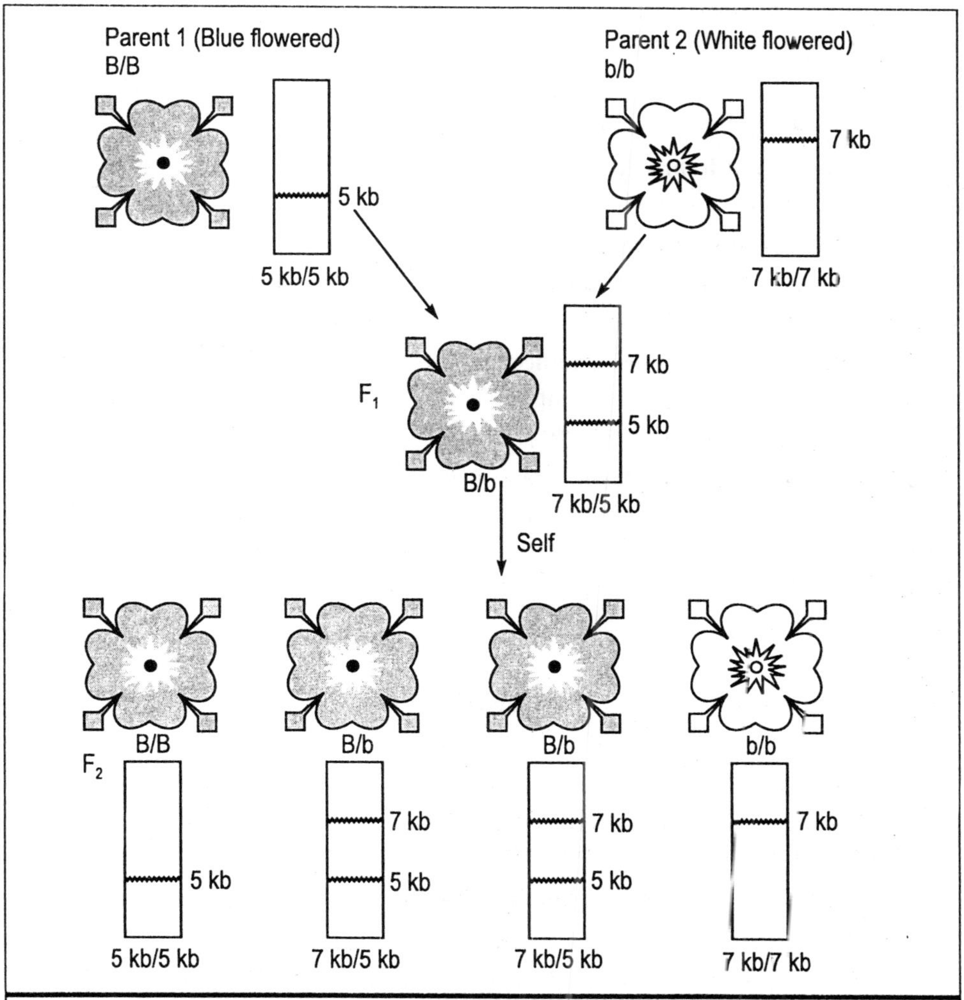

Fig. 7.2: Use of molecular marker (RFLP) in plant breeding based on phenotypic character

identify linked markers. Bulked segregate analysis (BSA) helps to detect the polymorphism between two species. Then the segregating population is tested, generally the polymorphism between the species is likely to be linked to genes for the trait.

(c) **Accelerated back-crossing**: Marker assisted selection accelerates the back crossing and selection for desired trait which help in earlier release of improved variety. Instead of several back-crossing and then selection which requires lots of time, molecular markers facilitate selection of individuals with more of the recurrent genome at each generation.

Applications of MAS

Markers have been used in the breeding of desirable disease resistance property against virus and fungal pathogen. For introducing this trait into cultivar variety, MAS helps in the following ways for the new breeding strategy.

Identification of the breeding line: Molecular markers are used to identify the breeding line among the large number of germplasm available.

Identification of hybridity: Using RAPD analysis the somatic hybrid or the hybrid nature of self-pollinated crops can be easily identified.

Purity of breeding lines: Cross contamination and seed harvestation may lead to contamination of breeding lines. Molecular markers can be used to assist establishment of pure breeding lines and check contamination of breeding.

Prediction of hybrid performance (Heterosis): Genetic distance between possible parents can be estimated by employing molecular markers. RFLP microsatellite markers are selected as useful marker for these predictions.

Identification of germplasm: Identification of several useful genetic resource of possible parents for use in breeding requires suitable molecular technique. RAPD marker is useful tool for the survey of germplasm. Survey of rice germplasm using RAPD shows linkage between the presence of specific marker and QTL for novel character.

QUESTIONS

1. Define mutation, mutagen and mutant. Discuss importance of mutation in plant breeding. Mention few achievements.
2. Define aneuploidy and polyploidy. Write down their significance in plant improvement. Give examples.
3. Define biotechnology. What are the recent techniques adopted in crop improvement.

8 CULTIVATION AND IMPROVEMENT OF SOME CROPS

Human is dependent on plants for the essentials of his existence. The three major necessities of human life, e.g., food, clothing and shelter are supplied mainly by plants. The number of plant species economically involved in human existence is estimated about 1,000–2,000. Hills in 1952 classified economically important plants as follows:

A. *Food Plants:* cereals, millets, pulses, vegetables, fruits;

B. *Food Adjuncts:* spices, condiments, beverages;

C. *Drug Plants:* medicinal plants, fumitories, masticatories;

D. *Industrial Plants:* plants yielding fibre, timber, rubber, gum and resin, tannin, dye, oil, sugar and starch.

In this chapter, cultivation and improvement of some of the crop plants yielding cereals (rice, wheat), fibres (cotton, jute) and beverages (tea) have been discussed.

8.1 RICE

Rice, *Oryza sativa*, the staple food of majority of the Indians, is grown practically all over India.

Cultivation area: The major states of rice cultivation are West Bengal, Bihar, Maharashtra, Orissa, Tamil nadu, Uttar Pradesh, Madhya Pradesh, Andhra Pradesh, Assam.

Soil: Rice grows well in clays or clay-loams. For normal growth of the crop soil pH between 5 to 8.5 is suitable.

Climate: Following climatic conditions are suitable for rice cultivation:

Temperature	21°C to 35°C
Rainfall	Sufficient rainfall
Humidity	High

Cropping season: Depending on the type of rice, cropping season vary:

Type	Season
Aus	Autumn
Aman	Winter
Boro	Summer

Rotation: Rice is grown in rotations with jute, wheat, potato, pulses, oil-seeds, vegetables, etc.

Preparation of land: The preparation of land depends upon whether the crop is to be grown under the 'dry', 'semi-dry' or the 'wet' system.

In *dry-system,* the land is ploughed well; weeds and stubbles are collected and burnt; manures are applied well in advance before sowing. In *wet-system,* the land is irrigated before sowing and then ploughed in standing water. Finally ladder is used to level the field.

Sowing: Under *dry-system* of cultivation improved, pure, viable seeds are sown directly by broadcasting or drilling in lines. For broadcasting, seed rate varies from 80-100 kg/ha and for line sowing from 60-70 kg/ha. In *wet-system* of cultivation, the crop is generally grown by transplantation. For transplanted crops, the seedlings are raised in nurseries. Generally 2-3 seedlings of 3-5 weeks age are transplanted in each hill. Normally early varieties are planted 20 cm from line to line and 10-15 cm from plant to plant, while medium and late varieties are planted with 20 cm × 15 cm and 23 cm × 23 cm spacings respectively. The sowing time for different types of rice is as follows:

Type	Sowing Season
Aus	May – June
Aman	June – July
Boro	November – December

Manuring: Chemical fertilizers and organic manures are equally important for rice cultivation. Nitrogen, either in organic or inorganic form, is the best manure for rice. Farmyards manure, compost, oil-cakes, fish manure and green twigs are commonly used in India. 10-15 tonnes of well rotted farmyard manure/ha is recommended at the time of land preparation.

After-care: In field, water is allowed to stand a depth of 1.2-2.5 cm until transplanted seedlings are well established. After this, 5 cm of water with frequent draining and reirrigation is maintained in the field till 1 or 2 weeks before harvesting. Few weedings should be given and soil near roots of plants should be stirred.

Disease and control: Important rice diseases and their controlling measures are as follows :

Name of disease	Caused by	Controlling measures
Blast	*Piricularia oryzae*	(i) Growing disease resistant varieties;
Bacterial leaf blight	*Xanthomonas oryzae*	(ii) Seed treatment with organo-mercurial compounds;
Brown spot	*Helminthosporium oryzae*	(iii) Use of selected systemic fungicides.

Pest and control: Important rice pests and their control are:

Name of the pest	Controlling measures
Paddy stem borer (*Tryporyza incertulus*)	(i) Use of insecticides; (ii) Use of fumigants, viz., methyl bromide;
Paddy bug (*Leptocorisa varicornis*)	(iii) Removal and burning of weeds and grasses.

Harvesting: The crop is harvested when the 'ears' are nearly ripe and the colour of the straw is still slightly green. Crop is cut with sickles, dried in the field for 2–4 days and stacked in the threshing yard. The harvesting time for different types of rice is as follows :

Type	Harvesting season
Aus	September–October
Aman	November–December
Boro	April–May

Threshing: The crop is then threshed either with sticks or by beating against a wooden log or by treading the entire crop under foot of bullocks or by pedal and power threshers. Finally husk is removed by pounding method or by rice hullers.

Storage: Cleaned unhusked paddy is dried and stored in jute bags and granaries.

Yield: The yield of paddy varies according to the type:

Type	Yield
Aus	1,249 kg/ha
Aman	1,121 kg/ha
Boro	2,164 kg/ha

The yield of high yielding variety may be upto 8,000 kg/ha

Improvement of crop through plant breeding:

(a) **Introduction of new varieties**: Many new varieties have been introduced from foreign countries and acclimatized, for example, IR-8, IR-20, Palman 579 – all are from Philippines, Intan from Indonesia, Taichung from China, etc.

(b) **Selection from promising local strain**: Sp_1, Sp_2, Type 1 (selected from Ramjeban of Dehradun), Type A 64 (selected from Hansaraj and Unnao).

(c) **Hybridization**: Hybridization between certain good varieties can combine many good qualities of different varieties into one variety. Then selection is done to screen out the best.

Name of the variety	Cross Product
Pankaj	Peta × Tongkai Rotan
Sita	IR-8 × IR-12
K-78-13	Shinei × Ch 971
Kalinga-2	Dunghansali × IR-8
Ratna	TKM-6 × IR-8
Rasi	IR-8 × N-22
Surekha	IR-8 × Siam 29
Kiran	N-22 × TN-1

(d) **Mutation breeding**: Chemical or physical mutagen application has the potential to get many good promising varieties, e.g., Jagannath (Mutant of T-141), K-84 (Mutant of T-65), Hybrid mutant 95 (Jhona 349 × TN 1), etc.

(e) **CMS line development for hybrid seed production**: Rice hybridization requires male sterile plants. Five commercial male sterile lines (cytoplasmic) are identified in rice which are being used extensively for hybrid seed production, e.g., Wild rice – wild aborted or WA type, *Oryza sativa* f. spontanea, Chinsurah Boro II (BT type), Gambiaca (Gam type), Shan-Tao-Bai.

(f) **Application of tissue culture**: The techniques of tissue culture can help the improvement of the crop in many ways. Specially anther or pollen culture has helped the crop to develop the varieties with higher yield as well as resistance against blast, tungro, leaf blight and also maturity within a very short period. Dihaploids of Indica × Japonica crosses were successfully raised through anther culture. Six good strains of rice have been selected from pollen plants of F1 anther culture of hybrid rice variety Shan You No. 2.

(g) **Transgenic rice plant development**: Introduction of the foreign gene is another attempt to develop the crop with disease or insect resistance property and good nutritional quality, etc.

Rice varieties have been developed by introducing Bt-gene to give the insect resistance property against yellow stem borer. 'Golden Rice' variety has been developed which bears the gene for carotene production.

Important improved varieties:

Early maturing	High yielding: Improved Local:	Rasi, Ratna, Kshitish Dular, NC 1626
Medium maturing	High yielding: Improved Local:	Prakash, Kunti, Sashyasru Bhasamanik, Rupsail, Latisail, Raghusail, Patnai-23

Late maturing	High yielding :	Pankaj, Mahsuri, Suresh, Biraj
	Improved Local :	Tilakkachari, Achra 108/1, Kumargore
	Deep water tolerant :	Jaladhi 1, Jaladhi 2
	Salt tolerant :	SR 26B, Patai 23
	Flood tolerant :	FR 13A, FR 43B
Disease resistant varieties	Bacterial blight resistant :	N-22, IR-20, IR-22, Ratna
	Bacterial Leaf streak resistant :	IR-20, Krishna, Jagannath
	Bunt resistant :	T-6, T-8, T-10, T-22, T-19, etc.
	Brown spot resistant :	Ch-13, Ch-45, T-141, Co-20, T-2112, T-2118, T-960
	Blast resistant :	ADT-10, Co-4, TKM-1, Co-29, 30

Research Stations:

1. Central Rice Research Institute, Cuttack.
2. Indian Agricultural Research Institute, New Delhi.
3. International Rice Research Institute, Manila, Philippines.

Table 8.1 : Popular high yielding varieties of rice released by Central Variety Release Committee (CVRC) and states (From H. K. Chaudhuri)

Variety	*Parentage*	*Year*	*Breeding method*
C.V.R.C.			
1. *IR 8*	DGWG × Peta	1966	Introduction from Philippines
2. *Jaya*	TN 1 × T 141	1968	Pedigree
3. *Pankaj*	Peta × Tongkai Rotan	1969	Selection from introduced segregating material
4. *Jagannath*	T 141 Mutant	1969	Mutant
5. *Sabarmati*	(TN1 × Basmati 370) × Basmati 370	1970	Back cross
6. *Ratna*	TKM 6 × IR8	1970	Pedigree
7. *IR 20*	IR 262 × TKM 6	1970	Introduction from Philippines
8. *Rasi*	IR 8 × N 22	1977	Pedigree
Andhra Pradesh			
1. *Tella Hamsa*	HR 12 × T (N) 1	1971	Pedigree
2. *Surekha*	IR 8 × Siam 29	1976	Pedigree
3. *Phalguna*	IR 8 × Siam 29	1977	Sister selection of Surekha
Bihar			
1. *Sita*	IR8 × IR 12-178-2-3	1972	Selection from introduced segregating material
2. *Kiran*	N 22 × T (N) 1	1976	Pedigree
3. *Rajendra Dhan 201*	IR8 × Tadukan	1979	Selection from introduced segregating material
Gujarat			
1. *GAUR 1*	Zinya 31 × IR9-60	1973	Pedigree

Contd.

Table 8.1 : Popular high yielding varieties of rice released by Central Variety Release Committee (CVRC) and states (From H. K. Chaudhuri)

Variety	*Parentage*	*Year*	*Breeding method*
Jammu & Kashmir			
1. *K 84*	T 65 mutant		Mutant
2. *K 78-13*	Shinei × Ch. 971		Cross of two introductions
Karnataka			
1. *Manila*	C4-63 selection	1970	Introduction from Phillippines
2. *Intan*	—	1975	Introduction from Indonesia
3. *Prakash*	T 90 × IR8	1977	Pedigree
Madhya Pradesh			
1. *Garima*	Cross 4 × T (N) 1	1975	Pedigree
2. *Jagriti*	Rikku 132 × R4	1976	Pedigree
3. *Kranti*	Cross 116 × IR8	1976	Pedigree
Maharashtra			
1. *Surya*	BG 79 × I R 400-28-4-5	1973	Pedigree
2. *Radhanagari 185-5*	Halwasal × T (N) 1	1974	Pedigree
3. *Karjat 14-7*	IR8 × Zinya	1975	Pedigree
4. *Ratnagir 68-1*	IR8 × Sigadis	1975	Pedigree
Orissa			
1. *Kalinga 2*	Dunghansali × IR8	1973	Cross with Hungarian introduction
2. *Kumar*	T 90 × IR 8	1974	Pedigree
3. *Suphala*	T 141 × T (N) 1	1976	Pedigree
Punjab			
1. *Hybrid Mutant 95*	Jhona 349 × T (N) 1	1972	Hybrid mutant
2. *Palman 579*	—	1972	Introduction from Philippines
3. *PR 106*	IR 8 × peta/5 × Belle Patna	1978	Multiple cross, introduced from Philippines
Rajasthan			
1. *Chambal*	IR8 × NP 130	1975	Pedigree
Tamil Nadu			
1. *Vaigai (CO 37)*	T (N) 1 × CO 29	1974	Pedigree
2. *Bhavani*	Peta × BPI 76	1976	Pedigree
3. *Raja Rajan (CO 40)*	CO 25 × IR8	1977	Pedigree
4. *TKM 9*	TKM 7 × IR8	1978	Pedigree
Uttar Pradesh			
1. *Saket 4*	TKM 6 × IR8	1971	Secondary selection from CR 44-35 of Cuttack
2. *IR 24*	IR8 × IR 127-2-2	1972	Selection from introduced segregating material
3. *Prasad*	IR 747 B2-6-3 × IR 579-48	1978	Pedigree

8.2 WHEAT

Wheat, *Triticum aestivum,* is an important staple food extensively cultivated in India.

Cultivation area: It is widely cultivated in Uttar Pradesh, Bihar, Punjab, Madhya Pradesh, Maharashtra, Haryana and Rajasthan. Indo-Gangetic Plain is the best wheat producing region.

Climate: The following climatic conditions favour crop growth:

Temperature	Cool climate
Rainfall	130-1,000 mm/annum
Humidity	Relatively dry weather

Soil: Easily irrigable and drainable loamy or clay-loamy soil is suitable for cultivation of wheat.

Rotation: Wheat is rotated with a kharip crop like jowar, bajra, cotton or with leguminous crop. In West Bengal rice-jute-wheat rotation is generally practised.

Preparation of land: Several ploughings, repeated harrowing in the rainy season followed by 3 or 4 cultivations and planking immediately before sowing produce a good seed bed for unirrigated crop on alluvial soil. For irrigated crop, the land is irrigated before sowing and number of ploughings is reduced.

Sowing: Viable and disease resistant seeds are shown by broadcasting, drilling or dibbling at a rate of 75-120 kg/ha. Sowing time is November for medium long duration variety and second fortnight of November for short duration variety.

Irrigation: For irrigated wheat, first irrigation should be given 20-25 days after sowing. Irrigation at tillering, heading and grain filling stages are essential.

After care: Weeding may be done by spraying 2, 4-D or hand weeding is practised.

Manuring: The unirrigated wheat crop is not normally manured although application of well-rotted farmyard manure or compost or oil-cake at the rate of 20-30 quintal/hectare about 6 weeks before sowing is desirable. For the irrigated wheat crop, though actual dose depends upon soil fertility, fertilizer mixture of 60-120 kg/ha N, 30-60 kg/ha P_2O_5 and 30-60 kg/ha K_2O is recommended.

Disease and control: Important diseases of wheat and their control are as follows:

Name of the disease	Causal organism	Controlling measures
Black stem rust	*Puccinia graminis tritici*	Growing resistant varieties
Loose smut	*Ustilago nuda var. tritici*	Hot water treatment of seed

Pest and control: Important wheat pests and their controlling measures are as follows:

Name of the pest	Controlling measures
Stem borer (*Sesamia inferens*)	Spraying endosulfan or dusting BHC
Termites (*Microtermes obesi*)	Adding aldrin or chlordane dust in soil

Harvesting: When the grain is dead-ripe and the straw is golden yellow and brittle, harvesting is done. Generally it begins in mid February and continues till April, depending on variety.

Threshing: Harvested grains are threshed by treading under the feet of cattle on a threshing floor or by pedal power threshers. Next winnowing is done with winnowing baskets.

Storage: Thoroughly dried grains should be stored in moisture proof and fumigated store rooms.

Yield: The yield of unirrigated crop is about 300-400 kg/ha and the yield under irrigated condition with high yielding varieties followed by correct production technology is 8 tones/ha.

Improvement of crop through plant breeding: In India the crop improvement in wheat can be divided into three phases:

First phase : Development of pure line varieties like NP4, Pb 8, K-13, Bansi, etc.

Second phase : Introduction of different characters through hybridization to give varieties like NP 52, C 518, C 519, Niphad, etc.

Third phase : Production of disease resistant variety like NP-700, NP-800, etc.

(a) **Introduction of new varieties**: Sonora 64 and Lerma Rojo 64 are the two varieties which have been introduced from Mexico to India as they bear the disease resistance property.

(b) **Hybridization with related species**: To diversify the genetic basis of Indian "durum" wheat, the interspecific hybridization programme has been taken to introduce genes from five "*Triticum*" species, namely *T. dicoccum, T. carthlicum, T. turgidum, T. timopheevi* and *T. aestivum* for higher yield and disease resistance property. Resistance to several diseases has been incorporated into cultivated wheat from closely related species of *Triticum, Aegilops* and *Agropyron.*

Name of the hybrid	Cross between the varieties
Arjun	Lerma Rojo 64A × Nai 60
Shekhar	K 812 × Kalyan Sona
Janak	E 557 × HD 845
Girija	E 5165 × E 417

There are some varieties which have been developed through selection process after hybridization.

Kalyan Sona → Penjamo 62 Sib × Gabo 55 — Selection from S 227.

Sonalika → II 54-388 × NIOB × Lerma Rojo — Selection from S-308.

(c) **Multiline varieties**: Many lines have been hybridized together to develop the varieties like Bithur, KSML 3 and MLKS 11 and are released. Discovery of Norin-10 (dwarf gene) and its introduction by hybridization has yielded many new varieties.

(d) **Mutation breeding**: Mutagen treatment to any existing good variety may yield with new character, like *Sharbati Sonora* has been developed from a mutant line of Sonora-64.

(e) **Cytosterile line development**: The hybridization between *T. vulgare* × *Aegilops caudata* and *T. durum* × *Aegilops ovata* named as *Aegilo-triticum* are cytosterile in nature. *T. timopheevi* is the best cytoplasm which is interactive with *Triticum* nucleus to give male sterility. 15 different cytoplasm have been recognised which can induce male sterility in common wheat and form the basis of alternative system for hybrid production.

(f) **Application of tissue culture**: Among different techniques in tissue culture, anther culture in wheat has developed more than 20 new varieties and strains in China. Spring wheat Huapei No. 1, winter wheat Jinghua No. 1, Jinghua No. 2 and 3 – all are the good varieties for plant breeding bearing the characters like vigorous tillering, cold resistance, early maturity and good grain quality. 'Florin' is the doubled haploid variety developed in France which has been used in breeding to develop the new strain Jingdan 2288 that bear the characters like large spike, plenty of grains, vigorous tillering, stripe rust and powdery mildew resistant and resistance to lodging.

Some Important Hybrid Varieties:

Grain characters	Variety
Soft and white	Dudhia or Pissi or Dudhya
Hard and red	Lal kanak
Soft and red	Desi, Lal, Lalpissi, Baija
Hard and amber	Sharbati

Disease Resistance Varieties:

Rust resistant	S 227, S 307, S 308, S 331, Lerma Rojo, Sonora 63, 64
Bunt resistant	Kalyan Sona, S 227, PV-18
Smut resistant	NP 817, 865, 871, 886, 888
Karnal Bunt resistant	HD 1907, HI 358, HP 743, L 176, L 191, M 137-A

Research Station: Indian Agricultural Research Institute; New Delhi.

Table 8.2: Some released and popular varieties of wheat (From H.K. Chaudhuri)

Name	Parentage	Year	Remark
1. *Sonora 64*	Yaktana 54 × Yaqui	1965	Introduction
2. *Kalyan Sona*	Penjamo 62 sib × Gabo 55	1967	Selection from S 227
3. *Sharbati Sonora*	Sonora 64 Mutant	1968	Mutant
4. *Sonalika*	II 54-388-An (Yt. 54) × NIOB × Lerma Rajo	1967	Selection from S 308
5. *Janak*	E 5557 × HD 845	1973	
6. *Deshratna*	S 503 × NNP 835	1974	
7. *Arjun*	Ler. Rojo 64 A × Nai 60	1974	
8. *UP 262*	5308 × Bayio 66	1975	
9. *UP 115*	(NP 887 × E 4870) × UP 302	1976	
10. *WL 711*	(S 308 × chris) × Kalyan sona	1976	
11. *Bithur*	—	1978	First multiline released in India for U.P.
12. *Shekhar* (K. Purvi)	K 812 × Kalyan sona	1979	
13. *Malvia-12*	NP 876 × Ciano 66	1979	
14. *Girija*	E 5165 × E 4717	1979	

8.3 COTTON

Cotton is the world's greatest industrial crop, the chief fibre yielding plant. All the cultivated cottons fall under 4 species; 2 belonging to old world, *Gossypium arboreum* and *G. herbaceum* which are diploid in nature; another 2 belonging to new world, *Gossypium hirsutum* and *G. barbadense* which are tetraploid in nature.

Areas of cultivation: In India cotton is mainly cultivated in the states of South Western India – Gujrat, Maharashtra, Karnataka, Tamil Nadu, Kerala, Madhya Pradesh, Punjab, Rajasthan, Uttar Pradesh, Haryana.

Climate: Cotton grows well in sub-tropical areas with following climatic conditions:

Temperature	21°–27°C
Humidity	Less
Rainfall	50 cm

Soil: Cotton grows best in well drained loamy soils. Indo-Gangetic alluvial soil, black soil of Central and Southern India and also red laterite soil are good for cotton production.

Rotation: Cotton is commonly rotated with wheat, legume, jowar, bajra, sesamum, etc.

Preparation of land: Before sowing the land is given one or two ploughings in South and some parts of North India, whereas in Central and Western India the land is harrowed three or four times, and ploughing is done once.

Sowing: Seeds are generally sown by broadcasting or drilling method at the rate of 15-25 kg/ha for new world cotton and 10-18 kg/ha for old world cotton during April to July. Mechanical sowing is done keeping the space 45-60 cm in between rows and 30-45 cm within the rows. Before sowing, cotton seeds are treated with different types of chemical reagents or mud or cowdung due to the fuzziness of seeds.

Manuring: Generally cotton is not manured sufficiently. Green manure, i.e., leguminous crop may be grown before cotton. Fertiliser dose depends on the variety and location of cotton crop grown. Nitrogen @ 40-100 kg/ha and phosphate @ 20-80 kg/ha are applied for irrigated cotton and half of this dose is required for rainfed crop.

After care: Weeds are removed with the help of hand hoes in some places. Cotton is mainly unirrigated crop, watering is required at the time of flowering and fruiting stage. For maturation of bolls, sufficient soil moisture is required.

Disease and control: Common diseases of cotton and their control are as follows:

Name of the disease	Causal organism	Control
Anthracnose	*Glomerella gossypi* *Colletotrichum indicum*	(a) Treating seeds with organomercurials (b) Spraying the crop with copper fungicides
Black arm and Angular leaf spot	*Xanthomonas malvacearum*	(a) Growing resistant varieties (b) Treating seeds with organomercurials or agrimycin
Powdery mildew	*Leveillula taurica*	Dusting crop with sulphur
Wilt	*Fusarium vasinfectum*	Growing resistant varieties

Pest and control: Common cotton pests and their control measures are as follows:

Name of the pest	Control
Cotton spotted bollworm	Removing and destroying attacked shoots and bolls, spraying with 0.1% carbaryl and 0.03% monocrotophos at 15 days interval
Cotton stem borer	Controlled by burning and uprooting withered plants
Pink bollworm	Exposing the seeds in hot sun for 2-3 days before sowing, clean cultivation practices, spraying with insecticides like spotted bollworm

Harvesting: Harvesting is done during October to March. It is harvested in 3 or 4 pickings after the maturity of bolls. Seed cotton is removed from the bolls in the field; in closed-boll cottons, partially open bolls are collected from the plants in the field.

Yield: The average yield of seed-cotton is 400-600 kg/ha. The lint yield is 150-200 kg/ha.

Crop improvement through plant breeding: The All India Coordinated Cotton Improvement Project started in 1967 and now there are 30 centres in various agroclimatic zones.

(a) **Hybrid variety production**: *`Hybrid 4'* is the breakthrough in cotton production, which is the first commercial hybrid in India. *'Varalakshmi'* is another hybrid variety developed next. *`Sujata'* and *`Suvin'* are the two high spinning cotton varieties.

(b) **Improvement through selection**: Examples of few cotton varieties after *pure line selection* are : Co-2, Co-7, A-10, A-231R, C520, etc. *Mass selection* has yielded the varieties like – Dodahatti local, Dharwar American, Combodian cotton, etc.

(c) **Mutation breeding**: Mutagenic treatment in existing varieties developed some new useful characters better than the parental variety. MCU-7 is the mutant variety obtained from L114 EE variety.

(d) **Cytoplasmic male sterile line development**: Introduction of cytoplasmic male sterility to cultivars from wild species is another potential use of *Gossypium* cytoplasm. The nucleus of *G. hirsutum* has been incorporated into a number of cytoplasms of wild species. The male fertility restorer genes have been identified in several cases.

(e) **Improvement through amphidiploid production**: For improvement, wild cotton species can be utilized. The F1 plants are treated with colchicine and the amphidiploids produced, are crossed with cultivated tetraploid species. For example, (*G. thurberi* × *G. arboreum*) × *G. hirsutum* and back-crossed to *G. hirsutum,* developed lines with superior lint quality but otherwise poor quality would be there.

Some important improved varieties:

Superlong staple	Amaravati, Hybrid 4, MCU-9, Varalakshmi, CRS 156. Sujata, Suvin
Long staple	Suman, MCU-7, JK Hy 1
Medium staple	AKH-4, Khandwa-2, SH-131
Short staple	LD-133

Research Station: Central Cotton Research Institute, Nagpur.

Table 8.3: Some cotton varieties released (From H.K. Chaudhuri)

Variety	*Parentage*	*Year*	*Remark*
1. *MCU 9*		1978	Superlong staple of 60s counts
2. *Amaravati*		1978	Superlong staple of 50s counts
3. *Hybrid 4*	hybrid	1970	Superlong staple of 50s counts

Contd.

Table 8.3: Some cotton varieties released (From H.K. Chaudhuri)

Variety	Parentage	Year	Remark
4. *Varalakshmi*	hybrid	1972	Superlong staple of 80s counts
5. *CRS 156*	hybrid of Acala × SB 1085	1974	Superlong staple of 100s counts
6. *Sujata*		1969	Superlong staple of 100s counts
7. *Suvin*		1974	Superlong staple of 120s counts
8. *H 565-C*		1978	Long staple 40s counts
9. *Suman*	CP 15/2	1976	Long staple 40s counts
10. *MCU 7*	L 114 EE mutant	1972	Long staple 40s counts
11. *JK Hy 1*		1976	Long staple 44s counts
12. *AKH 4*		1975	Medium staple of 20s counts
13. *Khandwa-2*		1971	Medium staple of 36s counts
14. *SH 131*		1976	Medium staple of 30s counts
15. *LD 133*		1978	Short staple of 16s counts

8.4 JUTE

Next to cotton, jute forms the most important Indian fibre crop used more extensively for various purposes. Two species, *Corchorus capsularis* and *C. olitorius* are cultivated.

Areas of cultivation: In India, jute is grown mostly in the states of North-East India – West Bengal, Bihar, U.P., Orissa, Assam, Tripura and Meghalaya. The crop is cultivated annually on about 8 lakh hectares in India.

Climate: Jute grows well in warm and humid climate.

Temperature	24°-35°C
Humidity	90%
Rainfall	Sufficient

Soil: Jute grows well on the soil which is neither very heavy clay nor very sandy. Loamy alluvial soil with deposits of fertile silts is best for jute cultivation. The optimum soil pH is about 6.4.

Preparation of land: As jute seeds are very small, it requires a clean and clod-free seed bed. Therefore, the land is ploughed and cross ploughed for at least 5-6 times. After each operation, the land is laddered for breaking up the clods.

Manuring: Additional manuring is generally not required in soil receiving fresh silt. In other soils, 40-80 q/ha of well rotten cattle manure or compost and 450 kg/ha of weed ash are used during preparation of land.

Sowing: After treatment with organo-mercurial compounds or Bavistin 50%, the seeds are generally sown by broadcasting at the rate of 5-8 kg/ha. Nowadays, line sowing with seed drills has been greatly adopted. In this practice, seeds are sown in rows with a spacing of 20-22 cm. between the rows and 5-7 cm within the row. Sowing time varies according to species:

Species	Sowing time
C. capsularis	Mid February-April
C. olitorius	Mid March-May

After-care: When the seedlings attain a height of 7-20 cm, the crop is then lightly hoed with 'bida' or 'achra'; subsequently two or three hand weedings are given and the crop is gradually thinned to a spacing of 9-10 cm between plants in broadcasted fields.

Disease and control: Common diseases of jute and their control measures are as follows:

Name of the disease	Causal organism	Controlled by
Stem/root rot	*Macrophomina phaseoli*	Seed treatment with organo-mercurial (e.g. agrosan) or non-mercurial (e.g. captan) compounds
Wilt	*Macrophomina phaseoli* (Primary pathogen) *Fusarium solani* and *Pseudomonas sp.* (Secondary pathogen)	Crop rotation and proper drainage of the field.
Soft rot	*Sclerotium rolfssi*	Clean cultivation, deep ploughing and copper oxychloride spray

Pest and control: Common jute pests and their control measures are as follows:

Pest	Control
Jute semilooper (*Anomis subulifera*)	Removing the larvae; spraying with 0.05% endosulfan or 0.1% carbanyl.
Hairy caterpillar (*Dicrieia oblique*)	Plucking the infested leaves in the early stage; spraying with parathion (0.03-0.4%) or endosulfan (0.04%)

Crop rotation: Jute is rotated with paddy and pulses. On lowlands, jute follow paddy (aman) constitutes the normal rotation. On highlands, the crop is grown in rotation with pulses (gram, pea, etc.), potato, wheat, barley, oat, mustard.

Harvesting: The crop is to be harvested 120-135 days after planting when pods begin to appear. Depending on the variety grown and the time of sowing, the harvesting period varies from June to October.

At the time of harvesting, plants are cut near the ground level or in case of flooded land, plants are uprooted. The harvested plants are usually left on dry land for 3-4 days for the shedding and the drying up of leaves, then tied in bundles and kept covered with green leaves, weeds or earth for 3-4 days.

Extraction of fibre (Processing of Jute):

(a) **Retting or steeping**: Bundles are retted or steeped in gently flowing, fairly deep, clean, soft water for 10-15 days in July–September to rot out the softer gummy tissues. Sometimes retting may take 25 days or more specially when the temperature of water is low. In this process complete submergence of stalks is essential to ensure proper retting of the root end. The bundles with their root ends are generally kept in water for 2-4 days before entire bundles are fully submerged.

(b) **Stripping**: After retting is complete, the stalks are stripped by "beat-break-jerk" process. In this operation 8-10 stalks are taken at a time, their root ends beaten with a wooden or bamboo mallet until the fibres are loosened. Loosened fibres are then wrapped around the fingers and the stalks or stems are jerked forward and backward over the surface of water till the fibre is completely separated. The separated fibre is then rinsed, washed, cleaned, wrung and piled on the ground. Finally, strands are opened out and dried in the sun for 2-4 days and tied into bales, ready for the market. The fibre content of the plant varies from 4.5–7.5%.

Yield: The yield varies according to species and variety:

capsularis: 9.0-13.5 quintals of dry fibres/hectare

olitorius: 13.5-18.8 quintals of dry fibres/hectare

Recently recommended improved varieties have yield potential of 40 q./ha.

Improvement of the crop through plant breeding: Improvement of jute is being done through following breeding practices:

(a) **Selection**: JRC-321, a variety suitable for lowland cultivation, is the product of selection from local variety Hewti.

(b) **Hybridization**: JRO-3690, a high yielding variety, is an example of crossing of two low yielding mutants of *C. olitorius*.

(c) **Mutation breeding**: Improved varieties JRO-514, JRO-412 are the result of mutation breeding of *C. olitorius*.

Improved varieties:

C. capsularis	Sonali (JRC-321), Sabuj Sona (JRC-212), Shyamali (JRC-7447)
C. olitorius	Chaitali (RRO 878), Basudev (JRO-7835), Baisakhi (JRO-632), Navin (JRO-524)

Research Station: Jute Agricultural Research Institute (JARI), Nilganj, Barrackpore, West Bengal, established in 1953 has two sections —

(i) Research Station in Assam, (ii) Multiplication farm at Panagarh.

Jute Technological Research Laboratories (1938) was transferred to ICAR in 1965 which work on: (i) Technological Research on Jute, (ii) Evaluation of fibre quality of new strains.

8.5 TEA

Tea, *Camellia sinensis*, is one of the most popular caffeine containing beverage, obtained from tender leaves of the tea plants after processing. Tea contains 13-18% tannin, 2-5% theine and a volatile oil. There are two main varieties, e.g., *bohea* (Chinese tea) and *assamica* (Indian, i.e., Assam tea).

Cultivation area: In India, tea is cultivated largely in Assam, West Bengal (Darjeeling & Jalpaiguri districts), Kerala and Tamil Nadu.

Climate: Following climatic condition, are suitable for tea plant:

Temperature	20°C to 32°C
Rainfall	175-380 cm/annum
Humidity	Wet

Soil: Loamy or forest land rich in organic matter with acidic pH is suitable for tea cultivation. Hill slopes favour the growth of tea or any other place where water doesn't become stagnant.

Preparation of land: Firstly the area under cultivation should be cleaned of forest growth. The land should be prepared carefully considering the three important aspects:

(i) Protection of land from soil erosion;

(ii) Cultivation primarily for weed suppression and better root development;

(iii) Use of shade plants to protect soil and young plants.

Planting methods: There are two methods of planting, viz., 'Square' and 'Rectangular' systems. In both cases, pits measuring 30.4-45.2 cm in depth, 22.8 cm in diameter and 1.2-1.5 m apart in either direction are dug and filled with organic matter containing surface soil.

Preparation of seedling and its transfer: Tea is normally propagated by seeds but propagation by means of cuttings, budding, etc., is also successful. Seeds are first sown in germination bed, nextly young seedlings are transferred to 'jungle' or 'basket' nurseries. Then, either 6-18 months old 'basket' seedlings or 1-2 years old 'jungle' seedlings are planted in April-may or September-October respectively in pits already dug in the 'square' or 'rectangular' system.

Plantation of shade trees: Plants like *Albizzia* spp., *Dalbergia anamica, Derris robusta, Leucaena glauca,* etc., are planted to provide shade to tea bushes.

Manuring: Frequent nitrogenous manuring is essential for better growth of leaves. Application of compost, green manure and fertilizer mixtures supplying 27 kg N, 13.5 kg P_2O_5 and 13.5 kg K_2O per acre are essential.

Disease and control: Common diseases of tea and their control are as follows:

Name of the disease	Causal agent	Controlled by
Red rust and algal red spot	*Cephaleuros parasiticus*	Proper manuring and spraying with Bordeux mixture or lime sulphur.
Blister blight	*Exobasidium* sp.	Spraying with perenox, blitox

Pest and control: Common tea pests and their control are as follows:

Name of the pest	Controlled by
Red spider (*Tetranychus bioculatus*)	Spraying with lime sulphur
Tea mosquito (*Helopeltis* spp.)	Spraying with kerosenised water and DDT

Pruning: Tea bushes are pruned regularly and the best pruning practice is —

(i) **1st year after planting**: The plant is decentered by cutting the centre of growing stem at 20-30.4 cm from the ground.

(ii) **2nd year after planting**: Plants are left unpruned.

(iii) **3rd year after planting**: Plants are pruned across at 45.5 cm removing strong growing centre shoot leaving lateral branches.

(iv) **4th and following years**: Pruned to leave 1.2 cm of new wood.

Plucking: Tea leaves are plucked by hand or with scissors. The first plucking is usually made when the plants are five years old and is restricted to the terminal bud, tender leaves and three expanded leaves.

Yield: The average yield/acre of tea from seed borne plants is 326 to 372 kg. and vegetatively propagated plants of about 1006 kg.

Grades: Generally there are eight grades of tea:

(i) Broken Orange Pekoe, (ii) Orange Pekoe, (iii) Orange Fannings : made from the bud, first leaf and the softer stalk; (iv) Broken Pekoe, (v) Pekoe, (vi) Broken Pekoe Souchong, (vii) Pekoe Fannings : made from the larger first, second, third leaves and the intervening stalks; (viii) Dust : Consisting merely of grains of tea.

Processing of tea: Processing of tea leaves varies for different commercial varieties, e.g., Black Tea, Green Tea, Brick Tea and Oolong Tea.

(a) **Black Tea**: Black tea, common in India, is manufactured through following operations —

(i) **Withering**: The plucked leaves are taken to withering house where these are spread out either on slopping wire netting racks or on horizontal hessian cloth 'chung' for 18-24 hours to eliminate water content of leaves.

(ii) **Rolling**: Withered leaves are taken to rolling rooms where they are subjected to three rolls, each of 30 mins. duration by machine. This process curls the leaves and burises the cells of the leaves, so that their sap is exposed to action of oxygen in air. Room temperature is maintained at 24°-26°C.

(iii) **Fermenting**: Rolled leaves are spread on a cement floor of fermenting bed for 2.5 to 4.5 hours at about 24°-26.5°C. During this operation, tannins of leaves acted on by enzymes causing the change of leaf colour from light red to brown.

(iv) **Firing**: Fermented leaves are dried in a current of hot air by 'pressure' type drier machine at about 60°-65°C for 30-40 minutes in order to arrest fermentation and for slow desiccation of leaves.

(v) **Sorting and grading**: The dried leaves are passed through sieves of different meshes and sorted out which are then graded into different grades and packed in separate plywood tea chests lined with aluminium foil.

(b) **Brick Tea**: Brick tea, mainly consumed in Tibet and USSR, is manufactured through following operations —

(i) **Panning**: Leaves are subjected to heat treatment at a temperature of 71°-94°C in a large iron vessel built into a brick stove for 10 minutes. This operation makes the leaves a little soft and turn olive green.

(ii) **Rolling**: Leaves are then taken to rolling machine to roll for 30 minutes by which the fibr. of the leaves well lacerated and broken.

(iii) **Fermenting**: Leaves are placed in a heap upon a cement floor, covered with tarpaulin for 5 days. This operation helps in fermentation process by fungus to cause typical flavour.

(iv) **Drying**: Fermented leaves are dried either in sun or on brick furnace fired by charcoal.

(v) **Moulding**: Bulk of dried leaves is rammed as tight as possible into the moulds (23.0 cm × 11.3 cm × 1.2 m) which hold materials for four bricks of 1.8 kg each. After three days, the bricks have settled to allow of the moulds opening. The bricks are then packed in paper and put in the sun.

(c) **Green Tea**: Green tea, mainly produced in China and Japan, is manufactured in the following ways:

(i) **Panning**: Same as brick tea.

(ii) **Steaming and rolling**: After panning, leaves are heat treated at 120°C by steaming to make the leaves soft so that these are rolled by hand on a bamboo mat.

(iii) **Drying**: Rolled leaves are subjected to brick furnace to dry upon trays over charcoal fires until they are perfectly crisp.

(d) **Oolong Tea**: Oolong tea with colour of black tea and flavour of green tea, is mainly produced in Formosa.

(e) **C.T.C. Tea**: Now-a-days tea processing method has been reduced, e.g., (i) Crushing, (ii) Tearing and (iii) Curling, which is called C.T.C. method.

Crop improvement through plant breeding: Chinese type tea was introduced in India, Assamese type tea is grown all over world. As tea is cross-pollinated crop, seed setting is poor, mainly clonal propagation by stem cutting is done. Clonal selection, varietal crosses, mutation, etc., are the ways of crop improvement.

Introduced and selected varieties: TV-1, and TV-2.

Hybridized and selected varieties: TV17 (TV-1 × St. 202)

TV 24 (Chinese × Assamese type)

Variety developed by polyclonal cross: Gaurisankar (St. 203)

Variety developed by biclonal F_1 hybrid: Nandadevi (St. 378), F_1 hybrid-St. 397.

Application of tissue culture: As the crop is mainly clonally propagated, so the micropropagation technique is very much helpful to maintain the stock of raised hybrid variety. And as the crop is cross-pollinated, so the hybrid seed production is also difficult to maintain the line. Tissue culture procedure helps to get the selected clones in multiple amount.

Research Station: Tocklai Experimental Station, Jorhat runs in association with Tea Research Association.

Table 8.4: Tea Varieties released (From H.K. Chaudhuri)

Variety	Parentage	Year	Remark
A. *Clonal Varieties :*			
TV 1	19/29/13	1949	Introduction in 1918 and selection
TV 2	Stock 14	1959	Introduction in 1914 from China and selected
TV 17	TV 1 × St 202	1968	F_1 hybrid
TV 22	12/41/42	1976	Open pollinated variety
TV 24		1979	F_1 hybrid of Chinese and Assamese type
B. *Seed varieties :*			
Gaurisankar	(4/16, 6/16/, 8/7, 8/17, 10/18,	1954	Polyclonal cross of 7 clones
(St. 203)	12/15 and 12/18		
Nandadevi	(14/5/35) × (14/16/28)	1968	Biclonal F_1 hybrid
St. 397	(19/29/13) × (19/35/2)	1976	F_1 hybrid

QUESTIONS

1. Describe the cultivation procedure in rice. State the improvement done in this crop through plant breeding practice. Mention few improved varieties.
2. Discuss the method of cultivation of wheat in India. State the scope of improvement of this crop. Mention the varieties recommended for cultivation.
3. Write down the cultivation practice of cotton. Mention the improvement done in this crop. Give examples of improved varieties.
4. Give an account of jute cultivation in West Bengal. State the methods of extraction of jute fibre. Write the scope of improvement of jute. Name one jute research centre in India.
5. Describe cultivation of tea. State the processing of tea-leaves. Write down its scope of improvement.

BIOMETRY

9 BIOMETRY: AN INTRODUCTION

Statistics is a branch of science which deals with methods of collection, classification and analysis, i.e., drawing inferences from collected data; then testing of hypothesis and making comment on these.

The statistics can be applied in all scientific branches like social, physical and biological. The use of statistics in Biology is known as **Biostatistics** or **Biometry**. It deals with the application of different statistical methods and analysis of data collected from biological system.

9.1 APPLICATION OF BIOMETRY

In biological system, it is very difficult to draw any concrete conclusion about any happening, as there are much differences among the individuals. Such as in the same plant, the leaves are all different; again in a field, all the same kind of plants are not identical; in the same place, the same crop does not respond equally in different plots. So it is very difficult to reach at a definite and reliable conclusion unless we take the help of statistical analysis of the observed data.

In case of plant breeding, the application of biometry has immense importance, as the statistical analysis only helps to conclude about a population of plant; such as —

(a) How the character, which is acquired by the population, is distributed, how the character is being inherited, etc.

(b) Whether two characters in a population are correlated or not, if correlated, how much they are correlated.

(c) In case of plant breeding whether Mendelian Genetics is being followed or not, can be tested by chi-square test.

(d) The application of fertilizer or irrigation or any kind of cultural practice can be effectively justified by application of test of significance.

(e) The laws of probability is of importance in genetics for forecasting the chance of obtaining certain result from a cross and elucidating the operation of genetic principle.

(f) The effects of two or more different fertilizer application on the same field can be analysed by analysis of variance.

(g) Study of alleles of genes in populations and the factors which maintain or change the frequencies of particular genotype in populations can be studied.

9.2 LIMITATION OF BIOMETRY

(a) All the conclusions about statistical analysis depend on the availability of sample data. If the sampling is biased, the analysis will be eronotic.

(b) Statistics can be used and applied only on collective data, not on individual data.

(c) Statistical laws are always true in case of large population on the average, not on the small sample.

9.3 POPULATION AND SAMPLE

In statistical investigation, the **population** can be defined to any well defined group of individuals or of observations of a particular type available in a particular space or area which is being studied for statistical analysis. For example, all the rice plants in a field can be called as population.

The population may be finite or infinite. If a population consists of a fixed number of values then it is said to be **finite,** such as number of plants of same species in a particular boundary area, or the total number of flowers in a plant. On the other hand, population in which it is not possible to count the number of individuals then it is called **infinite,** such as number of cells in a callus mass, number of phytoplanktons in a pond, etc.

In statistical analysis, a fraction of population is considered or studied to analyse or comment on the whole population, i.e., representative of the entire population is called **sample**. Any sample should represent the population in unbiased manner and sample should be specific and of convenient size used in statistical tests. If all the pods in a garden is considered as population, then few pods from that garden will be the sample.

9.4 VARIABLE AND VARIATE

Any qualitative or quantitative character which show variation from one individual to the other in the same population is known as **variable**. An individual observation of any variable is known as **variate**.

Depending on the attribute or character, the variable may be of qualitative or quantitative type.

Qualitative variable is the one which describes the qualitative character of the sample, such as colour of flowers, quality of seed surface, i.e., smooth or wrinkled, etc.

Quantitative variable is the measurable character of the variate or individual, such as height of the plant, tillering number of plant, weight of seed grain/plant, etc.

But depending on the continuity or break in the represented data, the variables are of two types:

Discrete variable: When the variable is measured in unit number then it is called discrete variable, such as number of pods/plant, here the number cannot be in decimel or in fraction.

Continuous variable: If the variable has the fractional value then it is called continuous variable, such as length of plant, yield of seeds, etc. Here the data must have fractional value like 17.2 cm or 17.2 kg/ha.

9.5 SAMPLING

In biological experiment, it is not possible to collect complete information about a population. If the number of pods/plant is to be collected from a field then it is really time consuming and also rarely possible to do it. Then few plants are taken into account for studying the whole population of plant in that field.

The method by which only few items are selected from the population in such a way so that they will represent the population in unbiased way is called **sampling**.

The **size of sample** is an important factor in statistical analysis which depends on the number of sampling units selected from a population for investigation. The size should not be too big or too small. Taking only 10 values from a plot of 1000 plants will give erroneous result and also it is difficult to handle with 1000 number of values.

Why sampling is essential ?

A. Sampling saves time, the data can be collected and summarised more quickly with a sample than a complete count of the whole population.

B. In case of infinite population, sampling is the only method for statistical analysis.

C. Sampling reduces the cost of experiment because only a few selected items are studied in sampling.

Limitations of Sampling

A. If the sampling is not done properly, i.e., if it is biased then it misleads which results in false, inaccurate interpretation.

B. There may be personal biasness during sampling or choice of method of sampling which may also lead to erroneous interpretation.

Criteria for Good Sampling

A. Selected samples from the population should be homogenous and should not have any differences when compared with the population.

B. Reasonable number of items are to be included in the sample to make the result more reliable.

C. The selected sample should have the similar characteristics as the original population from which it has been selected.

D. The individual items composing the sample should be independent from each other.

E. The number of observations included in a sample should be more to make the results more reliable.

Methods of Sampling

The proper method of selection of samples and the relation between the sample and population is the matter which determines the method of sampling. There are various methods of sampling, it totally depends upon a statistician which method will be applicable for proper selection of method of sampling (Fig. 9.1).

A. Simple Random Sampling

This method is followed where each item of the population has an equal chance of being included in the sample. Random sampling suggests that selection should be done without any biasness.

To ensure the randomness of selection one can adopt either lottery method or refer to table of random numbers.

Lottery method is the simplest and most popular method, all the items are numbered and slips of identical size and shape are made. All are shuffled together and selection is done blindly.

Table of random numbers can be used in place of blind selection. For this purpose, Random Number Table (5 digit) prepared by Snedecor and Cochran (1988) can be used either horizontally or vertically for selection of sample without biasness.

This method of sampling is more scientific because there are less chances of personal biasness in sampling from the population and chances of selection of every item is equal.

B. Systematic Sampling

This method is applied when the population is large, scattered and not homogeneous. Systematic procedure follows to choose a sample by taking every K th individual, where K refers the sample interval calculated by the formula:

$$K = \frac{\text{Total population}}{\text{Sample size desired}}$$

Example 1: 20% sample to be taken from 1000 individual of a population,

$$K = \frac{1{,}000}{20\% \text{ of } 1{,}000} = 5.$$

So the, First sample will be the 5th individual,
Second sample will be the 10th individual,
Third sample will be the 15th individual,
and like this way.

C. Stratified Sampling

This method of sampling is followed when the population is not homogeneous, hence the population is first divided into several homogeneous groups or strata and the sample is drawn from each stratum at random. This method is useful as it represents the proportionate representative sample from each group and it gives greater accuracy.

D. Cluster Sampling

A cluster is randomly selected group. This method is used when the units of population are natural groups such as school, hospitals, etc. The technique of cluster sampling allows small number of target population and the data provided statistically valid at 95% confidence limits.

E. Non-random Sampling

This method is called as judgement sampling. In this method the selection process of sample is somewhat subjective. The choice of sample items depends exclusively on the judgement of the investigator. The investigator exercises his judgement in the choice and includes those items in the sample which he considers most typical for his investigation.

For example, from a rice field one investigator may select only the healthy plants for artificial inoculation of a pathogen.

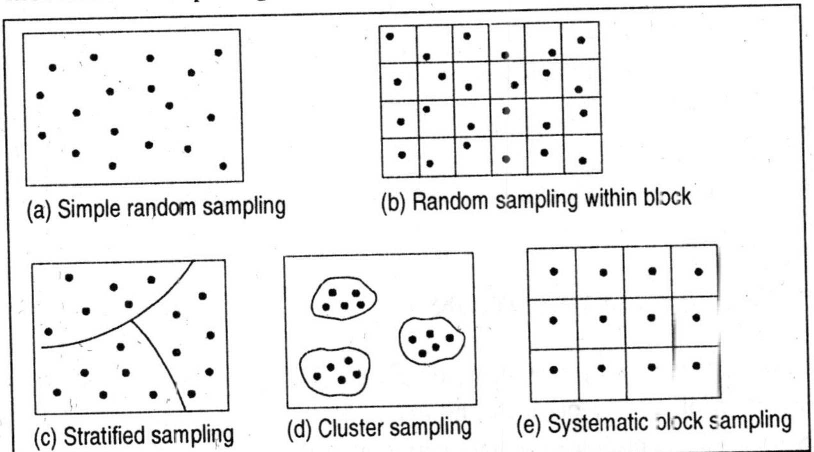

Fig. 9.1: Different sampling methods

9.6 DATA AND ITS TYPES

A set of values recorded for an event is called **data**. The data collected for statistical analysis do not consist of observations that are identical, they are variable in nature.

When a set of data is collected by personal investigation from the original source or by performing some experiment is called **primary data**.

When the data collected by some other sources and is used after some processing for further analysis then the data is called **secondary data**.

Classification of data

The data may be of two types such as:

(a) **Qualitative data**: The character of a population which cannot be numerically expressed such as colour of flower, nature of seed coat, etc., are called qualitative data.

(b) **Quantitative data**: The magnitude of any character which can be numerically measured is called quantitative data.

The quantitative data may be of two types:

(a) **Continuous** — When the values of variate do not exhibit any break, such as height of plant during growth.

(b) **Discrete** — When the values of variate vary by positive or negative jumps, such as the yield of rice in different plots after application of particular treatment or fertilizer application.

Main objectives of classification of data

1. To make proper use of raw data.
2. To make the study of data and comparison easier.
3. To put up the collected material to statistical treatment.
4. To simplify the complexities of raw data and make it possible to draw statistical inferences.
5. To give importance to the main information gathered while keeping aside the unnecessary information.

9.7 FREQUENCY DISTRIBUTION

For any statistical analysis, the handling with raw data requires some treatment, i.e., the classification of data to organise the available values in a more compact way. The frequency distribution presents the data very concisely indicating how frequently a variable occurs in a group of study.

Construction of frequency distribution table

If there are repetitions in individual values or items of any investigation, suitable frequency table can be formed. These frequency table may be discrete or may be continuous in nature.

The available **raw data** at first should be converted into arrayed data. For biostatistics the raw data are arranged in ascending order to make it **arrayed data** (Example-2).

Then for **discrete frequency distribution table,** the values of variables are written in one column and the repetition of that value is written against it which is the frequency (Table 9.1, 9.2).

But for **continuous frequency distribution table,** the values are grouped in fixed interval and then the frequency within that interval is observed and noted.

The **number of classes** or the range of **class interval** is an important factor for making this kind of frequency distribution table. There is no fixed rule for how many **classes** can be formed, generally it depends on observation of available data. minimum 3 classes and maximum 20 classes can be formed.

The size of class interval also depends on the range of data and the number of classes, it is equal to the difference between highest and lowest value divided by the number of classes,

$$i = \frac{H-L}{K},$$

where i = class interval

H = Highest value

L = Lowest value

K = number of classes.

Overlapping frequency distribution table

Values of variables are grouped in such a fashion that the upper limit of one class interval is represented in next class interval. In an example (Example 2), number of pods ranges from 15 to 41, the classes may be 15-17, 17-19, 19-21, etc. (Table 9.3).

Non-overlapping frequency distribution table

Values of variables are grouped in such a fashion that the upper level of one class interval do not overlap the preceeding class interval. In the above example, number of pods ranges from 15 to 41, the classes can be prepared like 15-17, 18-20, 21-23, etc (Table 9.4).

Cumulative frequency distribution table

Cumulative frequency is determined by adding the frequency of a class interval with the frequency of the preceeding class interval. The cumulative frequency table can be prepared from both the overlapping and non-overlapping frequency distribution table (Table 9.5, 9.6).

Relative frequency

This is calculated from the cumulative frequency against the total population or sample.

$$\text{Relative frequency of a class} = \frac{\text{Cumulative frequency of that class}}{\text{Total no. of sample}}$$

Class limit

It is defined as two boundaries of a class, i.e., the highest and lowest values of a class, which can be represented by L_1 and L_2.

Mid-value of class-interval

The central point of a class interval is called its mid-value or mid-point of that class, which is obtained by using the following formula.

$$\text{Mid - value of a class} = \frac{L_1 + L_2}{2}$$

Example 2 : Following raw data is obtained in an investigation. 100 pea plants bore pods ranging from 15 to 41 in a garden of pea plants (see Raw Data Table A):

Raw Data Table A:

33	31	28	15	17	17	16	18	16	18	20	22	24
25	31	27	30	29	33	28	20	22	23	25	41	39
30	36	37	27	33	28	31	29	32	31	29	34	19
22	25	40	19	21	24	30	26	37	27	28	32	32
31	29	34	21	23	25	40	26	38	27	26	33	28
34	29	30	30	35	29	23	29	26	38	27	32	28
34	35	29	30	33	32	35	29	24	26	38	27	36
28	34	29	35	30	33	32	36	37				

Arrayed Data Table B:

15	16	16	17	17	18	18	19	19	20	20	21	21
22	22	22	23	23	23	24	24	24	25	25	25	25
26	26	26	26	26	27	27	27	27	27	27	28	28
28	28	28	28	28	29	29	29	29	29	29	29	29
29	30	30	30	30	30	30	30	31	31	31	31	31
32	32	32	32	32	32	33	33	33	33	33	33	34
34	34	34	34	35	35	35	35	36	36	36	37	37
37	38	38	38	39	39	40	40	41				

Table 9.1 : Determination of frequency of each variable		
No. of Pods Variable	Tally mark	Repetition number of Plants or Frequency
15	I	1
16	II	2
17	II	2
18	II	2
19	II	2
20	II	2
21	II	2
22	III	3
23	III	3
24	III	3
25	IIII	4
26	卌	5
27	卌 I	6
28	卌 II	7
29	卌 IIII	9
30	卌 II	7
31	卌	5
32	卌 I	6
33	卌 I	6
34	卌	5
35	IIII	4
36	III	3
37	III	3
38	III	3
39	II	2
40	II	2
41	I	1

Table 9.2 : Variables and respective frequency			
No. of Pods (Variables)	No. of Plants (Frequency)	No. of Pods (Variables)	No. of Plants (Frequency)
15	1	29	9
16	2	30	7
17	2	31	5

Contd.

Table 9.2 : Variables and respective frequency

No. of Pods (Variables)	No. of Plants (Frequency)	No. of Pods (Variables)	No. of Plants (Frequency)
18	2	32	6
19	2	33	6
20	2	34	5
21	2	35	4
22	3	36	3
23	3	37	3
24	3	38	3
25	4	39	2
26	5	40	2
27	6	41	1
28	7		$\Sigma f = 100$

Table 9.3 : Overlapping frequency table

No. of Pods in class interval	No. of plants in frequency
15–17	3
17–19	4
19–21	4
21–23	5
23–25	6
25–27	9
27–29	13
29–31	16
31–33	11
33–35	11
35–37	7
37–39	6
39–41	4
	$\Sigma f = 100$

Table 9.4 : Non-overlapping frequency table

No. of Pods in class interval	No. of plants in frequency
15–17	5
18–20	6
21–23	8
24–26	12

Contd.

Table 9.4 : Non-overlapping frequency table

No. of Pods in class interval	No. of plants in frequency
27–29	22
30–32	18
33–35	15
36–38	9
39–41	5
	$\Sigma f = 100$

Table 9.5 : Overlapping cumulative frequency distribution table

Class interval	Frequency	Cumulative frequency
15–17	3	3
17–19	4	3 + 4 = 7
19–21	4	7 + 4 = 11
21–23	5	11 + 5 = 16
23–25	6	16 + 6 = 22
25–27	9	22 + 9 = 31
27–29	13	31 + 13 = 44
29–31	16	44 + 16 = 60
31–33	11	60 + 11 = 71
33–35	11	71 + 11 = 82
35–37	7	82 + 7 = 89
37–39	6	89 + 6 = 95
39–41	5	95 + 5 = 100

Table 9.6 : Non-overlapping cumulative frequency distribution table

Class interval	Frequency	Cumulative frequency
15–17	5	5
18–20	6	5 + 6 = 11
21–23	8	11 + 8 = 19
24–26	12	19 + 12 = 31
27–29	22	31 + 22 = 53
30–32	18	53 + 18 = 71
33–35	15	71 + 15 = 86
36–38	9	86 + 9 = 95
39–41	5	95 + 5 = 100

9.8 REPRESENTATION OF DATA

A. Tabular representation of data

Tabulation is a process of orderly arrangement of data into series or rows or columns, where they can be read at a glance. This process may be called as summarisation of data in an orderly manner within a limited space. It helps in simplifying the raw data, comparison can be done easily, reveals the pattern of distribution of any attribute, detects the omissions and errors.

Types of Table

(a) **Simple Table**: In simple table only one character is considered such as length of papaya plant in a field.

Length of plant (cm)	6–10	11–15	16–20	21–25	26–30
No. of plants	5	10	11	9	3

(b) **Complex Table**: In complex table more than one character is considered such as, length, sex of the plant, disease incidence, etc.

Length (cm)	No. of Papaya Plants			
	Male		Female	
	Infected	Healthy	Infected	Healthy
6–10	2	1	1	1
11–15	2	4	2	2
16–20	1	4	2	4
21–25	1	2	2	4
26–30	0	1	1	1

B. Graphic representation of Data

The quantitative and continuous data can be represented graphically by using the following diagrams:

1. Histogram
2. Frequency polygon
3. Frequency curve
4. Cumulative frequency curve
5. Dot diagram
6. Line diagram
7. Bar diagram
8. Pie-chart

Histogram

This graph is used for continuous frequency distribution. The width of class interval marked along with the X-axis. On these width, rectangles of areas proportional to the corresponding frequencies of the respective class intervals are erected.

If the class intervals are of equal lengths, then the heights of the rectangles are proportional to the corresponding frequencies and for unequal class intervals, the heights of the rectangles are proportional to the ratios of the frequencies to the width of the corresponding class (Fig. 9.2).

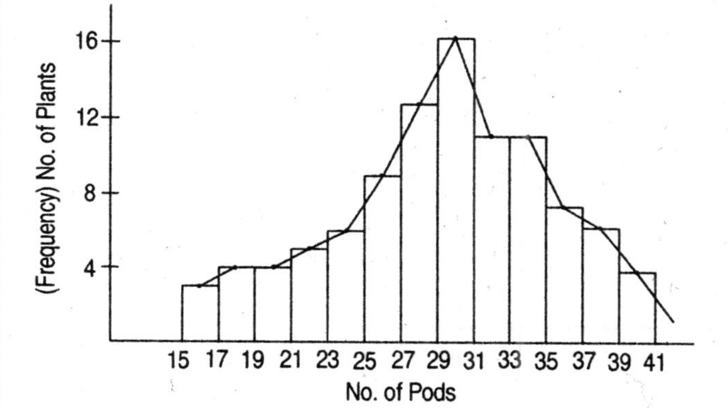

Fig. 9.2: Histogram and Frequency polygon (Example 2)

Frequency polygon and Frequency curve

The values of variables for an ungrouped data are taken on the X-axis and their frequencies are put on the Y-axis. Whereas for grouped data, the mid-point of each class-interval are put on the X-axis and frequency on the Y-axis are put and the dots are joined by straight line is called **frequency polygon** (Fig. 9.2), and the free hand curved drawing will represent the **frequency curve** (Fig. 9.3).

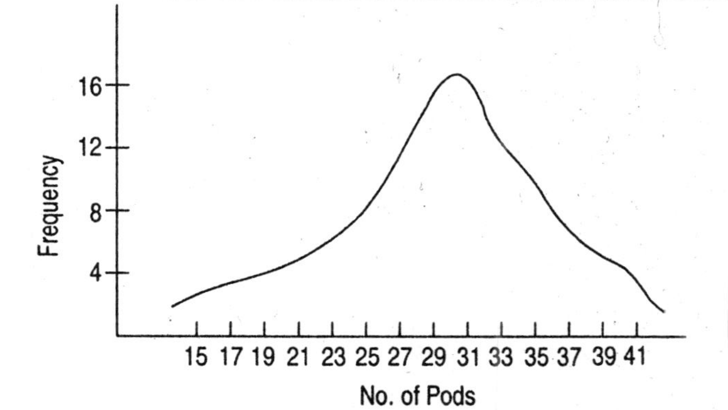

Fig. 9.3: Frequency curve (Example 2)

Cumulative frequency curve

This curve is drawn with the help of cumulative frequency distribution table. The mid-point of class interval (in case of grouped data), or the values of variable (in case of ungrouped data) are put on the X-axis and their cumulative frequency are put on the Y-axis which represents the **cumulative frequency curve** (Fig. 9.4).

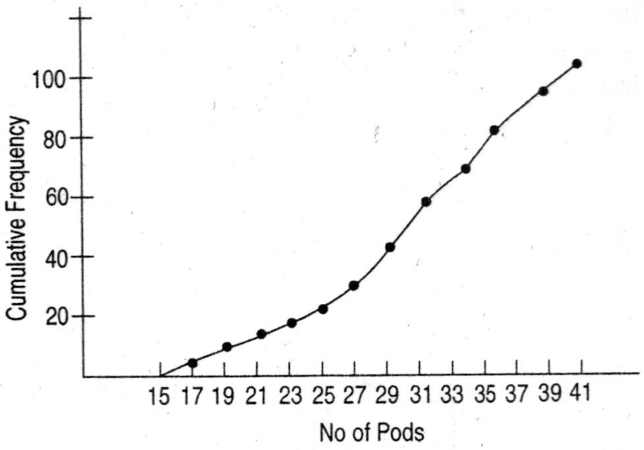

Fig. 9.4: Cumulative frequency curve (Example 2)

Dot diagram

This kind of diagram is prepared after cross tabulation in which frequencies of at least two variables have been cross classified. One variable being independent and the other variable being dependent. This type of graphic representation shows the nature of correlation between two variable characters, X and Y, of the same individual, such as colour and texture of seed coat of peas, length of pod and number of seeds/pod, etc. (Fig. 9.5).

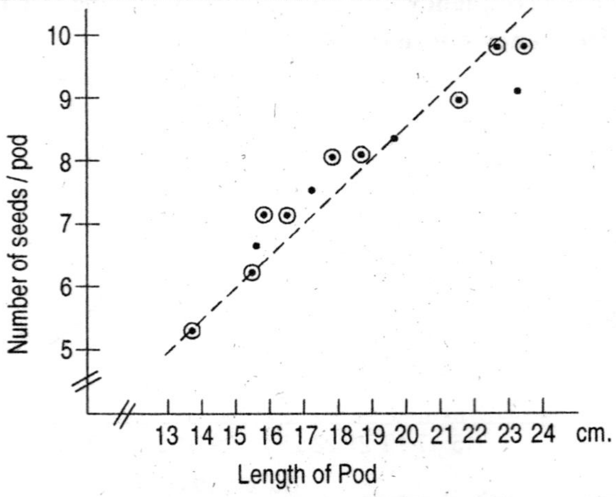

Fig. 9.5: Dot diagram (Example 3)

Example 3: Length of pods and number of seeds/pod are observed in 10 samples.

Length of pods (cm)	14	16	16	17	18	19	22	22	23	24
Number of seeds/pod	5	6	7	7	8	8	9	9	10	10

Line diagram

This is the simplest type of diagram where data is represented by the line only following the frequency distribution table. In case of ungrouped data, the values of variables are put on X-axis and the frequency is put on Y-axis, the straight lines are drawn proportional to the frequencies (Fig. 9.6).

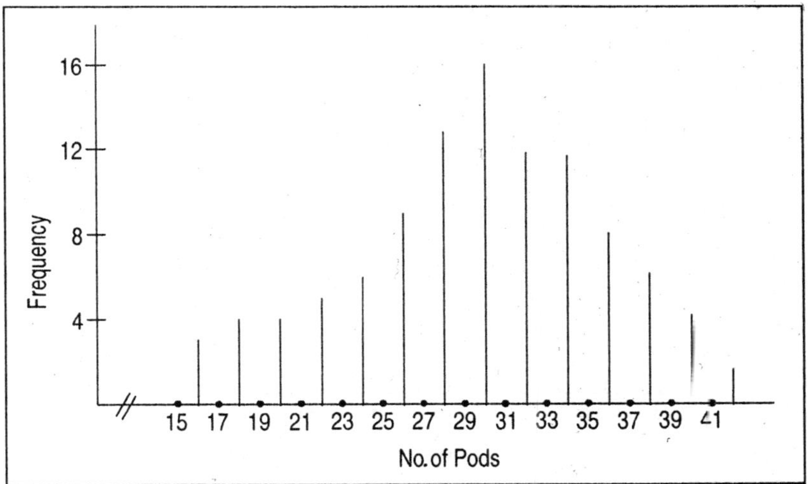

Fig. 9.6: Line diagram (Example 2)

Bar diagram

This is one dimensional diagram where bars of equal width are drawn either horizontally or vertically which represent the frequency of the variable. The width of bars should be uniform throughout the diagram.

Bar diagram may be of 4 types (Fig. 9.7):

Simple Bar diagram: This type of bar diagram is used to represent only one variable by one figure.

Divided Bar diagram: When the frequency is divided into different components then the diagrammatic representation is also called a divided bar diagram.

Percentage Bar diagram: The total length of bars corresponds to 100 and the divisions of the bar correspond to the percentage to different components.

Multiple Bar diagram: When a comparison between two or more related variables has to be made, then this type of bar diagram becomes essential.

Example 4: In an investigation, the total cereal crop production is noted in the following table. Make a divided Bar diagram and a percentage bar diagram from the data (Fig. 9.7b, c).

Years	Rice (Metric tons)	Wheat (Metric tons)	Maize (Metric tons)	Millet (Metric tons)	Total (Metric tons)
1993-94	1,383	634	513	400	2,930
1994-95	2,021	1,383	521	313	4,238
1995-96	1,914	1,413	551	900	4,578
1996-97	2,664	1,636	424	265	4,989

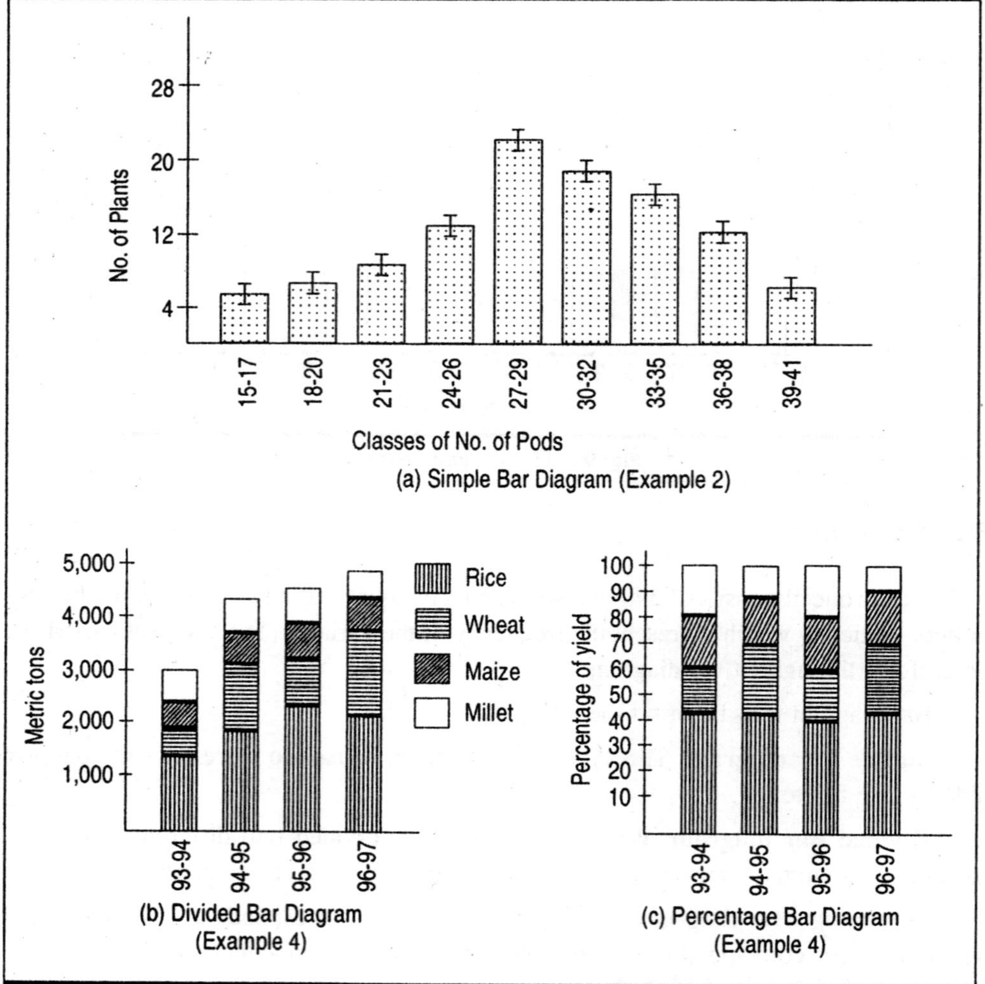

Fig. 9.7: Bar diagram and types

Pie diagram

It is an easy way of presenting discrete data of qualitative characters such as colour of flowers, colour of seed coat, texture of seed coat, etc. The frequencies are shown in a circle. Degree of angle denotes the frequency and area of the sector helps to compare at a glance.

Size of the angle is calculated according to following formula:

$$\text{Size of the angle} = \frac{\text{Class frequency}}{\text{Total observation}} \times 360^\circ$$

Example 5: In a hybridization experiment, among the F_2 hybrid seeds, the following results are observed.

Brown and large	—	54
Brown and small	—	18
White and large	—	18
White and small	—	6

Make a pie diagram of this observation (Fig. 9.8).

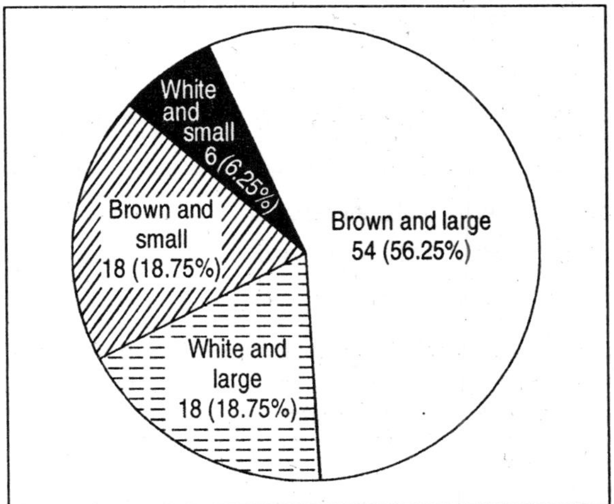

Fig. 9.8: Pie diagram (Example 5)

QUESTIONS

1. Define and mention the uses of biostatistics or biometry.
2. Define the following : (a) Data (b) Population (c) Sample (d) Variable (e) Primary data (f) Secondary data.
3. What do you mean by sample and sampling? Give some good methods of sampling.
4. What are the criteria for good sampling? Which sampling method is mostly followed in case of biostatistics?

5. Define the following : (a) Frequency distribution, (b) Cumulative frequency distribution, (c) Class interval.
6. The following data was found by a student in a grassland community :

   No. of seeds per *Indigofera* plant.

39	55	35	45	49	52	48	33	48	47	50	51	53	50	55
53	54	53	50	48	49	31	33	50	55	50	51	50	53	55
52	49	51	50	50	44	51	50	58	59	57	59	60	58	51

   (a) Make a simple frequency distribution table and also cumulative frequency distribution table.

   (b) Prepare overlapping and non-overlapping frequency distribution table.

   (c) Make histogram, frequency polygon, frequency curve and cumulative frequency curve with the above data.
7. Make simple bar diagram, divided bar diagram, percentage bar diagram from the following data :

   Plant community studied in an area by quadrat method shows following result in 5 quadrats.

	Quadrat 1	Quadrat 2	Quadrat 3	Quadrat 4	Quadrat 5
Herb	1,535	742	1,055	1,249	985
Shrub	20	18	15	17	25
Tree	2	5	3	3	4

8. Make a pie diagram with following data of a hybridization experiment :

F_2-seeds show the result :	Yellow and smooth seed	:	75
	Yellow and wrinkled seed	:	20
	Green and smooth seed	:	25
	Green and wrinkled seed	:	10

9. Prepare frequency distribution table, overlapping and non-overlapping frequency distribution table, histogram, frequency polygon, cumulative frequency curve and simple bar diagram from following data.

   No. of seeds/fruit :

12	14	13	6	7	21	17	15	13	12	7	8	11	20	21	11	19	18	9
10	13	15	12	14	13	11	9	8	16	14	18	9	10	16	15	16	10	10
11	13	14	12	14	13	14	15	12	17									

10. Prepare frequency distribution table, overlapping and non-overlapping frequency distribution table, histogram, frequency polygon, cumulative frequency curve and simple bar diagram from following data.

    Weight of seeds in gram :

2.1	1.8	1.7	1.5	1.6	2.3	2.0	1.1	1.2	1.8	1.6	1.5	1.6	1.7	1.3	1.4
1.7	1.3	1.4	1.7	1.9	1.8	1.6,	1.8	1.4	1.3	1.5	1.7	1.8	2.0	1.9	2.1
2.3	2.6	1.9	1.8	1.5	1.8	1.2	1.4	1.6	1.7	1.8	2.0	1.9	2.2	2.5	1.9
2.0	1.9	2.1	2.2	2.4											

10 MEASURES OF CENTRAL TENDENCY AND DISPERSION

I. CENTRAL TENDENCY

Generally it is found that the values of the variable tend to concentrate around some central value of observations of an investigation, which can be taken as a representative for the whole data. The tendency of the distribution is known as **central tendency** and the measures devised to consider this tendency are known as **measures of central tendency**. This value helps to understand the characteristics of entire mass of statistical data.

This measurement of central tendency should be based on all data available, it should be least affected by the fluctuations of sampling and should not be affected by the extreme values of the data.

The central tendency can be measured by position and also numerically. The **mathematical average** is called 'mean' and **positional average** can be expressed by 'median' and 'mode'. There are three different types of measurement of mean: (a) Arithmetic Mean, (b) Geometric Mean, (c) Harmonic Mean.

A. MATHEMATICAL AVERAGE

10.1 MEAN

(a) Arithmetic Mean

It is most commonly used of all the averages. It is the value which we get by dividing the aggregate of various items of the same series by the total number of observations.

Calculation for ungrouped data

When observations are denoted by x values showing x_1, x_2, x_3, ..., x_n; the total number of observations is calculated by summing up the observations and dividing the sum by the total number of observations (n)

$$\bar{x} = \frac{x_1 + x_2 + x_3 + \cdots + x_n}{n}$$

Example 1: The pod length of ten pods of a plant shows following data:

5.2 cm, 5.3 cm, 5.6 cm, 5.7 cm, 5.4 cm,

5.2 cm, 5.3 cm, 5.3 cm, 5.4 cm, 5.2 cm.

Find out the average pod length of the plant.

$$\bar{x} = \frac{\Sigma x}{n} = \frac{5.2+5.3+5.6+5.7+5.4+5.2+5.3+5.3+5.4+5.2}{10} \text{cm}$$

$$= \frac{53.6}{10} \text{cm} = 5.36 \text{ cm}$$

Calculation for grouped data

When the series is discrete, each value of the variable is multiplied by their respective frequencies, sum of all values is divided by total number of frequencies. Variable x has the values like x_1, x_2, x_3, ..., x_n and their frequencies are f_1, f_2, f_3, ..., f_n respectively.

Then **arithmetic mean**

$$\bar{x} = \frac{f_1x_1 + f_2x_2 + f_3x_3 + \cdots + f_nx_n}{f_1 + f_2 + f_3 + \cdots + f_n} = \frac{\Sigma fx}{\Sigma f}$$

When the series is continuous, the arithmetic mean is calculated after taking the mid point value of class intervals.

$$\bar{x} = \frac{\Sigma fm}{\Sigma f}$$

where, $\bar{x}$ = Arithmetic mean

$\Sigma f \cdot m$ = Sum values of mid point value multiplied by their frequencies

Σf = Sum of frequencies

m = Mid points of various class intervals.

Example 2: An observation on 32 Balsam plants shows the following data. Calculate the arithmetic mean.

No. of flowers/plant (x)	4	5	6	7	8	9
No. of plants (f)	3	5	6	9	5	4

No. of flowers / plant (x)	No. of plants (f)	f × x
4	3	12
5	5	25
6	6	36
7	9	63
8	5	40
9	4	36
	Σf = 32	Σfx = 212

$$\bar{x} = \frac{\Sigma fx}{\Sigma f} = \frac{212}{32} = 6.62 \text{ (approx.)}$$

The average number of flowers / plant is 6.62.

Data of Example 2 of Chapter 9 is used for calculation of mean number of pods. The data arranged in class intervals of 3 are used here.

No. of pods/plant	Mid points of class (m)	No. of plants frequency (f)	m f.
15-17	16	5	80
18-20	19	6	114
21-23	22	8	176
24-26	25	12	300
27-29	28	22	616
30-32	31	18	558
33-35	34	15	510
36-38	37	9	333
39-41	40	5	200
		Σf = 100	Σmf = 2,887

$$\text{Arithmetic Mean} = \bar{x} = \frac{\Sigma mf}{\Sigma f} = \frac{2{,}887}{100} = 28.87.$$

Merits, Demerits and Uses of Arithmetic Mean

Merits

1. It has the simplest formula to calculate and it is easily understood.
2. It is rigidly defined mathematical formula, the same result will come on repeated calculations.
3. The calculation is based on all the observations.
4. It is least affected by sampling fluctuation.

5. The arithmetic mean balances the value on either side.
6. It is the best measure to compare two or more series.
7. Arithmetic mean is totally dependent on values not on the position.

Demerits

1. It cannot be calculated if all the values are not known.
2. The extreme values have greater effect on mean.
3. The qualitative data cannot be measured in this way.

Uses

1. The arithmetic mean is mostly used in practical statistics.
2. Mean helps to calculate many other estimates in statistics.
3. The arithmetic mean is most popular method of any measurement used by common people to get the average of any data.

(b) Geometric Mean

The geometric mean is defined as the n-th root of the product of n observations.

$$\textbf{Geometric Mean (GM)} = \sqrt[n]{x_1 \cdot x_2 \cdot x_3 \cdots x_n}$$

Where n = number of observations; $x_1, x_2, x_3, \ldots, x_n$ = variable values.

When n is small then the above formula can be applied but in case of large 'n' number the logarithms are used to find out the GM

$$\text{GM} = \text{Anti log}\frac{\log x_1 + \log x_2 + \log x_3 + \cdots + \log x_n}{n} = \text{Anti log}\frac{\log x}{n}$$

Example 3: Find out the geometric mean of the following seeds. x denotes the weight of each seed in mg.

5 mg, 7 mg, 8 mg, 6 mg and 4 mg.

x	log x
5 mg	0.70
7 mg	0.85
8 mg	0.90
6 mg	0.78
4 mg	0.60
	Σ log x = 3.83

$$\frac{\Sigma \log x}{5} = \frac{3.83}{5} = 0.77$$

Antilog of $\frac{\Sigma \log x}{5}$, i.e., antilog of 0.77 = 5.89 mg

So the geometric mean of seed weight = 5.89 mg.

This mean is based on all observations, rigidly defined, less affected by extreme values. This mean is difficult to understand, compute and interpret.

This mean is mostly helpful in averaging ratios, percentage and determining ratio of change. This mean is important in construction of index number.

(c) Harmonic Mean

When the variables are expressed in ratios or rates, the proper average to be calculated through harmonic mean. The harmonic mean is defined as the reciprocal of arithmetic mean of the reciprocal of the given values.

The harmonic mean is applicable only in restricted field such as oxygen consumption/hour, calorie requirement/hour, CO_2 evolution/hour, flow of sap/min, etc.

$$\textbf{Harmonic mean (HM)} = \frac{n}{\frac{1}{x_1}+\frac{1}{x_2}+\frac{1}{x_3}+\cdots+\frac{1}{x_n}} = \frac{n}{\Sigma\left(\frac{1}{x}\right)}$$

Where n = Total number of observation; x_1, x_2, x_3 are the values of variables.

Example 4: In a particular experiment, 5 different sets of *Hydrilla* plants showed O_2 evolution/hour, was recorded.

2.5 c.c./hour, 1.8 c.c./hour, 2.0 c.c./hour

2.2 c.c./hour, 2.4 c.c./hour.

$$HM = \frac{5}{\frac{1}{2.5}+\frac{1}{1.8}+\frac{1}{2.0}+\frac{1}{2.2}+\frac{1}{2.4}} = \frac{5}{.4+.55+.5+.45+.41} = \frac{5}{2.3} = 2.17 \text{ c.c./hour.}$$

So, harmonic mean of the observation is 2.17 c.c./hour. This HM determination is based on all the observations of a series. It gives more weightage to the smaller items and also not much affected by sample fluctuation.

It is not very easy to calculate and also the positive and negative, both values, cannot be computed.

B. AVERAGE OF POSITION

From the data of any observation, one can find a peak in the middle with higher and lower values distributed more or less symmetrically towards both sides of the peak.

10.2 MODE

In a frequency distribution, **'mode'** is defined as "the value of the variable for which the frequency is maximum". From the definition it is clear that mode cannot be determined from a series of individual observation, always depends on the frequency of occurrence of any item.

When the concentration of data gives only one peak then the distribution is **unimodal,** but if the data concentrates at two or more points on a scale of values, then the series is called **bimodal** or **multimodal.**

In the **Example 2,** we find the maximum frequency in case of variable value 7. So the mode value of this observation is 7. This type of distribution is called **unimodal** distribution.

In the **Example 2** of chapter 9, the maximum frequency (22) is observed in case of class value 27-29 (Table 9.4), the mid value of this class is 28. So, the mode value of this observation is **28.**

Example 5: In another observation on 30 Balsam plants shows the following data.

No. of flowers/plant (x)	3	4	5	6	7	8	9	10
No. of plants (f)	1	3	2	8	5	8	2	1

Here the mode value cannot be calculated by mere inspection, as the maximum frequency is observed in case of two values of variable 6 and 8. So to determine the modal class, the data is grouped. If we take 2 values together then the grouped data can be arranged in following ways:

Class value	Mid value (m)	frequency
3-4	3.5	4
5-6	5.5	10
7-8	7.5	13
9-10	9.5	3

Here the modal class is 7-8, where mid value is 7.5, so the mode-value of this distribution is 7.5. This type of distribution is called **bimodal** distribution.

Merits and Demerits of Mode

Merits

The mode value avoids the effects of extreme items. The value is got by mere inspection of datas. All values need not to be known, it refers to a measurement which is most usual and most likely variate. The bimodal or multimodal distribution give good indication of the heterogeneity of the population.

Demerits

This value does not need any kind of computation. It becomes difficult to comment on bimodal or multimodal distribution. This value is less dependable as all observations in a series do not have any influence on the value.

10.3 MEDIAN

The median of a distribution is defined as the value of that variable which divides the total frequency into two equal parts when the series is arranged in ascending or descending order of magnitude. So in a distribution, half of the values remain below median value and half of the values remain above the median value.

Median value for ungrouped data

Median value is the value of the $\frac{n+1}{2}$th item. But this formula is applicable straightly when item number is odd. But when the item number is even, the median value is calculated by the mean value of $\frac{n}{2}$th and $\left(\frac{n}{2}+1\right)$th items.

$$\therefore \text{ Median} = \frac{\frac{n}{2}\text{th value} + \left(\frac{n}{2}+1\right)\text{th value}}{2}$$

Example 6: Calculate the median number of flowers in the following observation obtained from garden plants.

Plant no.	1	2	3	4	5	6	7	8
No. of flowers	20	17	25	18	23	21	16	26

Item no.	No. of flowers/plant Ascending	No. of flowers/plant Descending
1	16	26 *
2	17	25
3	18	23
4	20	21
5	21	20
6	23	18
7	25	17
8	26 *	16

The observations are arranged in both ascending and decending order. In case of observation of 7 plants the * marked item no. should not be considered.

If we take 7 observations, then the median value will be value of $\frac{7+1}{2}$th, i.e., 4th item, i.e., **20.**

If we take 8 observations, then the median value will be the mean of $\frac{8}{2}$ th and $\frac{8}{2}+1$ th item.

∴ Mean of 4th and 5th item, i.e., mean of 20 and 21 which is **20.5.**

Median value for grouped data

For grouped data, the classes are arranged according to the ascending order and respective frequencies are written against them. The frequencies are then cumulated and position of the median is calculated by the same formula. The median value is the mid value of the class in which the median item value is placed.

From **Example 2** of **Chapter 9** :

Class interval	Mid value	Frequency	Cumulative frequency
15-17	16	5	5
18-20	19	6	11
21-23	22	8	19
24-26	25	12	31
27-29	28	22	53
30-32	31	18	71
33-35	34	15	86
36-38	37	9	95
39-41	40	5	100

As the total number of variables is 100, so the median value will be the value which is in between the value of 50th and 51st item value.

50th and 51st item value is in the class interval 27-29 (no. of pods).

The median value is **28** of this observation.

Merits and Demerits of Median

Merits

In normal distribution, the median value is near the mean value which is more easier to calculate. This value eliminates the effect of extreme items, since they are not taken into account for its calculation, only the middle items are required to be known.

Demerits

When the distribution is irregular then the median value is not at all the true representative of the series. In case of grouped data also, the precision is not there, this value is not very useful for further analysis, as it is difficult to handle mathematically.

Relationship between Mean, Median and Mode

If the distribution is symmetrical then the Mean, Mode and Median value coincide, otherwise the distance between the Mode and Median is usually twice the distance between the Median and the Mean. Thus

Mode – Median = 2 (Median – Mean)

or

Mode – Mean = 3 (Median – Mean)

In the **Example 2** of **Chapter 9,** the calculated arithmetic mean is 28.87, the mode value is 28, and the median value is also 28. Coincidence of the Mean, Mode and Median value reflects the symmetrical distribution of the observed data.

10.4 DISTRIBUTION PATTERNS

In the previous sections, the frequency distribution and their graphical representation like frequency polygon, histogram, bar diagram have been discussed. But sometimes the knowledge of distribution pattern is required to comment on the population on the basis of observation of sample.

Normal distribution

If we observe in any population any attribute is distributed mostly near the mean value and equally distributed to the higher and lesser value gradually in decreasing order then the distribution pattern is called **normal distribution.**

When this kind of normally distributed attribute is plotted graphically with the help of available data, the normal distribution pattern gives a bell shaped symmetrical curve which is called **'normal distribution curve'**. In this curve the mean value lies in the peak of the curve.

Properties of normal distribution curve

1. It is a continuous bell shaped curve which is associated with continuous variable.
2. There is only one maximum peak (unimodal). The normal curve is symmetrical (Fig. 10.1a) and asympotic (touches at infinity).
3. The height of normal curve is maximum at its Mean. Mean, Median and Mode coincides in normal curve.
4. The peak divides the distribution in two equal halves.
5. Most of the observations are clustered around the Mean and there are relatively a few observations at the extremes.
6. The normal distribution curve has a fixed mathematical characteristic feature independent of the scale (unit of measurement) of magnitude.

Skewness and Kurtosis

In normal distribution, most of the cases fall in the middle but there are cases in which central tendency do not exhibit normal behaviour.

There are two types of **divergence** from normal distribution — (i) **Skewness** and (ii) **Kurtosis**.

(i) **Skewness** means that the curve is not symmetrical. In a skewed distribution, the Mean, Median and Mode do not coincide, it pulls the Median and Mean away from Mode either left or right. In a skewed distribution, the frequency curve is not bell shaped and values do not lie on both sides of measure of central tendency equally. Here Mean, Median and Mode fall at different points.

In **symmetrical distribution curve,** Mode coincides with Mean and Median.

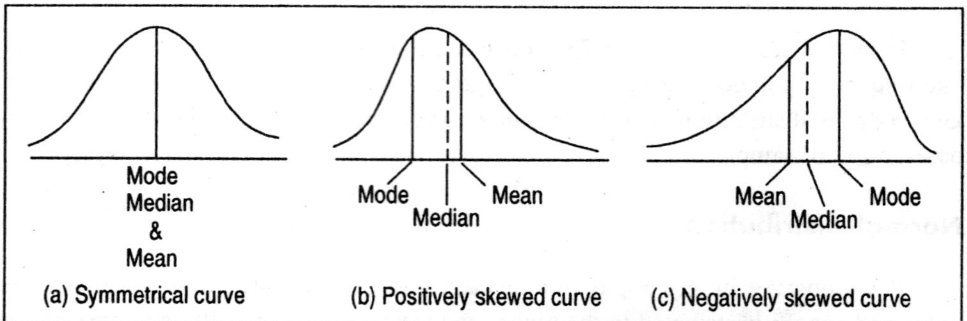

Fig. 10.1: Symmetrical and skewed curve

In **positively skewed curve**, the value of Mean and Median lie away from Mode values (right hand), the values are greater than Mode.

In **negatively skewed curve,** the value of Mean and Median lie left hand to Mode value, the values are lesser than the Mode value (Fig. 10.1).

(ii) **Kurtosis** means the bulginess, it measures the degree of peakedness of a frequency distribution. In case of unimodal normal distribution, the peak has a flat top while the others may have peaked top. The degree of peakedness of a distribution is called as kurtosis.

We may have **three kinds of kurtosis** (Fig. 10.2):

1. **Leptokurtic curve** where the curve has high peak and long tails.
2. **Mesokurtic curve** which has moderately high peak comparable to the normal curve, peak is neither high nor flat.
3. **Platykurtic curve** when the curve has very flat top for its peak.

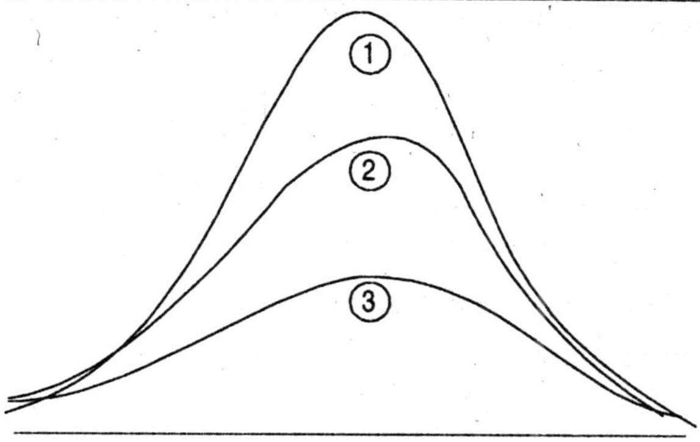

(1) Leptokurtic, (2) Mesokurtic, (3) Platykurtic curve

Fig. 10.2: Kurtosis

II. DISPERSION

In statistics, dispersion is commonly used to mean the scattering of data, deviation, fluctuation, spread or variability of data. The term **'dispersion'** can be defined as the degree to which the individual values of the variate scatter away from the average or the central value.

Variability is a normal biological phenomenon, and it is important to measure this variability. This measurement helps us to find out how individual observations are dispersed around the mean of a large series. This is also named as **measures of dispersion**.

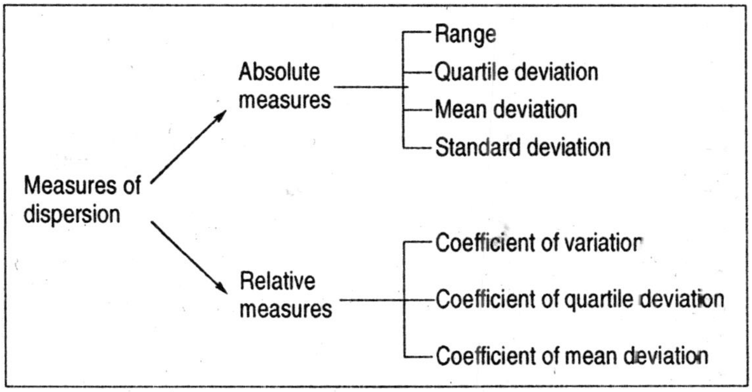

10.5 MEAN DEVIATION

It is the easier way of measuring dispersion with little statistical importance. When the value of variables differ in its value from its mean by an amount, the difference is known as the deviation. Deviations above the mean are negative and below the mean are positive.

Mean deviation (δ) is the average of the absolute values of the deviation from the mean.

$$\text{MD or } \delta = \frac{\Sigma|d|}{n}$$

where, $|d| = x - \bar{x}$ = deviation from actual mean, not considering the '+' and '–' sign.

n = total number of observations.

For **grouped data**, the mean deviation is calculated following the formula:

$$\delta = \frac{\Sigma|f \cdot d|}{\Sigma f}$$

where, $|f \cdot d|$ = multiplication of frequency with each class deviation, not considering the '+' or '–' sign.

Σf = sum of frequencies.

Calculation of the mean deviation from the Example 2 of Chapter 9:

A. When the data has not been arranged in classified manner.

No. of Pods x	Frequency f	fx	Deviation $\|d\|$	$f\|d\|$
15	1	15	14	14
16	2	32	13	26
17	2	34	12	24
18	2	36	11	22
19	2	38	10	20
20	2	40	9	18
21	2	42	8	16
22	3	66	7	21
23	3	69	6	18
24	3	72	5	15
25	4	100	4	16
26	5	130	3	15
27	6	162	2	12
28	7	196	1	7
29	9	261	0	0
30	7	210	1	7
31	5	155	2	10
32	6	192	3	18
33	6	198	4	24

Contd.

No. of Pods x	Frequency f	fx	Deviation |d|	f|d|
34	5	170	5	25
35	4	140	6	24
36	3	108	7	21
37	3	111	8	24
38	3	114	9	27
39	2	78	10	20
40	2	80	11	22
41	1	41	12	12
	Σf = 100	Σfx = 2,890		Σf|d| = 478

$$\overline{x} = \frac{\Sigma fx}{\Sigma f} = \frac{2,890}{100} = 28.90 \text{ or } 29$$

$$\text{Mean deviation} = \frac{\Sigma f|d|}{\Sigma f} = \frac{478}{100} = 4.78$$

B. **When the same data has been arranged in classified form**

No. of Pods per Plant	Midpoints m	No. of Plants f	Deviation |d|	|f × d|
15-17	16	5	13	65
18-20	19	6	10	60
21-23	22	8	7	56
24-26	25	12	4	48
27-29	28	22	1	22
30-32	31	18	2	36
33-35	34	15	5	75
36-38	37	9	8	72
39-41	40	5	11	55
	Total	Σf = 100		Σ|fd| = 489

Here deviations are shown as positive values since all negative values are treated as positives. Thus Σ|fd| comes to 489 and the mean deviation $= \frac{\Sigma|fd|}{\Sigma f} = \frac{489}{100} = 4.89$

Significance: Mean deviation is easy to calculate but it has less mathematical value, rarely applied for biostatistics. As the negative values are ignored, so the mean deviation is less meaningful.

10.6. VARIANCE

Variance depends on the deviations where the squared deviations are summed up and then divided by the number of observations to get the sample variance. It has a distinct advantage over mean deviation as the squaring is done for the deviated values, as a result all values become positive.

$$\textbf{Sample variance } (s^2) = \frac{\Sigma(x-\overline{x})^2}{n} = \frac{\Sigma d^2}{n}$$

where, $\overline{x}$ = Mean of the sample

n = Total no. of observation.

In case of grouped data, due to frequency distribution, the variance $s^2 = \frac{\Sigma f \cdot d^2}{\Sigma f}$

Since the sample is the part of the population with only n number of items, to get the estimate of population variance the formula for sample variance is transformed to —

$$\textbf{Population Variance} = \sigma^2 = \frac{\Sigma d^2}{n-1} \quad \text{or} \quad \frac{\Sigma f \cdot d^2}{\Sigma f - 1}$$

Calculation of variance from the data observed in Example 2 of Chapter 9:

No. of Pods (x)	Frequency (f)	Deviation (|d|)	Deviation^2 (d^2)	f $\times d^2$
15	1	14	196	196
16	2	13	169	338
17	2	12	144	288
18	2	11	121	242
19	2	10	100	200
20	2	9	81	162
21	2	8	64	128
22	3	7	49	147
23	3	6	36	108
24	3	5	25	75
25	4	4	16	64
26	5	3	9	45
27	6	2	4	24
28	7	1	1	7
29	9	0	0	0
30	7	1	1	7

Contd.

No. of Pods (x)	Frequency (f)	Deviation (|d|)	Deviation² (d^2)	$f \times d^2$
31	5	2	4	20
32	6	3	9	54
33	6	4	16	96
34	5	5	25	125
35	4	6	36	144
36	3	7	49	147
37	3	8	64	192
38	3	9	81	243
39	2	10	100	200
40	2	11	121	242
41	1	12	144	144
	$\Sigma f = 100$			$\Sigma fd^2 = 3{,}638$

$$\bar{x} = 28.9 \text{ or } 29;\ \Sigma fd^2 = 3{,}638$$

$$\Sigma f = 100;\ \text{variance, } s^2 = \frac{\Sigma fd^2}{\Sigma f} = \frac{3{,}638}{100} = 36.38$$

Significance: Variance is a quantitative mathematical expression representing squared units. It has two major difficulties:

1. When the deviation and number of observation are more, then variance becomes a large number which is difficult to be expressed numerically.
2. The unit in which the variance is expressed is not in the same unit of the observation, such as, if the observations are made in cm, then the variance is expressed in sq. cm.

10.7 STANDARD DEVIATION

This is most useful method of measurement of dispersion of a series where the values deviated from mean are squared and summed up and then expressed as square root of the summed up value divided by no. of observations. Thus the standard deviation is defined as the square root of the variance.

$$\text{Sample Standard Deviation or } s = \sqrt{s^2} \text{ or } \sqrt{\frac{\Sigma d^2}{n}} \text{ or } \sqrt{\frac{\Sigma f \cdot d^2}{\Sigma f}}$$

where, d = deviation from mean

n = total no. of observations

f = frequency of each class

When the population standard deviation is estimated then the formula is:

$$\text{Population standard deviation or } \sigma = \sqrt{\frac{\Sigma d^2}{n-1}} \text{ or } \sqrt{\frac{\Sigma f \cdot d^2}{\Sigma f - 1}}$$

Why n – 1 is used to calculate the standard deviation when sample size is small?

When the sample size is large enough (e.g., 1,000) to reach the population size, i.e., if 'n' approaches towards 'ν' (n = sample size, ν = population size), then the value of variance will show negligible difference whether divided by n or n – 1. But when the sample size is small, e.g. 30, then the value of population variance should be obtained by dividing with n – 1, which gives more appropriate estimate of population variance.

n – 1 denotes number of degrees of freedom, i.e., number of comparisons that can be made between any one observation and the rest number of observations taking them in pairs.

Computation of standard deviation needs 7 steps:

Step I : Mid-point calculation of each class (m).

Step II : Calculation of mean of total observations $(\bar{x})$.

Step III : Deviation from mean of each mid value of each class $d = (\bar{x} - m)$.

Step IV : Squaring of each deviation (d^2).

Step V : Multiplication of squared deviation with frequency $(f \cdot d^2)$.

Step VI : Summation of all $f \cdot d^2$ and then division by $(\Sigma f - 1)$.

Step VII : Square root of the total value, $\sqrt{\frac{\Sigma fd^2}{\Sigma f - 1}}$.

Calculation of standard deviation from the data observed in Example 2 of chapter 9:

Variance (s^2) calculated previously in this example is 36.38.

∴ Standard deviation or $s = \sqrt{36.38} = 6.03$

Calculation of standard deviation when the data is classified or grouped from Example 2 of Chapter 9:

No. of Pods/Plant	Mid point (m)	No. of Plants (f)	Deviation (|d|)	d^2	$f \cdot d^2$
15-17	16	5	13	169	845
18-20	19	6	10	100	600
21-23	22	8	7	49	392
24-26	25	12	4	16	192
27-29	28	22	1	1	22
30-32	31	18	2	4	72
33-35	34	15	5	25	375
36-38	37	9	8	64	576
39-41	40	5	11	121	605
		$\Sigma f = 100$			$\Sigma fd^2 = 3{,}679$

From previous calculation the data shows the mean $(\overline{x}) = 29$.

Deviation is calculated using the formula $d = \overline{x} - m$

$$\text{Standard deviation or } s = \sqrt{\frac{3{,}679}{100}} = \sqrt{36.79} = 6.06.$$

Merits and Demerits of Standard Deviation

Merits

1. The calculation is based on all observations.
2. It is more rigidly defined.
3. Less affected by fluctuations of sampling compared to other measures of dispersion.
4. It summarises the deviation of large number of observations from mean and is expressed as one unit of variation.

Demerits

1. It requires a lengthy calculation, i.e., squaring of deviations and then again square root of summed up values.
2. Not very simple to understand.
3. The calculation gives more weightage to extreme values.

Uses of Standard Deviation

1. It helps in correlating and comparing of different samples.
2. It helps in finding the suitable size of sample for valid conclusion.
3. It helps in finding the standard error which determines whether the difference between means of two similar samples by chance or real.
4. The value of mean and standard deviation help to comment on the population on the basis of observation of sample (Fig. 10.3).

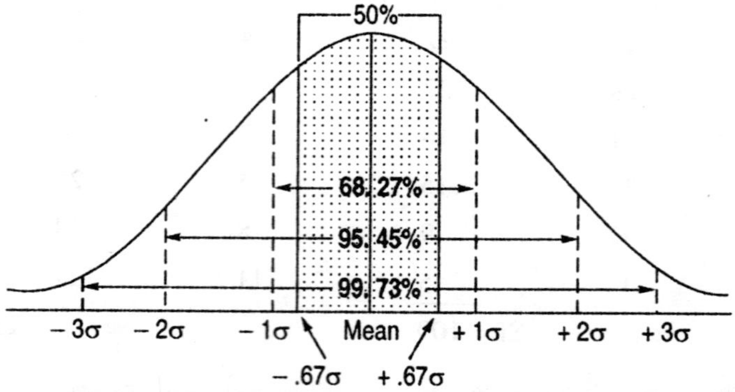

Fig. 10.3: Relationship between Normal Distribution Curve and Standard Deviation

(a) 50% of total observations lie in an area bounded by a distance of 0.6745 σ on both side of the mean.

(b) Mean ± 1σ covers the 68.27% area of the curve.

(c) Mean ± 2σ covers the 95.45% area of the curve.

(d) Mean ± 3σ covers the 99.73% area of the curve.

10.8 COEFFICIENT OF VARIATION

In measurement of dispersion, we use the units which are used for observation. But if we want to compare the dispersion of two different characters in the same population, then the calculation of coefficient of variation is needed. This measurement is expressed in percentage ignoring the units. For example, from the same population if we like to study the length of pod and number of seeds/pod and to compare the dispersion of both the two characters, we must calculate the **coefficient of variation.**

$$\text{Coefficient of variation} = \frac{\text{Standard deviation}}{\text{Mean}} \times 100$$

or, $\text{c.v.} = \frac{\sigma}{\bar{x}} \times 100$ or $\frac{s}{\bar{x}} \times 100$ in case of samples or population respectively.

This coefficient of variation is also helpful to get an idea or compare two different populations about the dispersion of a character. More the c.v., more is the inconsistency about the dispersion of character.

Example 7: In two different populations (Batch I and Batch II) the seed number/fruit is calculated:

Batch I	7, 9, 6, 8, 6, 5, 7, 8, 6, 8
Batch II	10, 8, 9, 10, 11, 10, 5, 6, 4, 7

Batch I

Seed number/fruit (x)	Frequency (f)	f × x	Deviation (|d|)	Deviation² (d²)	f × d²
5	1	5	2	4	4
6	3	18	1	1	3
7	2	14	0	0	0
8	3	24	1	1	3
9	1	9	2	4	4
	Σf = 10	Σfx = 70			Σfd² = 14

$$\overline{x} = \frac{\sum fx}{n} = \frac{70}{10} = 7$$

$$\sigma = \sqrt{\frac{\sum fd^2}{n-1}} = \sqrt{\frac{14}{9}} = 1.25$$

$$\text{c.v.} = \frac{\sigma}{\overline{x}} = \frac{1.25}{7} \times 100 = 17.8\%$$

Batch II

Seed number/fruit (x)	Frequency (f)	f × x	Deviation (|d|)	Deviation² (d²)	f × d²
4	1	4	4	16	16
5	1	5	3	9	9
6	1	6	2	4	4
7	1	7	1	1	1
8	1	8	0	0	0
9	1	9	1	1	1
10	3	30	2	4	12
11	1	11	3	9	9
	Σf = n = 10	Σfx = 80			Σfd² = 52

$$\overline{x} = \frac{\sum fx}{n} = \frac{80}{10} = 8$$

$$\sigma = \sqrt{\frac{\Sigma fd^2}{n-1}} = \sqrt{\frac{52}{9}} = 2.4$$

$$\text{c.v.} = \frac{\sigma}{\bar{x}} = \frac{2.4}{8} \times 100 = 30\%$$

The calculation shows :

Batch I	Batch II
$n_1 = 10$	$n_2 = 10$
$\Sigma x_1 = 70$	$\Sigma x_2 = 80$
$\bar{x}_1 = 7$	$\bar{x}_2 = 8$
$\sigma_1 = 1.25$	$\sigma_2 = 2.4$
$\text{c.v.}_1 = 17.8\%$,	$\text{c.v.}_2 = 30\%$

From the above result, we can conclude that the Second Batch of fruits has relatively more number of seeds/fruit and also this Second Batch has inconsistent number of seeds per fruit, i.e., variation of seed number is more.

10.9 STANDARD ERROR

Standard error of a statistics provides a measure of accuracy of that statistics. Standard error is the **index number** and theoretically the difference between statistics and parameter should be zero.

Standard error of the mean (SE_M) is an estimate of the range by which the means of a number of sets deviate about the mean of those means. It is calculated by the formula :

$$SE_M = \frac{s}{\sqrt{n}} \text{ or } \frac{\sigma}{\sqrt{n}}$$

s = Sample standard deviation

σ = Population standard deviation

n = Total no. of observations

Calculation of SE_M from Example 2 of Chapter 9:

The standard deviation calculated in this example previously is, s = 6.06 and n = 100.

$$\therefore \text{the } SE_M = \frac{s}{\sqrt{n}} = \frac{6.06}{\sqrt{100}} = \frac{6.06}{10} = 0.606$$

In **Example 7,** the SE_M value calculated in both the batches:

$$SE_M \text{ of Batch I} = \frac{1.25}{\sqrt{10}} = \frac{1.25}{3.16} = 0.39$$

$$SE_M \text{ of Batch II} = \frac{2.4}{\sqrt{10}} = \frac{2.4}{3.16} = 0.76$$

SE_M value is more reduced than Standard deviation value, it becomes further reduced when number of observations are more. So the magnitude of SE_M depends upon the number of individuals (n) of the sample. This value is directly related with population standard deviation and inversely related to square root of sample size.

QUESTIONS

1. How will you calculate the arithmetic mean from individual data or grouped data? Write down the merits and demerits of arithmetic mean.
2. Define and explain mean, median and mode. Which one is more acceptable in statistical analysis and why?
3. Define geometric mean and harmonic mean. When will you take the help of this type of mean calculations?
4. Define normal distribution curve. Mention its properties. Discuss the divergence from normal distribution.
5. What do you mean by dispersion or variability? How will you measure the dispersion of a series of observations.
6. What is standard deviation? Why measurement of SD is more applicable than MD in Biostatistics. Show the steps of calculation of standard deviation. What are its merits and demerits?
7. Write short notes on :

   (a) Variance; (b) Co-efficient of variation; (c) Standard error.
8. Classify the data and calculate mean and standard deviation.

   No. of seeds/fruit

   12 14 13 6 7 21 17 15 13 12 7 8 11 20 11 19 18 9 10

   13 13 15 12 14 13 11 9 8 16 14 18 9 10 16 15 16 10 10

   13 11 14 12 14 13 11 13 14 15 12 and 17
9. No. of *Indigofera* plants/m^2

   20 24 23 21 22 40 39 39 40 38 37 38 37 25

   30 29 28 26 25 26 25 34 36 31 35 36 31 35

   36 35 32 35 33 35 and 34.

   (i) Calculate the mean number of plants/m^2

   (ii) Calculate the standard error of mean.
10. Grain lengths of two varieties of rice are given below. Calculate the mean, SEM and c.v. of grain length of the two varieties. Write down your comments on comparison.

Variety A		Variety B	
Grain length (mm)	No. of grains	Grain length (mm)	No. of grains
9-11	3	9-11	0
12-14	5	12-14	8
15-17	9	15-17	8
18-20	3	18-20	4

11 TEST OF SIGNIFICANCE AND ANALYSIS OF VARIANCE

I. TEST OF SIGNIFICANCE

In biological research when we compare any character of two samples, we calculate the significance of difference in the mean and variance to draw a meaningful conclusion. For example, if we are interested to compare the protein content of two seed varieties of maize or yield of two crop varieties, then it is necessary to assess the significance of the difference in mean or variance. **Test of significance** helps us in determining whether the difference between the two samples are actually due to chance factor or the difference is really significant among the samples.

11.1 NULL HYPOTHESIS AND ALTERNATIVE HYPOTHESIS

Testing of hypothesis is the procedure which approaches the comparison between means. In biological science statistical test of hypothesis plays an important role. The test of a sample mean $(\overline{x})$ is done against a population mean (μ) or between the two means of two samples $(\overline{x}_1$ and $\overline{x}_2)$. The hypothesis or a statement is that either $\mu = \overline{x}$ or $\overline{x}_1 = \overline{x}_2$, i.e., examination should be done to determine whether there is no difference between μ and $\overline{x}$ or between $\overline{x}_1$ and $\overline{x}_2$.

The hypothesis is that μ and $\overline{x}$ are equal or $\overline{x}_1$ and $\overline{x}_2$ are equal which is called **null hypothesis**. It is expressed in the following way

$$H_0 : \overline{x}_1 = \overline{x}_2 \qquad \text{or} \qquad H_0 : \mu = \overline{x}$$

$\overline{x}_1$ = Mean of sample 1 $\qquad \mu$ = Population mean,

$\overline{x}_2$ = Mean of sample 2 $\qquad \overline{x}$ = Sample mean

Null hypothesis (H_0) can be defined as the hypothesis which is a test for possible rejection under the assumption that it is true, i.e., it is accepted.

Alternative hypothesis (H_1 or H_A) can be defined as the hypothesis which is complementary to the null hypothesis. Rejection of null hypothesis leads to the acceptance of alternative hypothesis.

Generally it can be written as,

$$H_A \text{ or } H_1 : \mu \neq \overline{x} \quad \text{or} \quad \overline{x}_1 \neq \overline{x}_2$$

which states that $\overline{x}$ is not equal to population mean (μ) or there is difference between two sample means, i.e., either $\overline{x}_1 > \overline{x}_2$ or $\overline{x}_1 < \overline{x}_2$.

11.2 STUDENT'S 't'-TEST

To test the significance of difference of means of two samples, W.S. Gosset applied a statistical tool called **'t'-test**. According to nick name of Gosset, the test has been named as **Student's 't'-test**. In this test we make a choice between two alternatives:

1. To accept the null hypothesis (no difference between the two means);
2. To reject the null hypothesis (the difference between the means of two samples significant).

The test is applied to small sample (n<30), and samples must be drawn randomly from normal population. The 't' is defined as quantity representing the difference between the sample mean or true mean or population mean expressed in terms of the standard error.

$$t = \frac{\text{Difference between sample means}}{\text{Standard error of the difference between means}}$$

$$t = \frac{\overline{x}_1 - \overline{x}_2}{SE_D}$$

$\overline{x}_1$ and $\overline{x}_2$ = Mean of two samples

SE_D = Standard error of the difference between means

Degree of Freedom

To calculate the standard error of the difference, the **degree of freedom** can be obtained by one less than the number of observations of sample. In case of unpaired 't' test, i.e., when the comparison between two samples is done, degree of freedom is calculated by the formula

$(n_1 + n_2) - 2$, where n_1 = no. of observations in sample I

n_2 = no. of observations in sample II.

Whereas, in case of paired 't' test, the degree of freedom is (n – 1), as there is the same number of observations.

11 TEST OF SIGNIFICANCE AND ANALYSIS OF VARIANCE

I. TEST OF SIGNIFICANCE

In biological research when we compare any character of two samples, we calculate the significance of difference in the mean and variance to draw a meaningful conclusion. For example, if we are interested to compare the protein content of two seed varieties of maize or yield of two crop varieties, then it is necessary to assess the significance of the difference in mean or variance. **Test of significance** helps us in determining whether the difference between the two samples are actually due to chance factor or the difference is really significant among the samples.

11.1 NULL HYPOTHESIS AND ALTERNATIVE HYPOTHESIS

Testing of hypothesis is the procedure which approaches the comparison between means. In biological science statistical test of hypothesis plays an important role. The test of a sample mean $(\overline{x})$ is done against a population mean (μ) or between the two means of two samples $(\overline{x}_1$ and $\overline{x}_2)$. The hypothesis or a statement is that either $\mu = \overline{x}$ or $\overline{x}_1 = \overline{x}_2$, i.e., examination should be done to determine whether there is no difference between μ and $\overline{x}$ or between $\overline{x}_1$ and $\overline{x}_2$.

The hypothesis is that μ and $\overline{x}$ are equal or $\overline{x}_1$ and $\overline{x}_2$ are equal which is called **null hypothesis**. It is expressed in the following way

$$H_0 : \overline{x}_1 = \overline{x}_2 \qquad \text{or} \qquad H_0 : \mu = \overline{x}$$

$\overline{x}_1$ = Mean of sample 1 $\qquad$ μ = Population mean,

$\overline{x}_2$ = Mean of sample 2 $\qquad$ $\overline{x}$ = Sample mean

Null hypothesis (H_0) can be defined as the hypothesis which is a test for possible rejection under the assumption that it is true, i.e., it is accepted.

Alternative hypothesis (H_1 or H_A) can be defined as the hypothesis which is complementary to the null hypothesis. Rejection of null hypothesis leads to the acceptance of alternative hypothesis.

Generally it can be written as,

$$H_A \text{ or } H_1 : \mu \neq \bar{x} \quad \text{or} \quad \bar{x}_1 \neq \bar{x}_2$$

which states that $\bar{x}$ is not equal to population mean (μ) or there is difference between two sample means, i.e., either $\bar{x}_1 > \bar{x}_2$ or $\bar{x}_1 < \bar{x}_2$.

11.2 STUDENT'S 't'-TEST

To test the significance of difference of means of two samples, W.S. Gosset applied a statistical tool called **'t'-test**. According to nick name of Gosset, the test has been named as **Student's 't'-test**. In this test we make a choice between two alternatives:

1. To accept the null hypothesis (no difference between the two means);
2. To reject the null hypothesis (the difference between the means of two samples significant).

The test is applied to small sample (n<30), and samples must be drawn randomly from normal population. The 't' is defined as quantity representing the difference between the sample mean or true mean or population mean expressed in terms of the standard error.

$$t = \frac{\text{Difference between sample means}}{\text{Standard error of the difference between means}}$$

$$t = \frac{\bar{x}_1 - \bar{x}_2}{SE_D}$$

$\bar{x}_1$ and $\bar{x}_2$ = Mean of two samples

SE_D = Standard error of the difference between means

Degree of Freedom

To calculate the standard error of the difference, the **degree of freedom** can be obtained by one less than the number of observations of sample. In case of unpaired 't' test, i.e., when the comparison between two samples is done, degree of freedom is calculated by the formula

$(n_1 + n_2) - 2$, where n_1 = no. of observations in sample I

n_2 = no. of observations in sample II.

Whereas, in case of paired 't' test, the degree of freedom is $(n - 1)$, as there is the same number of observations.

Example 1: In an experiment to find out the effect of a hormone spray on the seed yield of dwarf french bean, the following results were obtained. Analyse the data using the 't' test for inference on the effect of hormonal spray on the seed yield.

No. of seeds/plant

Control : 30 35 31 36 38 32 25 39 26 27 40 41 35 38 32
31 38 30 35 36 25 28 29 37 36 36 30 31 34 33
35 40 30 32 28 26 29 30 30 35 36 37 38 39 30
31 39 40 35 36

Treated : 35 38 40 45 40 42 46 48 49 50 35 38 45 47 45
46 50 51 46 41 42 48 49 51 52 50 53 54 55 43
45 50 44 51 52 53 52 48 50 49 47 49 50 49 51
52 50 51 52 40

Control (x_1)					**← No. of seeds/plant →**			**Treated (x_2)**	
x_1	f	fx_1	x_1^2	fx_1^2	x_2	f	fx_2	x_2^2	fx_2^2
25	2	50	625	1250	35	2	70	1225	2450
26	2	52	676	1352	38	2	76	1444	2888
27	1	27	729	729	40	3	120	1600	4800
28	2	56	784	1568	41	1	41	1681	1681
29	2	58	841	1682	42	2	84	1764	3528
30	7	210	900	6300	43	1	43	1869	1849
31	4	124	961	3844	44	1	44	1936	1936
32	3	96	1024	3072	45	4	180	2025	8100
33	1	33	1089	1089	46	3	138	2116	6348
34	1	34	1156	1156	47	2	94	2209	4418
35	6	210	1225	7350	48	3	144	2304	6912
36	6	216	1296	7776	49	5	245	2401	12005
37	2	74	1369	2738	50	7	350	2500	17500
38	4	152	1444	5776	51	5	255	2601	13005
39	3	117	1521	4563	52	5	260	2704	13520
40	3	120	1600	4880	53	2	106	2809	5618
41	1	41	1681	1681	54	1	54	2916	2916
					55	1	55	3025	3025
	$\Sigma f=50$	$\Sigma fx_1=1670$		$\Sigma fx_1^2=56726$		$\Sigma f=50$	$\Sigma fx_1=2359$		$\Sigma fx_2^2=112499$

$$\bar{x}_1 = \frac{\Sigma fx_1}{\Sigma f} = \frac{1670}{50} = 33.4$$

$$\text{Variance, } \sigma_1^2 = \frac{\Sigma fx_1^2 - \frac{(\Sigma fx_1)^2}{\Sigma f}}{\Sigma f - 1}$$

$$= \frac{56726 - \frac{(1670)^2}{50}}{49}$$

$$= \frac{56726 - 55778}{49} = \frac{948}{49} = 19.347$$

$$\bar{x}_2 = \frac{\Sigma fx_2}{\Sigma f} = \frac{2359}{50} = 47.18$$

$$\text{Variance, } \sigma_2^2 = \frac{\Sigma fx_2^2 - \frac{(\Sigma fx_2)^2}{\Sigma f}}{\Sigma f - 1}$$

$$= \frac{1112499 - \frac{(2359)^2}{50}}{49}$$

$$= \frac{1112499 - 111297.62}{49} = \frac{1201.38}{49} = 24.518$$

$$t = \frac{|\bar{x}_1 - \bar{x}_2|}{\sigma_d} = \frac{|\bar{x}_1 - \bar{x}_2|}{\sqrt{\frac{\sigma_1^2}{n_1} + \frac{\sigma_2^2}{n_2}}} = \frac{|133.4 - 47.281|}{\sqrt{\frac{19.347}{50} + \frac{24.518}{50}}} = \frac{13.88}{\sqrt{0.387 + 0.49}} = \frac{13.88}{\sqrt{0.877}} = \frac{13.88}{0.936} = 14.829$$

Degrees of freedom = $n_1 + n_2 - 2 = 50 + 50 - 2 = 98$

Calculated t = 14.829

Tabulated t = 3.37 (p = 0.001)

Since the calculated t (14.829) far exceeds the tabulated t (3.37) with 98 d.f. at p = 0.001, the null hypothesis stating that there is no difference in the two sample means, is rejected. Alternatively there is a highly significant difference between the two sample means, i.e., the hormonal spray has a very significant effect on the seed yield.

Example 2: Two varieties of potato plants (A and B) yielded tubers as shown in the following table. Does the mean weight of tubers of the variety A significantly differ from that of variety B.

Variety A :	2.2	2.5	1.9	2.6	2.3	1.8	2.0	2.1	2.4	2.3	kg/plant
Variety B :	2.8	2.5	2.7	3.0	3.1	2.3	2.4	3.2	2.5	2.9	kg/plant

Variety A		**Variety B**	
x_A	x_A^2	x_B	x_B^2
1.8	3.24	2.3	5.29
1.9	3.61	2.4	5.76
2.0	4.00	2.5	6.25
2.1	4.41	2.5	6.25
2.2	4.84	2.7	7.29
2.3	5.29	2.8	7.84
2.3	5.29	2.9	8.41
2.4	5.76	3.0	9.00
2.5	6.25	3.1	9.61
2.6	6.76	3.2	10.24
$\Sigma x_A = 22.10$	$\Sigma x_A^2 = 49.45$	$\Sigma x_B = 27.40$	$\Sigma x_B^2 = 75.94$

$$\bar{x}_A = \frac{22.10}{10} = 2.21 \qquad \bar{x}_B = \frac{27.40}{10} = 2.74$$

$$\Sigma x_A^2 - \frac{(\Sigma x_A)^2}{n_A} = 49.45 - \frac{(22.1)^2}{10} = 49.45 - 48.84 = 0.61$$

$$\Sigma x_B^2 - \frac{(\Sigma x_B)^2}{n_B} = 75.94 - \frac{(27.4)^2}{10} = 75.94 - 75.08 = 0.86$$

$$n_A + n_B - 2 = 10 + 10 - 2 = 20 - 2 = 18$$

Pooled variance,

$$\sigma_d^2 = \frac{0.61 + 0.86}{18} = \frac{1.47}{18} = 0.082$$

Pooled standard deviation,

$$\sigma_d = \sqrt{\sigma_d^2} = \sqrt{0.082} = 0.286$$

$$\therefore \quad t = \frac{|\bar{x}_A - \bar{x}_B|}{\sigma_d\sqrt{\frac{1}{10} + \frac{1}{10}}} = \frac{2.74 - 2.21}{0.286\sqrt{\frac{1}{5}}} = \frac{0.53}{0.286 \times 0.45} = \frac{0.53}{0.129} = 4.108$$

Calculated t = 4.108

Tabulated t = 3.92 (0.001 p) for 18 d.f.

As the calculated t (4.108) for the difference of the two sample means is greater than the tabulated t (3.92) for 18 d.f. at $p \le 0.001$, the difference between the two means is therefore highly significant. The null hypothesis stating that the two sample means are not different is rejected at $p \le 0.001$.

Example 3: Application of fertilizer (NPK) was tested on the yield of rice grown in 10 plots. Another set of 10 plots of similar size and conditions was taken as the control. Determine whether the fertilizer has any significant effect on the yield of rice.

NPK applied		**NPK not applied (control)**	
Plot Nos.	Yield of rice (x_1) (Quintal/hectare)	Plot Nos.	Yield of rice (x_2) (Quintal/hectare)
1	16	1	10
2	14	2	12
3	18	3	11
4	15	4	9
5	13	5	13
6	17	6	13
7	16	7	12
8	15	8	14
9	14	9	13
10	13	10	11

NPK applied			**Control**		
Plot Nos.	Rice yield (x_1) (Quintal/hectare)	x_1^2	Plot Nos.	Rice yield (x_2) (Quintal/hectare)	x_2^2
1	16	256	1	10	100
2	14	196	2	12	144
3	18	324	3	11	121

Contd.

NPK applied			Control		
Plot Nos.	Rice yield (x_1) (Quintal/hectare)	x_1^2	Plot Nos.	Rice yield (x_2) (Quintal/hectare)	x_2^2
4	15	225	4	9	81
5	13	169	5	13	169
6	17	289	6	13	169
7	16	256	7	12	144
8	15	225	8	14	196
9	14	196	9	13	169
10	13	169	10	11	121
Total	$\Sigma x_1 = 151$	$\Sigma x_1^2 = 2305$	Total	$\Sigma x_2 = 118$	$\Sigma x_2^2 = 1414$

$$\therefore\ \bar{x}_1 = \frac{\Sigma x_1}{n_1} = \frac{151}{10} = 15.1 \qquad \therefore\ \bar{x}_2 = \frac{\Sigma x_2}{n_2} = \frac{118}{10} = 11.8$$

$$\Sigma x_1^2 - \frac{(\Sigma x_1)^2}{n_1} = 2305 - \frac{(151)^2}{10} \qquad \Sigma x_2^2 - \frac{(\Sigma x_2)^2}{n_2} = 1414 - \frac{(118)^2}{10}$$

$$= 2305 - 2280.1 = 24.9 \qquad = 1414 - 1392.4 = 21.6$$

Pooled variance,

$$\sigma_d^2 = \frac{\Sigma x_1^2 - \frac{(\Sigma x_1)^2}{n_1} + \Sigma x_2^2 - \frac{(\Sigma x_2)^2}{n_2}}{n_1 + n_2 - 2} = \frac{24.9 + 21.6}{10 + 10 - 2} = \frac{46.5}{18} = 2.58$$

$$\sigma_d = \sqrt{2.58} = 1.61;\ |\bar{x}_1 - \bar{x}_2| = |15.1 - 11.8| = 3.30$$

$$t = \frac{\bar{x}_1 - \bar{x}_2}{\sigma_d\sqrt{\frac{1}{n_1} + \frac{1}{n_2}}} = \frac{3.30}{1.61\sqrt{\frac{1}{10} + \frac{1}{10}}} = \frac{3.30}{1.61\sqrt{0.2}} = \frac{3.30}{1.61 \times 0.45} = \frac{3.3}{0.72} = 4.58$$

Tabulated t = 3.92 ($p = 0.001$) for 18 d.f. is found to be less than the calculated t (4.58). Thus the difference between the means of two samples is highly significant. The null hypothesis stating that the two means have no difference is rejected at $p = 0.001$.

11.4 PAIRED 't'-TEST

Paired 't'-test is applied when each individual gives a pair of observations. Here the paired data of independent observation from one sample only to be compared. This kind of observations are made available in biological sciences, such as:

(a) To study the effect of fertilizer, pesticide, drug on plants.

(b) To compare the effect of two different fertilizers or drugs.

(c) To compare the result of two techniques or the accuracy of two different instruments.

Steps to be followed for calculation of 't'

1. The difference in each set of paired observations are made.

$$\bar{x}_1 - \bar{x}_2 = D$$

2. The mean of difference $\bar{D}$ is calculated, $\bar{D} = \frac{\Sigma D}{n}$.

3. Standard deviation of the difference (S_D) and then standard error of mean from the same is calculated,

$$SE_D = \frac{S_D}{\sqrt{n}}, \quad \text{where} \quad S_D = \sqrt{\frac{\Sigma D^2 - \frac{(\Sigma D)^2}{n}}{n-1}}$$

4. The 't' value is determined by substituting the above values in the formula,

$$t = \frac{\bar{D}}{\text{Standard error of difference } (SE_D)} \quad \text{or} \quad t = \frac{\bar{D}}{\frac{S_D}{\sqrt{n}}}$$

5. The calculated value of 't' is compared with 't'-table to find out the significance at that particular degree of freedom.

Example 4: Determine the effect of a new growth regulator (x) on the callus growth of *Nicotiana* in laboratory condition. The observations are made on particular 10 calli.

Callus no.	**1**	**2**	**3**	**4**	**5**	**6**	**7**	**8**	**9**	**10**
Initial wt. of callus (g)	2.9	2.7	2.8	2.8	2.9	2.9	2.8	3.0	2.6	2.8
Final wt. of that callus (g)	3.0	2.8	2.9	2.8	3.2	3.1	3.0	3.2	2.7	2.9
Difference (D)	0.1	0.1	0.1	0	0.3	0.2	0.2	0.2	0.1	0.1
D^2	0.01	0.01	0.01	0	0.09	0.04	0.04	0.04	0.01	0.01

$$\Sigma D = 1.4 \qquad \Sigma D^2 = 0.26 \qquad n = 10$$

$$\bar{D} = \frac{1.4}{10} = 0.14$$

Standard deviation of the difference $= \sqrt{\dfrac{\Sigma D^2 - \dfrac{(\Sigma D)^2}{n}}{n-1}}$

$$= \sqrt{\dfrac{0.26 - \dfrac{1.96}{10}}{9}} = \sqrt{\dfrac{.06}{9}} = \sqrt{.0066} = .08$$

$$SE_D = \frac{.08}{\sqrt{10}} = \frac{.08}{3.16} = .025$$

$$t = \frac{\overline{D}}{SE_D} = \frac{0.14}{.025} = 5.6.$$

Tabulated t = 4.78 at p .001 with 9 degrees of freedom.

The calculated 't' far exceeds the tabulated 't' at p = 0.001 with 9 degrees of freedom. So the result is highly significant, i.e., the hormone has the positive effect on callus growth.

II. ANALYSIS OF VARIANCE

The classical technique for comparing the two sample means is based on student's 't' test, but a different statistical technique, **analysis of variance,** is required when analysis of data of three or more experimental or field samples are involved.

For example, a plant physiologist investigating the effect of IBA on emergence of roots from detached leaves. He used three different concentrations of IBA in five replicates. The aim of the investigator is to find out if there is any differential effect of IBA concentration on root emergence or not. The 't' test may be employed here to compare the means of treatment effects taking two at a time, i.e., comparison can be done between any two concentrations. For three different concentrations (A, B, C) altogether three tests (A & B, B & C, A & C) are required and which will make the procedure unnecessary lengthy. If in an experiment 7 different effects are to be studied then 21 't'-tests will be required. Also each time 't'-test is concerned with Type-1 error, i.e., the probability of false rejection of null hypothesis may be much higher than it appears to be.

In order to get certain positive results with greater efficiency, a different test is applied involving many samples which is called **analysis of variance** (ANOVA). Though the name of the test is somewhat misleading as it does not test the variance, rather the test is aimed at comparing means. The analysis is directed to variations between the treatments and variations among the individuals of particular treatment.

11.5 F-STATISTICS

The analysis of variance is based on **F-statistics** which is a ratio of variances. The ratio of variances due to treatments and variances due to random differences within the treatment is calculated and the estimate, known as the 'F', is then used for comparison. To find out the variance ratio 'F', the deviations are dealt in two parts : (a) deviation due to treatment effect and (b) deviation due to random differences in the individual items within treatments.

For calculation of 'F' value, the deviations due to all observations, deviations due to treatments and deviations due to random differences within treatments are needed.

Steps to be followed for calculation of F-statistics

1. The data should be arranged in tabular form showing the treatments in columns and replicates in rows.

Treatments

Replicates	1	2	3	
1	x_{11}	x_{21}	x_{31}	Σx_{r1}
2	x_{12}	x_{22}	x_{32}	Σx_{r2}
3	x_{13}	x_{23}	x_{33}	Σx_{r3}
4	x_{14}	x_{24}	x_{34}	Σx_{r4}
	Σx_{c1}	Σx_{c2}	Σx_{c3}	

* $\Sigma x_{c1}, \Sigma x_{c2}, \Sigma x_{c3}$ are the columnwise total

* $\Sigma x_{r1}, \Sigma x_{r2}, \Sigma x_{r3}\ \Sigma x_{r4}$ are rowwise total.

2. Column totals are summed up to get the grand total G.

$$G = \Sigma x_{c1} + \Sigma x_{c2} + \Sigma x_{c3}$$

3. The totals of the squares of each observations columnwise are summed up to get the value of A.

$$x_{11}^2 + x_{12}^2 + x_{13}^2 + x_{14}^2 = a_1$$

$$x_{21}^2 + x_{22}^2 + x_{23}^2 + x_{24}^2 = a_2$$

$$x_{31}^2 + x_{32}^2 + x_{33}^2 + x_{34}^2 = a_3$$

Therefore, $A = a_1 + a_2 + a_3$.

4. The squared column totals divided by the no. of observations are summed up to get the value of B.

$$\frac{(\Sigma x_{c1})^2}{n_{c1}} = b_1, \quad \frac{(\Sigma x_{c2})^2}{n_{c2}} = b_2, \quad \frac{(\Sigma x_{c3})^2}{n_{c3}} = b_3$$

Therefore, $B = b_1 + b_2 + b_3$.

5. The squared row totals divided by the no. of observations in row are summed up to get the value of C.

$$\frac{(\Sigma x_{r1})^2}{n_{r1}} = c_1, \quad \frac{(\Sigma x_{r2})^2}{n_{r2}} = c_2, \quad \frac{(\Sigma x_{r3})^2}{n_{r3}} = c_3, \quad \frac{(\Sigma x_{r4})^2}{n_{r4}} = c_4$$

Therefore, $C = c_1 + c_2 + c_3 + c_4$.

6. The squared grand total (G^2) is divided by the total no of observations made, which is the value of D.

$$D = \frac{(G^2)}{n} \qquad n = n_{c1} + n_{c2} + n_{c3}$$

7. Tabulation of all calculations for analysis of variance:

Source of variation	Sum of squares (SS)	Degree of freedom	Mean squares
Between treatments (column)	B – D	Column number - 1 (u – 1)	$\frac{B-D}{u-1}$
Between replicates (row)	C – D	Row number - 1 (v – 1)	$\frac{C-D}{v-1}$
Residual	$(A-D)-\{(B-D)+(C-D)\}$	$(u-1)(v-1)$	

8. Calculation of F-value between treatment variation and between replicate variation.

$$F \text{ (between treatments)} = \frac{\text{Mean square of variation between treatments}}{\text{Residual mean square}}$$

$$= \frac{\dfrac{(B-D)}{u-1}}{\dfrac{(A-D)-\{(B-D)+(C-D)\}}{(u-1)(v-1)}}$$

$$F \text{ (between replicates)} = \frac{\text{Mean square of variation between replicates}}{\text{Residual mean square}}$$

$$= \frac{\dfrac{(C-D)}{v-1}}{\dfrac{(A-D)-\{(B-D)+(C-D)\}}{(u-1)(v-1)}}$$

9. Table for comparison of F-value.

Source of variation	degree of freedom	Calculated F	Tabulated F	Probability	Significance level
Between treatments	$n_1 = u - 1$ $n_2 = (u-1)(v-1)$				
Between replicates	$n_1 = v - 1$ $n_2 = (u-1)(v-1)$				

10. When the calculated F-value is less than tabulated F-value then the treatment has no significant effect, whereas if calculated F-value is more than tabulated value then the treatment has significant effect.

Practical sheet for analysis of variance (one way classification)

Replicates	Treatments 1	2	3	4	Total
1	x_{11}	x_{21}	x_{31}	x_{41}	
2	x_{12}	x_{22}	x_{32}	x_{42}	
3	x_{13}	x_{23}	x_{33}	x_{43}	
4	x_{14}	x_{24}	x_{34}	x_{44}	
5	x_{15}	x_{25}	x_{35}	x_{45}	

Total

Σx $\quad \Sigma x_{c_1} + \Sigma x_{c_2} + \Sigma x_{c_3} + \Sigma x_{c_4} = G$ (Grand total)

Σx^2 $\quad \Sigma x_{c_1}^2 + \Sigma x_{c_2}^2 + \Sigma x_{c_3}^2 + \Sigma x_{c_4}^2 = A$

$$\frac{(\Sigma x)^2}{n_c} \qquad \frac{(\Sigma x_1)^2}{n_{c_1}} + \frac{(\Sigma x_2)^2}{n_{c_2}} + \frac{(\Sigma x_3)^2}{n_{c_3}} + \frac{(\Sigma x_4)^2}{n_{c_4}} = B$$

n_c = number of observations in the columns

$$\frac{(\Sigma x)^2}{n} \text{ or } \frac{G^2}{\text{Total number of observations}} = \frac{G^2}{n} = D$$

where c and r stand for the number of columns and rows, respectively.

Total SS = A – D $\qquad$ u (number of columns)

Between treatment SS = B – D $\qquad$ v (number of rows)

Residual SS = A – B

Analysis of variance table:

Source of variation	Sum of squares (SS)	Degrees of freedom	Mean squares
Between treatments	B – D	$u - 1 = n_1$	$\frac{B-D}{u-1}$
Residual	A–B = (A–D) – (B–D)	$u(v - 1) = n_2$	$\frac{A-B}{u(v-1)}$
Total	A – D	uv – 1	

$$F \text{ (Between treatments)} = \frac{\text{Between treatment mean square}}{\text{Residual mean square}} = \frac{B-D}{u-1} \Big/ \frac{A-B}{u(v-1)}$$

Inference table:

1	2	3	4	5	6
Source of variation	Degree of freedom	Calculated F	Tabulated F	Probability	Significance level

Practical sheet for analysis of variance (Two way classification)

Test subjects (Replicates)	Treatments 1	2	3	Total
1	x_{11}	x_{21}	x_{31}	$x_{11}+x_{21}+x_{31}=x_{r1}$
2	x_{12}	x_{22}	x_{32}	$x_{12}+x_{22}+x_{32}=x_{r2}$
3	x_{13}	x_{23}	x_{33}	$x_{13}+x_{23}+x_{33}=x_{r3}$
Total u = 3	$x_{11}+x_{12}+x_{13}$	$x_{21}+x_{22}+x_{23}$	$x_{31}+x_{32}+x_{33}$	= Grand total (G)
Σx	$(=\Sigma x_{c_1})$	$(=\Sigma x_{c_2})$	$(=\Sigma x_{c_3})$	
Σx^2	$x_{11}^2+x_{12}^2+x_{13}^2$ $(\Sigma x_{c_1}^2)$ $=a_1$	$x_{21}^2+x_{22}^2+x_{23}^2$ $(\Sigma x_{c_2}^2)$ $=a_2$	$x_{31}^2+x_{32}^2+x_{33}^2$ $(\Sigma x_{c_3}^2)$ $=a_3$	$(a_1+a_2+a_3)=(A)$
$\frac{(\Sigma x)^2}{n_c}$	$\frac{(\Sigma x_{c_1})^2}{n_{c1}}$ $=b_1$	$\frac{(\Sigma x_{c_2})^2}{n_{c2}}$ $=b_2$	$\frac{(\Sigma x_{c_3})^2}{n_{c3}}$ $=b_3$	$(b_1+b_2+b_3)=(B)$
$\frac{(\Sigma x)^2}{n_r}$	$\frac{(\Sigma x_{r1})^2}{n_{r1}}$ $=c_1$	$\frac{(\Sigma x_{r2})^2}{n_{r2}}$ $=c_2$	$\frac{(\Sigma x_{r3})^2}{n_{r3}}$ $=c_3$	$(c_1+c_2+c_3)=(C)$

$$\frac{(\Sigma x_{c_1}+\Sigma x_{c_2}+\Sigma x_{c3})^2}{n_{c_1}+n_{c_2}+n_{c3}}=\frac{G^2}{n}=D \qquad (D)$$

Total sum of squares = A – D

Between treatment sum of squares = B – D

Between rows sum of squares = C – D

Residual sum of squares = (A – D) – {(B – D) + (C – D)}

Table for analysis of variance:

Source of variation	Sum of squares	Degrees of freedom	Mean squares
1. Between treatments	B – D	u – 1	$\frac{B-D}{u-1}$
2. Between rows	C – D	v – 1	$\frac{C-D}{v-1}$
3. Residual	(A–D)–{(B–D)+(C–D)}	(u–1)(v–1)	$\frac{(A-D)-\{(B-D)+(C-D)\}}{(u-1)(v-1)}$
Total	A – D	uv – 1	

$$F\,(\text{Between treatments}) = \frac{\text{Between treatment mean square}}{\text{Residual mean square}} = \frac{\dfrac{B-D}{u-1}}{\dfrac{(A-D)-\{(B-D)+(C-D)\}}{(u-1)(v-1)}}$$

$$F\,(\text{Between replicates}) = \frac{\text{Between replicate mean square}}{\text{Residual mean square}} = \frac{\dfrac{C-D}{v-1}}{\dfrac{(A-D)-\{(B-D)+(C-D)\}}{(u-1)(v-1)}}$$

Inference table:

1	2	3	4	5	6
Source of variation	Degree of freedom	Calculated F	Tabulated F	Probability	Significance level

Example 5: A plant physiologist was investigating the effect of IBA on emergence of roots from detached leaves. He used three different concentrations in five replicates and got the following results. Calculate the 'F' value to study the analysis of variance.

Replicates	Conc. of IBA (ppm) 10	20	30		
1	5	7	4	$\Sigma x_{r1} = 16$	$C_1 = \frac{256}{3} = 85.3$
2	4	9	3	$\Sigma x_{r2} = 16$	$C_2 = \frac{256}{3} = 85.3$
3	6	8	5	$\Sigma x_{r3} = 19$	$C_3 = \frac{361}{3} = 120.3$
4	5	10	3	$\Sigma x_{r4} = 18$	$C_4 = \frac{324}{3} = 108.0$
5	6	9	4	$\Sigma x_{r5} = 19$	$C_5 = \frac{361}{3} = 120.3$
	$\Sigma x_1 = 26$	$\Sigma x_2 = 43$	$\Sigma x_3 = 19$	$G = 88$	$C = 519.2$
	$a_1=138$	$a_2=375$	$a_3=75$	$A=588$	$D = \frac{88^2}{15} = 516.3$
	$b_1 = \frac{26^2}{5} = 135.2$	$b_2 = \frac{43^2}{5} = 369.8$	$b_3 = \frac{19^2}{5} = 72.2$	$B = 577.2$	

$A - D = 71.7$

$B - D = 60.9$

$C - D = 2.9$

Source of Variation	SOS	D.f.	MS	F-Value	Tabulated F-Value		
					p = .05	p = .01	p = .001
Between Treatments	B – D = 60.93	u – 1 = 2	B–D/u–1 = 60.93/2 = 30.46	$\frac{30.46}{.984}$ = 30.95	4.5	8.7	18.5
Between Replicates	C – D = 2.9	v – 1 = 4	C–D/v–1 = 2.9/4 = 0.725	$\frac{.725}{.984}$ = 0.736	3.8	7.0	14.4
Residual	(A – D) – {(B – D) + (C – D)} = 71.7 – (60.93 + 2.9) = 7.87	(u – 1) (v – 1) = 2×4 = 8	$\frac{7.87}{8}$ = .984				

Inference:

The calculated F-value 30.95 for between treatments of three concentrations of IBA is found to be greater than the tabulated F value for $n_1 = 2$, $n_2 = 8$ degrees of freedom at .001 and .01 probability level. Thus the deviation is highly significant. Here the null hypothesis stating that the 3 different concentrations of IBA have the identical effect is rejected at 99.9% level.

The calculated F-value is 0.736 for between the replicates of each treatment is found far less than the tabulated F-value for n_1=4, n_2=8 degrees of freedom at 0.001 and 0.01 probability level. Thus the conclusion is there is no significant difference among the replicates. Here the null hypothesis is accepted, i.e., replicates do not have any difference.

QUESTIONS

1. What do you mean by test of significance? Give a brief account of laying down of hypothesis for the procedure of carrying out a significance test.
2. When will you perform the paired 't'-test and when will you follow unpaired 't'-test? Write down the formula for calculating the 't' value in both the processes.
3. What is student's 't' test? What is its main objective?

   In the following table two varieties of tomato plant (A) and (B) yield fruits as shown in the following table. Does the mean member of fruits (A) significantly differ from (B)?

Fruit yield/kg										
Variety A	2.2	2.5	1.9	2.6	2.3	1.8	2.0	2.1	2.4	2.3
Variety B	2.8	2.5	2.7	3.0	3.1	2.3	2.4	3.2	2.5	2.9

4. Two sets of potted plants were sprayed with 0.01 and 0.1 ppm dimethyl mercury salt solution respectively. The following results were obtained with regard to sterility percentage of seeds.

   (n_1=n_2=10 plants)

	% Sterile seeds									
Set I (0.01 ppm)	10	15	16	18	11	8	9	14	15	12
Set II (0.1 ppm)	15	18	19	17	12	14	13	18	19	20

Find out whether the two conc. of dimethyl mercury have the same effect on the percentage of sterility in seeds.

5. One growth regulator has been applied on 10 plants. After 15 days the increase in height was measured. Determine whether growth regulator activity has any significant effect on height.

Plant no.	1	2	3	4	5	6	7	8	9	10
Initial height (cm)	110	115	102	95	108	110	97	115	98	110
After 15 days										
Final height (cm)	110	116	102	98	112	112	98	120	102	111

6. What is analysis of variance? When will you adopt this method for analysing your data? What is the significance of this type of test?
7. What do you mean by F-statistic by one way analysis and two way analysis? Describe the method of calculation of F-value by both the methods.
8. In an investigation relating the effect of $HgCl_2$ of 3 different concentrations on the germination of rice grain, the following results were obtained.

Replicates	Control	10 mg/L	20 mg/L	30 mg/L
1	98	80	75	65
2	100	85	70	68
3	95	83	65	72
4	97	90	75	58
5	100	88	72	60

9. The following data shows no. of tillers in 4 varieties of rice. Give your comment by analysing the data by F-test.

Plot no.	No. of tillers/plant of the 4 varieties of rice			
	A	B	C	D
1	9	7	6	15
2	10	8	8	14
3	11	7	7	13
4	12	8	9	16
5	10	9	6	12

10. In an experiment number of grains per panicle due to 3 different doses of fertiliser and 3 different environmental conditions were observed. Analyse the data using test of variance to determine if there is any variation in the grain yield under different environmental condition or due to treatment.

	Doses of fertilizer		
	Low	Medium	High
Irrigated low land	120	150	110
Non-irrigated low land	95	120	95
Non-irrigated upland	70	80	85

12 PROBABILITY AND CHI-SQUARE TEST

I. PROBABILITY

The measurement of relative chance of occurrence of an event from among a set of alternatives can be defined as **probability**. When in an experiment there are chances of occurrence of many events then the question of probability arises. Such as, if a coin is tossed then either 'head' or 'tail' will happen; if a dice is thrown then there are possibility of getting 1 or 2 or 3 ... or 6.

Probability is a number expressed in a quantitative scale. When one event will not occur at all then the probability of that event is 0, and if there is any event which will happen positively without fail then the probability of that event is 1. But in biological science, mostly we find the probability of any event lies between impossibility to certainty i.e., the value ranges from 0 to 1.

Mathematically probability can be explained in the following way:

If an event can happen in 'a' number of ways, and fails to happen in 'b' number of ways, then the **probability of its happening** 'p' is written as.

$$p = \frac{a}{a+b} \quad \text{or} \quad p = \frac{\text{Number of events occurring}}{\text{Total number of events}}$$

Similarly **probability of failure** of any event is denoted as 'q', where,

$$q = \frac{b}{a+b} \quad \text{or} \quad q = \frac{\text{Number of events not happening}}{\text{Total number of events}}$$

Therefore, $p + q = \frac{a}{a+b} + \frac{b}{a+b} = 1$

So, if the probability of happening any event is 0.7, then the probability of not happening of that event is 0.3.

12.1 EVENTS

The results of any experiment in all possible forms are said to be **events** of that experiment. Such as, throwing of a dice has 6 possible outcomes, either 1 or 2 or 3 or 4 or 5 or 6. All these six outcomes are called events of that single experiment.

Null event

When there is no chance of getting an event is called **null** or **impossible event**. It is symbolically denoted by ϕ. Such as, survival of any human being forever is an impossible event or null event.

Sure event

If the likelihood of occurrence of any event is sure then the event is called **sure event.** Such as, the death of a human being is a sure event.

Equally likely events

If the likelihood of the occurrence of every event in an experiment is same then those are called as **equally likely events**. Such as, when a dice is thrown, there is no biasness, there are the possibility of coming any number 1 to 6 in equal probability, so the events are equally likely events.

Mutually exclusive events

If in an experiment the occurrence of one event prevents or rules out the happening of all other events, then these are called as **mutually exclusive events.** Such as, when a coin is tossed either 'head' or 'tail' will come. The occurrence of one event affects the occurrence of another event, both events cannot occur together, i.e., occurrence of 'head' rules out getting 'tail' in the same trial. Here the events are connected by the words 'either' or 'or'.

Simple event and compound event

Any event having only one sample point of a sample space is called **simple event** and if any event is decomposable into a number of simple events then it is called as **compound event.**

Such as, if a bag contains 4 white and 6 red balls, and if one ball is drawn then it is simple event, but if two balls are drawn together then the events will be — 'both the balls are white', 'both the balls are red', 'one ball is white and another ball is red' — these are compound events.

The compound events may be of two types:

Independent event

Two or more events are said to be **independent events** when the outcome of one event does not affect or is not affected by the other events . For example, if a coin is tossed twice, the result of second tossing would in no way be affected by the result of first tossing, so these are independent events.

Dependent event

The occurrence or non-occurrence of one event in one trail affects the probability of other events in other trails are called **dependent events**. For example, the probability of drawing a queen from a pack of 52 cards is 4/52, but if the card drawn for the first time (queen) is not replaced then the probability of second drawing of a queen is 3/51, as the pack now contains 3 queens and 51 cards.

12.2 ADDITION AND MULTIPLICATION RULES

Probability is estimated usually on the basis of following two rules of chances:

1. Addition rule
2. Multiplication rule

Addition Rule

This rule is applied when two events are mutually exclusive, i.e., both events cannot occur simultaneously. The birth of a male child excludes the birth of a female child in the same trial. The probability that one of several mutually exclusive events will occur is the sum of the probabilities of the individual events.

Mathematically, when two events A and B are **mutually exclusive**, the chance of occurrence or probability of occurrence of either A or B can be calculated from the following formula:

$$p\ (A \text{ or } B) = p(A) + p(B)$$

This rule is applicable to any number of mutually exclusive events as follows:

$$p(E_1 \text{ or } E_2 \text{ or } E_3 \dots E_n) = p(E_1) + p(E_2) + p(E_3) + \dots + p(E_n)$$

Example 1: If a dice is rolled, what is the probability of getting either 3 or 5?

Probability of getting 3 is $p(3) = \frac{1}{6}$

Probability of getting 5 is $p(5) = \frac{1}{6}$

$\therefore$ Probability of getting either 3 or 5 is $p(3 \text{ or } 5) = p(3) + p(5) = \frac{1}{6} + \frac{1}{6} = \frac{1}{3}$

Example 2: What is the probability of getting a king or a joker from a pack of 54 cards?

Probability of getting a king is $p(K) = \frac{4}{54} = \frac{2}{27}$

Probability of getting a joker is $p(J) = \frac{2}{54} = \frac{1}{27}$

So, the probability of either a king or a joker is

$$p(K \text{ or } J) = p(K) + p(J) = \frac{2}{27} + \frac{1}{27} = \frac{3}{27} = \frac{1}{9}$$

Addition rule changes when the events are **not mutually exclusive**, i.e., if two events A and B can occur simultaneously in few cases, then the rule becomes modified in the following way:

$$p(A \text{ or } B) = p(A) + p(B) - p(A \text{ and } B)$$

Example 3: What is the probability of getting a king or club from a pack of 52 cards?

In this example, getting a king and a club are not mutually exclusive events as there will be one king which is king of club. So the chance or probability of getting that event should be subtracted.

$$p(\text{King or Club}) = p(\text{King}) + p(\text{Club}) - p(\text{King and Club})$$

Probability of King = $p(\text{King}) = \frac{4}{52} = \frac{1}{13}$

Probability of Club = $p(\text{Club}) = \frac{13}{52} = \frac{1}{4}$

Probability of King and Club = $p\,(\text{King and Club}) = \frac{1}{52}$

So, $p(\text{King or Club}) = \left(\frac{1}{13} + \frac{1}{4}\right) - \frac{1}{52} = \frac{16}{52} = \frac{4}{13}$

Multiplication Rule

(a) **When the events are independent**: Probability of two or more independent events occurring together is the product of the probabilities of individual events. Symbolically, if p(A) and p(B) are the probabilities of two respective events A and B, and the happening of these two events are independent then the probability of happening both the events together can be calculated with the following formula:

$$p(A \text{ and } B) = p(A) \times p(B)$$

Thus the rule may be extended to any number of independent events like E_1, E_2, E_3, ..., E_n, and the formula will be as follows:

$$p(E_1 \text{ and } E_2 \text{ and } E_3 \ldots \text{ and } E_n) = p(E_1) \times p(E_2) \times p(E_3) \times \ldots \times p(E_n)$$

Example 4: If two dice are thrown simultaneously what is the probability of getting 3 in both the dice?

The probability of getting 3 in first dice is $p(A) = \frac{1}{6}$

The probability of getting 3 in 2nd dice is $p(B) = \frac{1}{6}$

So, the probability of getting 3 in both the dice is

$$p(A \text{ and } B) = p(A) \times p(B) = \frac{1}{6} \times \frac{1}{6} = \frac{1}{36}$$

(b) **When the events are dependent**: When the probability of happening one event is affected by the occurrence of another event then it is called **conditional probability**. Such as, conditional probability of happening A, when B has already happened, is denoted as p(A/B); conditional probability of B, and A has already happened, is denoted as p(B/A).

When the two events A and B are occurring simultaneously but any one event has conditional probability then the multiplication rule will be written as

$$p(AB) = p(A) \cdot p(B/A) \quad \text{or} \quad p(B) \cdot p(A/B)$$

where p(A/B) = Conditional probability of A given that B has happened
p(B/A) = Conditional probability of B given that A has happened

Example 5: Four cards are drawn consecutively four times from a pack of 52 cards. Find the chances of drawing an ace, a king, a queen and a jack. The cards are not replaced after each withdrawal.

Probability of drawing an ace = $p(A) = \frac{4}{52}$

Probability of drawing a king = $p(K) = \frac{4}{51}$

Probability of drawing a queen = $p(Q) = \frac{4}{50}$

Probability of drawing a jack = $p(J) = \frac{4}{49}$

So, the combined probability

$$p(\text{A and K and Q and J}) = p(A) \times p(K) \times p(Q) \times p(J)$$

$$= \frac{4}{52} \times \frac{4}{51} \times \frac{4}{40} \times \frac{4}{49} = 0.317$$

Example 6: Four cards are drawn in four consecutive drawals from a pack of 52 cards without replacing the cards after each drawal. What is the probability of drawing a king in each drawal?

The probability of getting a king in 1st drawal = $\frac{4}{52}$

The probability of getting a king in 2nd drawal = $\frac{3}{51}$

The probability of getting a king in 3rd drawal = $\frac{2}{50}$

The probability of getting a king in 4th drawal = $\frac{1}{49}$

So, the combined probability of getting a king in 4 consecutive drawals is

$$\frac{4}{52} \times \frac{3}{51} \times \frac{2}{50} \times \frac{1}{49} = \frac{1}{270725}$$

12.3 BINOMIAL DISTRIBUTION

The binomial distribution is one of the most widely used probability distributions of random discrete variates. This process is one where an experiment can result in only one or two mutually exclusive outcome such as success or failure, male or female, dead or alive, etc.

For example, in case of birth of a child there are two possible happenings, the male or female. So, the probability of male child is $p = \frac{1}{2}$ and also the probability of female child is $q = \frac{1}{2}$.

If the two deliveries of two ladies are considered then there are four possible outcomes.

1st lady	2nd lady	Probability	
M	M	$= p \times p = p^2$	
F	M	$= q \times p = pq$	2 pq
M	F	$= p \times q = pq$	
F	F	$= q \times q = q^2$	

We can get this probability distribution through the binomial expansion.

$$(p+q)^2 = p^2 + 2pq + q^2$$

In case of three births, we may get 3 males, 3 females, 2 males one female, one male two females, etc., which we can get through expansion of $(p + q)^3 = p^3 + 3p^2q + 3pq^2 + q^3$.

Therefore for 'n' number of events, the expected result may be expressed as $(p + q)^n$.

12.4 SIGNIFICANCE IN GENETICS

An understanding of the laws of probability is of great importance in genetics. Because it helps in —

(i) Forecasting the chance of obtaining certain results from a cross,

(ii) Elucidating the operation of genetic principles,

(iii) Assessment of goodness of fit of phenotypic ratio in relation to particular genetic principles.

The principles of probability can be applied to obtain the expected phenotypic ratios from multiple hybrid crosses by avoiding the use of complicated checker boards. However, the actual ratios of offspring obtained in a cross, very rarely, exactly tally with the expected results calculated by the principles of probability. Within certain limit this deviation may be attributed due to chance; but when the actual results deviate to a great extent, the factor other than chance is responsible. Considering the variability in biological materials, a probability of 0.05 is as significant.

II. CHI-SQUARE TEST

In biological experiments and field surveys, apart from quantitative data we get the qualitative data which is genetical character such as tall and short, colour of flower, seed coat character which do not have any numerical values. But the number of flowers or seeds having a particular colour falls under any category can be counted numerically.

This type of observation requires the calculation of the expected number of individuals under any category. Thus it becomes necessary to know whether there is any deviation in between the observed and expected frequencies. The measurement of this deviation is done with the help of a particular test which is called **Chi-square (χ^2) test**. The formula for Chi-square test is

$$\chi^2 = \Sigma \frac{(O-E)^2}{E}$$

where O = Observed value, E = Expected value.

12.5 APPLICATION OF χ^2-TEST

It is an alternative test to find significance of difference in two or more than two proportions:

(a) It can compare the values of two binomial samples when they are small.

(b) It can compare the frequencies of two multinomial samples.

(c) Chi-square measures the probability of association between two discrete attributes.

(d) The Chi-square test is applied as a test for **goodness of fit** which reveals the closeliness of observed frequency with those of the expected frequency. Thus it helps to answer whether physical or chemical factors did or did not have an effect.

(e) Occasionally it is desirable to compare one set of observations taken under particular conditions to those of a similar nature taken under different conditions. In this case there are no definite expected values, only the question is whether the results are dependent (contingent upon) or independent of conditions. Then the χ^2-test is called as test for independence or **contingency test.**

12.6 χ^2-TEST FOR GOODNESS OF FIT

χ^2-test is applied to a wide range of studies relating to experimental biology and field studies. The aim of this test is to test the closeliness of observed frequencies with those of the expected, i.e., how well the observed frequency curve fits into theoretical curve. If both the observed and expected frequency distribution are in complete agreement with each other then the χ^2-value will be zero. But in experimental observations there is always some degree of deviation. The critical χ^2-value will be exceeded due to sampling fluctuations.

For example, if a crossing experiment gives two different sizes of seeds in F_2 progeny then these two types seeds may segregate according to 3:1 (Mendelian Monohybrid), 1:1 (Monohybrid test cross), 9:7 (Complementary factor interaction), 13:3 (Inhibitory factor) or 15:1 (Duplicate gene interaction) ratio, etc.

Again if crossing experiment results in three types of seeds then these may be due to incomplete dominance (1:2:1), supplementary factor (9:3:4) or due to dominant epistasis (12:3:1) interaction, etc.

Likely, 4 types of seeds with 4 different combinations of two different characters may either follow the 9:3:3:1 (Mendelian dihybrid) or 1:1:1:1 (Dihybrid test cross) ratio for segregation.

Test for the goodness of fit is required in these above cases for studying the closeliness of observed data of the experiments with those of expected frequencies.

Steps to be followed to test the Goodness of Fit

1. Deviation between the observed and the expected results should be calculated.
2. Comparing the minimum deviation the null hypothesis should be selected for χ^2-test.

3. χ^2-value should be determined.
4. Comparing the calculated χ^2-value with tabulated χ^2-value the conclusion has to be made.

Example 7 → (3:1): Following number of seeds with the associated character is observed. Test the goodness of fit and comment.

Yellow seed : 428

Green seed : 152

Step I: Calculation of expected value for each ratio

(i) **3:1 ratio**

Total no. of samples = 428 + 152 = 580

$$\text{Expected yellow seeds} = \frac{3}{4}\times 580 = 435$$

$$\text{Expected green seeds} = \frac{1}{4}\times 580 = 145$$

(ii) **1:1 ratio**

$$\text{Expected yellow seeds and green seeds} = \frac{1}{2}\times 580 = 290$$

(iii) **9:7 ratio**

$$\text{Expected yellow seeds} = \frac{9}{16}\times 580 = 326.25$$

$$\text{Expected green seeds} = \frac{7}{16}\times 580 = 253.75$$

(iv) **13:3 ratio**

$$\text{Expected yellow seeds} = \frac{13}{16}\times 580 = 471.25$$

$$\text{Expected green seeds} = \frac{3}{16}\times 580 = 108.75$$

(v) **15:1 ratio**

$$\text{Expected yellow seeds} = \frac{15}{16}\times 580 = 543.75$$

$$\text{Expected green seeds} = \frac{1}{16}\times 580 = 36.25$$

Step II: Determination of expected segregation ratio

Table for calculation of minimum deviation

Observed value	Expected segregation ratios									
	3:1		1:1		9:7		13:3		15:1	
	Expected	Deviation	Expected	Deviation	Expected	Deviation	Expected	Deviation	Expected	Deviation
Yellow 428	435	7	290	138	326.25	101.75	471.25	43.25	543.75	115.75
Green 152	145	7	290	138	253.75	101.75	108.75	43.25	36.25	115.75
Total deviation		14		276		203.50		86.50		231.50

According to table, the deviation is minimum in 3:1 ratio, so the observed sample should fit with **3:1** ratio.

Step III: Calculation of χ^2-value

No. of classes	Character	Observed value (O)	Expected value (E)	(O – E)	$(O - E)^2$	$\frac{(O-E)^2}{E}$
1	Yellow	428	435	– 7	49	0.113
2	Green	152	145	+ 7	49	0.338

$$\chi^2 = \Sigma\frac{(O-E)^2}{E} = 0.113 + 0.338 = 0.451$$

Since, in this experiment the samples are of two classes, so degree of freedom = 2 – 1 = 1.

Step IV: Conclusion

The calculated Chi-square (χ^2) value is 0.451. The tabulated chi-square at 0.05 probability level with 1 degree of freedom is 3.841.

The calculated value is much less than the table value, so the deviation is insignificant, the observed deviation is due to chance factor only. It lies in the probability range 50-70%.

The observed result is in good fit with Mendelian Monohybrid ratio, i.e., **3:1**.

Step V: Comment

Comment: It is concluded that the experimental result shows **Mendelian monohybrid ratio** 3:1. So the assumed genotype of parents are :

P   Yellow   green

YY x yy

↓   ↓

gamete   Y x y

↓

F_1 Yy (selfed)

↓

F_2

♂ \ ♀	Y	y
Y	YY Yellow	Yy Yellow
y	Yy Yellow	yy green

YY Yellow }
Yy Yellow } 3
yy green 1

So the phenotypic ratio is 3:1.

Example 8 → (9:3:3:1): In an experiment on garden pea, we count 4 different kinds of plants,

Yellow cotyledon and inflated pod	—	555
Yellow cotyledon and constricted pod	—	185
Green cotyledon and inflated pod	—	195
Green cotyledon and constricted pod	—	65
Total		1000

Test the goodness of fit for this data and comment.

Step I: Calculation of expected value of assumed ratio

(i) **9:3:3:1 ratio**

Total number of plants = 1000

Yellow cotyledon and inflated pod = $\frac{9}{16} \times 1000 = 562.5$

Yellow cotyledon and constricted pod = $\frac{3}{16} \times 1000 = 187.5$

Green cotyledon and inflated pod = $\frac{3}{16} \times 1000 = 187.5$

Green cotyledon and constricted pod = $\frac{1}{16} \times 1000 = 62.5$

(ii) **1:1:1:1 ratio**

Yellow cotyledon and inflated pod = $\frac{1}{4} \times 1000 = 250$

Yellow cotyledon and constricted pod = $\frac{1}{4} \times 1000 = 250$

Green cotyledon and inflated pod $= \frac{1}{4} \times 1000 = 250$

Green cotyledon and constricted pod $= \frac{1}{4} \times 1000 = 250$

Step II: Determination of expected segregation ratio

	9:3:3:1		1:1:1:1	
Observed value	Expected value	Deviation |d|	Expected value	Deviation |d|
Yellow cotyledon inflated pod 555	562.5	7.5	250	305
Yellow cotyledon constricted pod 185	187.5	2.5	250	65
Green cotyledon inflated pod 195	187.5	7.5	250	55
Green cotyledon constricted pod 65	62.5	2.5	250	185
Total deviation		20		610

According to the table, we find that the deviation is minimum in case of 9:3:3:1 ratio. So it is assumed that the observed segregation pattern will be in good fit with this **9:3:3:1** ratio.

Step III: Determination of Chi-square value

No. of classes	Sample characters	Observed value (O)	Expected value (E)	(O–E)	$(O–E)^2$	$\frac{(O-E)^2}{E}$
1	Yellow, inflated	555	562.5	– 7.5	56.25	$\frac{56.25}{562.5} = 0.1$
2	Yellow, constricted	185	187.5	– 2.5	6.25	$\frac{6.25}{187.5} = 0.03$
3	Green, inflated	195	187.5	+ 7.5	56.25	$\frac{56.25}{187.5} = 0.3$
4	Green, constricted	65	62.5	+ 2.5	6.25	$\frac{6.25}{62.5} = 0.1$

$$\chi^2 = \Sigma \frac{(O-E)^2}{E} = 0.53$$

Since, in this experiment the samples are of 4 classes, so degrees of freedom = 4 – 1 = 3.

Step IV: Conclusion

The tabulated χ^2 value at 0.05 probability level for 3 degree of freedom is 7.81 which is more than the calculated value and it lies at 90-95% probability level. So the deviation found in the experiment is insignificant.

We can conclude that the observation is in good fit with Mendelian Dihybrid ratio **9:3:3:1**.

Step V: Comment

As the observation shows good fit with **Mendelian dihybrid ratio**, so the two genes for cotyledon colour and pod shape can independently assort.

P Yellow cotyledon, Inflated pod × Green cotyledon, constricted pod

YYII ↓ yyii ↓

gamete YI × yi

↓

F_1 YyIi

Yellow cotyledon, inflated pod

↓ (selfed)

F_2

Yellow inflated (Y-I-) = 9

Yellow constricted (Y-ii) = 3

Green inflated (yyI-) = 3

Green constricted (yyii) = 1

♂ \ ♀	YI	Yi	yI	yi
YI	YYII Yellow inflated	YYIi Yellow inflated	YyII Yellow inflated	YyIi Yellow inflated
Yi	YYIi Yellow inflated	YYii Yellow constricted	YyIi Yellow inflated	Yyii Yellow constricted
yI	YyII Yellow inflated	YyIi Yellow inflated	yyII green inflated	yyIi green inflated
yi	YyIi Yellow inflated	Yyii Yellow constricted	yyIi green inflated	yyii green constricted

Example 9 → (1:1): One yellow seeded pea plant when crossed with a green seeded pea plant, produced 50 yellow seeds and 46 green seeds in F_2. Write down your comment with the help of χ^2-analysis.

Sample character and sample size

No. of classes	Sample character	Sample size
1	Yellow seed	50
2	Green seed	46
	Total	96

Determination of expected segregation ratio (By estimating magnitude of deviation)

	3:1		1:1		9:7		13:3		15:1	
Observed value	Exp	|d|	Exp	|d|	Exp	|d|	Exp	|d|	Exp	|d|
Yellow seed 50	72	22	48	2	56	6	78	28	90	40
Green seed 46	24	22	48	2	42	6	18	28	6	40
Total deviation		44		4		12		56		80

According to the table, we find that the minimum deviation is in case of **1:1** ratio. So it is assumed that the observed sample should fit well with this ratio.

Determination of Chi-squre value:

No. of classes	Sample characters	Observed value (O)	Expected value (E)	(O–E)	$(O-E)^2$	$\frac{(O-E)^2}{E}$
1	Yellow	50	48	2	4	0.083
2	Green	46	48	2	4	0.083

$$\chi^2 = \Sigma\frac{(O-E)^2}{E} = 0.083 + 0.083 = 0.166$$

Conclusion: Since there are 2 classes, so the degree of freedom = 2 – 1 = 1. The calculated χ^2-value is 0.166 which is much less than the table value 3.841 for 1 degree of freedom at 0.05 probability level.

The χ^2 value 0.166 lies between 80-90% probability level. Therefore the deviation which is observed in the sample from the expected value is highly insignificant, and the observed ratio has a very good fit with the expected ratio, i.e., **1:1.**

Comment: As the experimental result shows good fit with 1:1 ratio, i.e., **Mendelian monohybrid test cross ratio,** so the assumed genotypes of the parents and the offsprings are as follows:

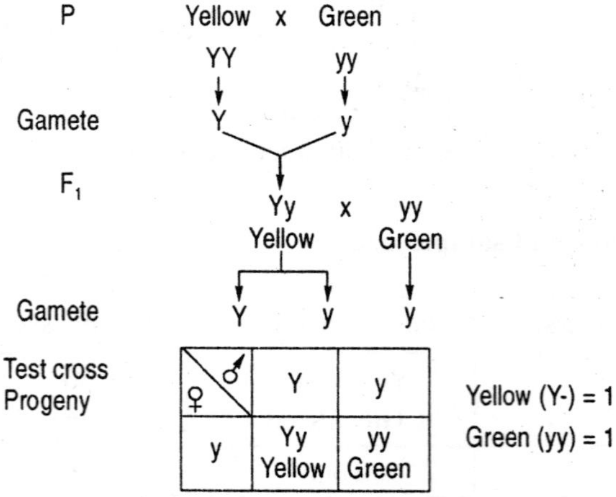

Example 10 → (9:7): Selfing of a hybrid plant produced a population with 120 pink flowers and 88 white flowers. Explain with χ^2-analysis, what does the result show?

Sample characters and sample size:

No. of classes	Sample character	Sample size
1	Pink flower	120
2	White flower	88

Total no. of samples = 208

Determination of deviation in different segregation ratio

	3:1		1:1		9:7		13:3		15:1	
Observed value	Exp	|d|	Exp	|d|	Exp	|d|	Exp	|d|	Exp	|d|
Pink 120	156	36	104	16	117	3	169	49	195	75
White 88	52	36	104	16	91	3	39	49	13	75
Total deviation		72		32		6		98		150

According to this table, we find that the deviation is minimum in case of **9:7** ratio. So, it is assumed that the observation should fit well with this ratio.

Determination of Chi-square value

No. of classes	Sample characters	Observed value (O)	Expected value (E)	Deviation (O–E)	$(O-E)^2$	$\frac{(O-E)^2}{E}$
1	Pink flower	120	117	3	9	0.076
2	White flower	88	91	3	9	0.098

$$\chi^2 = \Sigma\frac{(O-E)^2}{E} = .076 + 0.098 = 0.175$$

Conclusion: Since in the observation there are two classes, so the degree of freedom = 2 – 1 = 1

The calculated χ^2 value is 0.175, which is much less than the table value 3.84 for 1 degree of freedom at 0.05 probability level.

The χ^2 value 0.175 lies between 50-70% probability range. Therefore this deviation from the expected value is insignificant and the observed ratio is in good fit with **9:7** ratio.

Comment: It is concluded that the experimental result shows the characters with **complementary factor interaction** in F_2 generation.

As the observed sample are assumed to show complementary factor interaction, so the assumed genotypes are:

P_1 AAbb x aaBB

White flower White flower

↓ ↓

gamete Ab aB

F_1 ↓

AaBb

Pink flower

F_2

♂ \ ♀	AB	Ab	aB	ab
AB	AABB Pink	AABb Pink	AaBB Pink	AaBb Pink
Ab	AABb Pink	AAbb White	AaBb Pink	Aabb White
aB	AaBB Pink	AaBb Pink	aaBB White	aaBb White
ab	AaBb Pink	Aabb White	aaBb White	aabb White

A-B- = Pink 9

A-bb = White 3 }

aaB- = White 3 } 7

aabb = White 1 }

Here, the analysis shows that 2 pairs of factors control the same character and two dominant genes A and B are complementary to each other. Each of which has no effect on expression of character, but when in combination shows their effect, i.e., pink colour. Absence of any one of the them (A or B) leads to absence of pink colour and it can be concluded very easily that the colour character is controlled by two pairs of factors which are complementary to each other.

Example 11 → (13:3): From a plant after selfing total 96 seeds are harvested of which yellow seeds 79, brown seeds 17. Explain the result with χ^2-analysis.

Sample character and sample size

Sample character	Sample size
Yellow seed	79
Brown seed	17
Total	96

Determination of expected segregation ratio: (By estimating magnitude of deviation)

	3:1		1:1		9:7		13:3		15:1	
Observed value	Exp	|d|	Exp	|d|	Exp	|d|	Exp	|d|	Exp	|d|
Yellow seed 79	72	7	48	31	56	23	78	1	90	11
Brown seed 17	24	7	48	31	42	23	18	1	6	11
Total deviation		14		62		46		2		22

According to the table, we find that the minimum deviation is in case of **13:3** ratio. So it is assumed that the observed sample should fit well with this ratio.

Determination of Chi-square value

No. of classes	Sample characters	Observed value (O)	Expected value (E)	Deviation (O–E)	$(O-E)^2$	$\frac{(O-E)^2}{E}$
1	Yellow	79	78	1	1	0.0126
2	Brown	17	18	1	1	0.055

$$\chi^2 = \Sigma\frac{(O-E)^2}{E} = 0.0126 + 0.055 = 0.068$$

Conclusion: Since there are two classes, degree of freedom = 2 – 1 = 1. The calculated χ^2 value 0.068 is much less than the table value 3.841 for 1 degree of freedom at 0.05 probability level.

The χ^2-value 0.068 lies between 90-95% probability level. Therefore, this deviation from the expected value is highly insignificant and the observed ratio has a very good fit with the expected ratio.

Comment: As the experimental result shows good fit with **13:3** ratio. So it is assumed that the brown colour of seed coat is controlled by gene C, but the presence of another dominant gene D inhibits the effect of C, and the seed coat colour becomes yellow. The dominant gene D does not have its own expression but it inhibits the effect of another dominant gene C. So the effect of C is only expressed when there is recessive gene d. This phenomenon is called **Inhibitory Factor**.

P_1 Brown seed CCdd × Yellow seed ccDD

gamete Cd × cD

F_1 CcDd Yellow seeds

F_2

♀ \ ♂	CD	Cd	cD	cd
CD	CCDD Yellow	CCDd Yellow	CcDD Yellow	CcDd Yellow
Cd	CCDd Yellow	CCdd Brown	CcDd Yellow	Ccdd Brown
cD	CcDD Yellow	CcDd Yellow	ccDD Yellow	ccDd Yellow
cd	CcDd Yellow	Ccdd Brown	ccDd Yellow	ccdd Yellow

C-D- = Yellow = 9
ccD- = Yellow = 3
C-dd = Brown = 3
ccdd = Yellow = 1

So, the phenotypic ratio of Yellow seed : Brown seed = 13 : 3.

Example 12 → (15:1): In an experiment, the hybrid plant yielded 193 brown coloured seeds and 15 white seeds. Comment on the observed result with χ^2-analysis.

Sample character and sample size:

No. of classes	Sample characters	Sample size
1	Brown seed	193
2	White seed	15
	Total	208

Determination of expected segregation ratio (by estimating magnitude of deviation):

	3:1		1:1		9:7		13:3		15:1	
Observed value	Exp	|d|	Exp	|d|	Exp	|d|	Exp	|d|	Exp	|d|
Brown seed 193	156	37	104	89	117	76	169	24	195	2
White seed 15	52	37	104	89	91	76	39	24	13	2
Total deviation		74		178		152		48		4

According to the table, we find that the deviation is minimum in case of **15:1** ratio. So, it is assumed that the observed sample should fit with this ratio.

Determination of Chi-square value

No. of classes	Sample characters	Observed value (O)	Expected value (E)	Deviation (O–E)	$(O-E)^2$	$\frac{(O-E)^2}{E}$
1	Brown	193	195	2	4	0.0205
2	White	15	13	2	4	0.307

$$\chi^2 = \Sigma\frac{(O-E)^2}{E} = 0.0205 + .307 = 0.332$$

Conclusion: Since there are two classes, degree of freedom = 2 – 1 = 1. The calculated χ^2 value 0.332 is less than the table value 3.841 for 1 degree of freedom at 0.05 probability level.

The χ^2 value 0.332 lies between 75-90% probability level. Therefore this deviation from the expected value is insignificant and the observed ratio has good fit with the expected ratio.

Comment: As the experimental result shows good fit with 15:1 ratio, so it is assumed that the brown colour of seed is controlled by two pairs of factors or alleles. Presence of any one of them will give them brown seed colour. Whereas the recessive condition of both will give the white colour. This case may be represented as follows:

P_1 Brown seed x White seed

BBRR ↓ bbrr

Gamete BR ↓ br

F_1 BbRr
Brown seed
(selfed)

↓

F_2

♂ \ ♀	BR	Br	bR	br
BR	BBRR Brown	BBRr Brown	BbRR Brown	BbRr Brown
Br	BBRr Brown	BBrr Brown	BbRr Brown	Bbrr Brown
bR	BbRR Brown	BbRr Brown	bbRR Brown	bbRr Brown
br	BbRr Brown	Bbrr Brown	bbRr Brown	bbrr White

9 B-R- = Brown
3 B-rr = Brown } 15
3 bbR- = Brown :
1 bbrr = White 1

So the observed result shows the **duplicate factor** interaction in F_2 ratio.

Example 13 → (1:1:1:1): Test the goodness of fit in the following sample from an experiment with garden pea plant.

Round, yellow 31
Round, green 26
Wrinkled, yellow 27
Wrinkled, green 26

No. of classes	Sample character	Observed no.
1	Round, yellow	31
2	Round, green	26
3	Wrinkled, yellow	27
4	Wrinkled, green	26

Determination of expected segregation ratio:

	9:3:3:1		1:1:1:1	
Observed value	Expect.	|d|	Expect.	|d|
Round, yellow 31	61.875	30.875	27.5	3.5
Round, green 26	20.625	5.375	27.5	1.5
Wrinkled, yellow 27	20.625	6.375	27.5	0.5
Wrinkled, green 26	6.875	19.125	27.5	1.5
Total		61.75		7.0

According to the table, we find that the deviation is minimum in case of **1:1:1:1** ratio. So, it is assumed that the observed sample should fit with this ratio.

Determination of Chi-square value:

No. of classes	Sample characters	Observed no. (O)	Expected no. (E)	(O–E)	$(O-E)^2$	$\frac{(O-E)^2}{E}$
1	Round, yellow	31	27.5	3.5	12.25	$\frac{12.25}{27.5}=0.445$
2	Round, green	26	27.5	– 1.5	2.25	$\frac{2.25}{27.5}=0.082$
3	Wrinkled, yellow	27	27.5	– 0.5	0.25	$\frac{0.25}{27.5}=0.009$
4	Wrinkled, green	26	27.5	– 1.5	2.25	$\frac{2.25}{27.5}=0.082$

$$\chi^2 = \Sigma\frac{(O-E)^2}{E} = 0.618$$

Conclusion: Since there are 4 classes, degrees of freedom = 4 – 1 = 3. The calculated χ^2 value is 0.618, which is much less than the tabulated χ^2 value for 3 degree of freedom at 0.05 probability level.

The calculated value lies between 70-90% probability level. Therefore the deviation from the expected value is insignificant and the observed ratio has a good fit with the expected ratio.

Comment: As the experimental result shows good fit with 1:1:1:1 ratio, so it is assumed that the result is obtained from a **dihybrid test cross**. The dominant characters are yellow and round. Parental genotypes are assumed as

P_1 Yellow, Round x Green, Wrinkled
YYRR yyrr
↓ ↓
Gamete YR yr
↘ ↙
F_1 YyRr x yyrr
Yellow, Round Green, wrinkled
↓

♂ \ ♀	YR	Yr	yR	yr
yr	YyRr	Yyrr	yyRr	yyrr

YyRr → Yellow, round
Yyrr → Yellow wrinkled
yyRr → green, round
yyrr → green, wrinkled

Example 14 → (An example not in agreement with 1:2:1): A geneticist obtained 209 plants after selfing a hybrid, among which with red flower = 52 plants, with pink flower = 128 plants and with white flower = 29 plants. Show the goodness of fit with 1:2:1 ratio.

Character	No. of plants
Red	52
Pink	128
White	29
Total no. of plants	209

Determination of expected frequency of different plant:

$$\text{Red flower} = 209 \times \frac{1}{4} = 52.25$$

$$\text{Pink flower} = 209 \times \frac{2}{4} = 104.5$$

$$\text{White flower} = 209 \times \frac{1}{4} = 52.25$$

Calculation of Chi-square value:

Character	Observed value (O)	Expected value (E)	(O – E)	$(O-E)^2$	$\frac{(O-E)^2}{E}$	$\chi^2 = \frac{\Sigma(O-E)^2}{E}$
Red	52	52.25	– 0.25	0.0625	.00119	
Pink	128	104.5	23.5	552.25	5.28	15.626
White	29	52.25	– 23.25	540.56	10.345	

Conclusion: The calculated value of χ^2 = 15.626. Since there are 3 classes, so the degrees of freedom = (3 – 1) = 2. The tabulated χ^2 value for 2 degree of freedom at 0.05 probability level is 5.99. As the observed value is much higher than the tabulated value, so the result does not in agreement with 1:2:1 ratio, i.e., the observation **does not show goodness of fit** with the expected ratio.

In conclusion, the principle of incomplete dominance is not applicable here.

12.7 CHI-SQUARE TEST FOR ASSOCIATION OF ATTRIBUTES

By using 2 × 2 contingency table the χ^2 analysis is applied to test whether there is any association between two or more classifications, i.e., to test for independence of the two attributes.

Steps to be followed to calculate Chi-square

1. Null hypothesis should be set up, which is no association exists between the attributes.

   H_0 : No association exists between the attributes.

   H_1 : Association exists between the attributes.

2. Expected frequency (E) is calculated corresponding to each category by the formula.

$$E_{ij} = \frac{R_i \times C_j}{n}$$

   R_i = Sum total of row in which E_{ij} is present.

   C_j = Sum total of column in which E_{ij} is present.

   n = Total sample.

3. The Chi-square value is calculated according to formula

$$\text{Chi-square}\left(\chi^2\right) = \Sigma\frac{(O-E)^2}{E}$$

E = Expected value,
O = Observed value.

Degrees of freedom = (R – 1) (C – 1)

R = No. of rows,
C = No. of columns.

4. Table value is found out for a particular level of significance and for the calculated degree of freedom.
5. The calculated value and table value are compared, if the calculated value of $\chi^2 <$ the table value then the null hypothesis is accepted. But if the χ^2 value is larger than the tabulated value then null hypothesis is rejected.

Example 15: An experiment was carried out to see the effect of an organo-mercuric compound on the survival of seedling. Two different concentrations are applied to test whether the percentage of death in higher concentration is significantly different from that of lower concentration or both are independent.

	Conc. of Fungicide		Total
	10 ppm	30 ppm	
Dead	23	35	58
Alive	62	55	117
Total	85	90	175

In this case the null hypothesis will be – there is no significant difference, i.e., both the effects are independent.

H_0 : Both the effects are independent.

H_1 : Both the effects are significantly different.

Table showing expected results and deviation

	Low Conc.			High Conc.			
	Observed	Expected	(O–E)=d	Observed	Expected	(O–E)=d	Total
Dead	23	28.2	– 5.2	35	29.8	+ 5.2	58
Alive	62	56.8	+ 5.2	55	60.2	– 5.2	117
Total	85			90			175

$$\text{The Chi-square value } (\chi^2) = \Sigma\frac{(O-E)^2}{E}$$

In this example, as the (O–E) is always 5.2, so we can compute the values in following way.

$$\chi^2 = (\pm 5.2)^2\left(\frac{1}{28.2}+\frac{1}{29.8}+\frac{1}{56.8}+\frac{1}{60.2}\right)$$

$$= 27.04\times(0.035+.034+0.018+0.017)$$

$$= 2.812$$

The degree of freedom for this example is

$$(C-1)(R-1)=(2-1)\times(2-1)=1$$

The tabulated χ^2 value at degree of freedom 1, and at $p = 0.05$ level is 3.84.

As the calculated χ^2 value is much less than the tabulated χ^2 value so the null hypothesis is accepted, i.e., both the effects are independent, there is no relation of death percentage with low and high concentration of fungicide.

12.8 YATES CORRECTION

Yates correction is applied to increase the precision of χ^2-test, only when the degree of freedom is 1 and the expected classes are small (less than 30). In case of 1 degree of freedom, there is possibility of underestimating the probabilities listed in the table. This can be adjusted by subtracting the correction value from the deviation value.

For goodness of fit, Chi-square formula using Yates correction ($\frac{1}{2}$ or 0.5) will be:

$$\chi^2 = \Sigma \frac{\left\{|(O-E)|-\frac{1}{2}\right\}^2}{E}$$

In **Example 12,** $\chi^2 = \frac{(2-0.5)^2}{195} + \frac{(2-0.5)^2}{13} = 0.012 + 0.173 = 0.185.$

In case of contingency Chi-square, using Yates correction the Chi-square value is calculated as follows:

			Total
	a	b	R_1
	c	d	R_2
Total	C_1	C_2	N

$$\chi^2 = \frac{N\left\{(|ad-bc|)-\frac{N}{2}\right\}^2}{R_1 \times R_2 \times C_1 \times C_2}$$

where $\frac{N}{2}$ is the factor of Yates correction.

If we apply this formula in the **Example 15,** then

$$\chi^2 = \frac{175\left\{|(1265-2170)|-\frac{175}{2}\right\}^2}{58\times 117\times 85\times 90}$$

$$= \frac{175(905-875)^2}{51912900} = 2.253$$

where ad = 23 × 55 = 1265

bc = 35 × 62 = 2170

QUESTIONS

1. Express probability with the help of mathematical expression.
2. Describe the two rules of probability with suitable examples.
3. What is conditional probability? How do the rules get changed with conditional probability?
4. State and prove addition and multiplication rules in probability for three events.
5. A bag contains 7 green and 3 yellow coloured seeds. Two seeds are drawn at random one after the other without replacement. What will be the probability that both drawn seeds are yellow?
6. We obtained 52 plants of 1-52 cm. One plant is selected at random. What is the probability that it is either 12 cm or 40 cm of height (Here 4 plants are kept of equal size of each denominator).
7. Define χ^2 and mention its formula. What do you mean by goodness of fit and contingency table? Explain with examples.
8. Test the goodness of fit in the following samples :

	Sample character	Sample size		Sample chracter	Sample size
(A)	Brown, long seed	65	(C)	Flat pod	53
	Brown, small seed	35		Constricted pod	43
	White, long seed	45	(D)	Smooth seed	80
	White, small seed	55		Wrinkled seed	16
(B)	Red seed	35	(E)	Tall plant	92
	Brown seed	25		Dwarf plant	4

9. In a F_2 population of Ragi, the breeder observed 603 plants of grey coloured seeds and 217 plants of red coloured seeds. Test whether this proportion of plants is in agreement with Mendel's monohybrid ratio of 3:1.
10. In case of snapdragon we observed 70 red flowered and 27 pink flowered plants. What is your comment after χ^2-analysis?
11. In a field it was supposed to test that pink colour flower is associated with tall plant character. In the field 432 were short plants and 568 were tall plants, 305 produced white flower and 355 produced pink flower. Test the hypothesis whether the characters are associated.

13 CORRELATION AND REGRESSION

13.1 CORRELATION

In Biostatistics, sometimes we study two characters or variables on the same sample and try to find out the existance of any kind of relationship between these two characters. For example, different concentrations of pesticide and their effect on germination, panicle length and number of grains.

Definition

Correlation is the relationship which can reveal whether the change in one variable would cause change in the other or not. Such relationship between the two sets of characters or variables can be expressed quantitatively by the degree of relationship, called **Correlation Coefficient**.

Kinds of Correlation

There are 3 different kinds of correlations: **Positive, Zero** and **Negative** correlations.

(a) **Positive correlation**: When the values of two variables change together in the same direction then the relationship is called positive correlation.

This type of correlation may be perfect positive or moderately positive.

(i) **Perfect positive correlation**: When both the variables increase or decrease in the same proportion then it is perfect positive correlation.

(ii) **Moderately positive correlation**: In this case, two variables are positively correlated but the changes do not occur in the same proportion. The coefficient value lies between + 1 and 0.

(b) **Negative correlation**: If one variable increases (or decreases) and the other decreases (or increases) then the relationship is called negative correlation.

Such as size and number of fruits/plant are negatively correlated. This negative relation may also be of two kinds:

(i) **Perfect negative correlation**: This kind of relationship is really very rare in case of biological situation, such as increase in temperature decreases the lipid content of the cell.

(ii) **Moderately negative correlation**: In this relationship, the variables are negatively correlated but not very perfectly, such as increase in post harvesting period decreases the viability of seeds. Here also the coefficient value lies in between 0 and – 1.

(c) **Zero correlation**: If the two variables have no correlation, i.e., there is no consistency on value of observation, in such cases the two values of variables are called with zero correlation.

Coefficient of Correlation

When the two variables have any direct relationship then the degree of relationship between these two variables is expressed by quantitative expression which is called **Coefficient of Correlation**. This quantitative measure expresses the degree of closeness of the linear relationship between the two variables. The correlation coefficient is designated by the letter **'r'** and it is also called as **Karl Pearson's Coefficient of Correlation** which is calculated by the following formula:

$$r_{x,y} = \frac{\Sigma dx \cdot dy}{\sqrt{\Sigma dx^2 \cdot \Sigma dy^2}} \quad \text{or} \quad \frac{\Sigma dx \cdot dy}{n \cdot \delta x \cdot \delta y}$$

where
$dx = x_1 - \bar{x}$ [deviation of x variable]
$dy = y_1 - \bar{y}$ [deviation of y variable]
δx = S.D. of x variable
δy = S.D. of y variable.
n = total no. of observation.

Steps to calculate the value of 'r'

(a) Two series are made by x and y variable.

(b) Mean of both the series are calculated, $\bar{x}$ and $\bar{y}$.

(c) The deviation of each observation is calculated as dx and dy.

(d) Squaring of the deviations are noted.

(e) The deviations of both the variables are multiplied.

(f) All the data are summed up according to formula to calculate 'r'.

Properties of Correlation Coefficient (Fig. 13.1):

1. Correlation coefficient lies between – 1 and + 1, i.e., $-1 \le r \le +1$

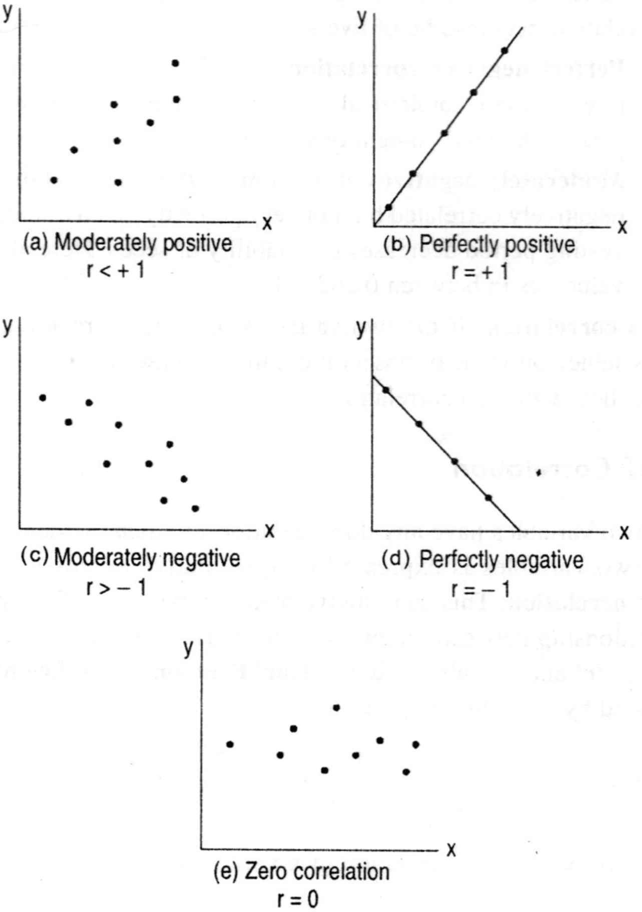

Fig. 13.1: Scatter diagrams showing different types and degrees of correlation

2. If r = + 1, the correlation is perfect and positive, if it is less than + 1 then moderately positive.
3. If r = – 1, the correlation is perfect and negative, if it is higher than – 1 then moderately negative.
4. If r = 0, there is no correlation between the variables.
5. The coefficient of correlation is not affected by change and scale of origin.

Example 1: Find out the correlation coefficient between the two attributes of 5 plants.

Height of the plant (x)	5 cm	9 cm	13 cm	17 cm	21 cm
Number of leaves per plant (y)	12	20	25	33	35

The data is arranged in the following table to calculate 'r':

x	dx	dx^2	y	dy	dy^2	dx.dy
5	5 – 13 = – 8	64	12	12 – 25 = – 13	169	104
9	9 – 13 = – 4	16	20	20 – 25 = – 5	25	20
13	13 – 13 = 0	0	25	25 – 25 = 0	0	0
17	17 – 13 = + 4	16	33	33 – 25 = + 8	64	32
21	21 – 13 = + 8	64	35	35 – 25 = + 10	100	80
$\Sigma x =$ 65 $\bar{x} = \frac{65}{5}$ =13		$\Sigma dx^2 =$ 160	$\Sigma y =$ 125 $\bar{y} = \frac{125}{5}$ = 25		$\Sigma dy^2 =$ 358	$\Sigma dx.dy =$ 236

Now,

$$r = \frac{\sum dx \cdot dy}{\sqrt{\sum dx^2 \cdot \sum dy^2}} = \frac{236}{\sqrt{160 \times 358}} = \frac{236}{\sqrt{57280}} = \frac{236}{239.33} = 0.986 \text{ (approx.)}$$

The value of r = 0.986 gives us the view that the two attributes like height of the plant and number of leaves are positively correlated and the value is as near about 1, so it can be concluded that the relationship is towards perfect positive relation.

Example 2: The effect of pesticide, 'Nuvan' is tested on germination of *Phaseolus* seeds. Find out the correlation coefficient.

Conc. of pesticide (ppm)	0	1	2.5	5	7.5	10	12.5	15	20
% germination	90	81	65	52	39	32	28	17	6

The data is arranged in the following table to calculate 'r':

x	dx	dx^2	y	dy	dy^2	dx.dy
0	– 8.16	66.58	90	+ 44.45	1975.8	– 362.71
1	– 7.16	51.26	81	+ 35.45	1256.7	– 253.82
2.5	– 5.66	32.03	65	+ 19.45	378.3	– 110.08
5	– 3.16	9.98	52	+ 6.45	41.6	– 20.38
7.5	– 0.66	0.436	39	– 6.55	42.9	+ 4.32
10	+ 1.84	3.38	32	– 13.55	183.6	– 24.93
12.5	+ 4.34	18.83	28	– 17.55	308.0	– 76.16
15	+ 6.84	46.78	17	– 28.55	815.1	– 195.28
20	+ 11.84	140.18	06	– 39.55	1564.2	– 468.27
$\Sigma x =$ 73.5 $\bar{x} = \frac{73.5}{9}$ = 8.16		$\Sigma dx^2 =$ 369.46	$\Sigma y =$ 410 $\bar{y} = \frac{410}{9}$ = 45.55		$\Sigma dy^2 =$ 6566.2	$\Sigma dx.dy =$ – 1507.31

$$r = \frac{\Sigma dx \cdot dy}{\sqrt{\Sigma dx^2 \cdot \Sigma dy^2}} = \frac{-1507.31}{\sqrt{369.46 \times 6566.2}} = \frac{-1507.31}{1557.55} = -0.9677$$

The value of r = – 0.9677 denotes that the two variables, i.e., pesticide concentration and germination percentage are negatively correlated, though it is not perfectly negative but approaching towards perfect negative correlation.

Significance of Correlation Coefficient

The calculated correlation coefficient should be checked from the correlation coefficient (r) table for the degree of freedom (number of pairs of observation minus one), i.e., (n – 1) at 0.05 to 0.001 probability level.

In **Example 1**, the 'r' value obtained is 0.986, the table r value at (10 – 1) = 9 degree of freedom at 0.001 probability level is 0.847. As the observed value is higher than the table value, so the 'r' value is highly significant.

In **Example 2**, the 'r' value is – 0.9677, the table 'r' value at (18 – 1) = 17 degree of freedom at 0.001 probability level is 0.693. As the observed value is much higher than the table value so, the 'r' value or the correlation is highly significant.

13.2 REGRESSION

Regression is used to denote the estimation and prediction of the average value of one variable for a specified value of the other variable. This estimation is done by deriving a suitable equation on the basis of available bivariate data. This equation is called **Regression equation** and its geometrical representation is called **Regression curve**.

The regression equation requires the **Regression coefficient**. The regression coefficient (b) is calculated in two different ways.

The regression coefficient of y on x is

$$b_{yx} = r\frac{\delta y}{\delta x}$$

and the regression coefficient of x on y is

$$b_{xy} = r\frac{\delta x}{\delta y}$$

δy = standard deviation of y

δx = standard deviation of x

Regression Line

When the bivariate data are plotted on graph paper, the concentration of points show certain pattern showing the relationship. When the trend points are found to be linear then by **least square method** we can obtain the **regression line**.

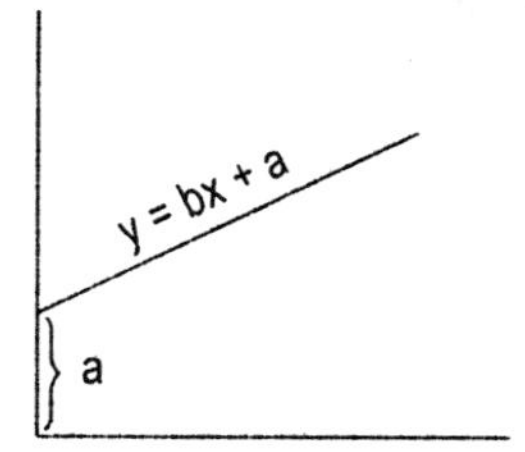

Fig. 13.2: Regression line showing the slope and constant

If two variables are linearly related then the relation can be expressed as y = bx + a, where 'b' is the slope of the line and 'a' is the intercept of that line.

If we put the values on regression equation then we can get the regression line equation easily. With the help of **Example 1,** we can discuss it.

The regression equation of x on y is

$$x - \bar{x} = b_{xy}(y - \bar{y})$$

where x = Value of x; y = Value of y;

$\bar{x}$ = Mean of x; $\bar{y}$ = Mean of y;

$$b_{xy} = r\frac{\delta x}{\delta y}$$

In example 1,

$$\bar{x} = 13, \quad \bar{y} = 25$$

$$\delta x = \sqrt{\frac{160}{5}} = \sqrt{32} = 5.65$$

$$\delta y = \sqrt{\frac{358}{5}} = \sqrt{71.6} = 8.46$$

$$b_{xy} = .986 \times \frac{5.65}{8.46} = .986 \times .66 = 0.658$$

So, the equation of x on y will be

$$x - 13 = 0.658(y - 25)$$

or $$x = 0.658y - 16.46 + 13$$

or $$x = 0.658y - 3.46$$

or $$y = \frac{x}{0.658} - \frac{3.46}{0.658} = 1.5x - 5.25$$

$$\therefore \quad y = 1.5x - 5.25$$

The regression equation of y on x is

$$y-\bar{y} = b_{yx}(x-\bar{x}) \qquad b_{yx} = .986\times\frac{8.46}{5.65}$$

$$\text{or} \quad y-25 = 1.476(x-13) \qquad = .986 \times 1.49$$

$$\text{or} \quad y = 1.476x-19.19+25 \qquad = 1.476$$

$$\text{or} \quad y = 1.476x+5.8$$

Significance

This equation will help us to get the estimate of one variable when the other is given or else we can predict the values of one variable when the other one is also assumed, i.e., extrapolation is possible.

QUESTIONS

1. Define and explain correlation and correlation-coefficient with examples.
2. With the help of scatter diagram define different types of correlation between two variables.
3. Define regression line. Derive the regression equation of x on y and y on x.
4. In an air pollution study a researcher obtained the following data with regard to number of lichen species and its respective distance from a chemical factory.

Distance away from factory (km)	0.5	1	2	3	4	5	7	10
No. of lichen species	1	2	5	8	11	16	19	28

   Comment on the correlation between the two variables.

5. The following data is observed in case of leguminous plant.

Number of seeds/pod	1	1	2	5	4	6	3	2	3	4
Length of pod (cm)	2	1.5	2	4.5	5	7	4	3	3	4.5

   Calculate the correlation coefficient and regression coefficient.

6. Length of panicle and number of grains were correlated in 10 randomly selected rice plants.

Length of panicle (cm)	10.0	12.5	15.0	12.0	16.0	11.0	14.5	17.0	14.0	10.5
No. of grains	65	102	105	85	115	72	85	110	92	75

   Calculate r and deduce regression equation.

14 MEASUREMENT OF GENE FREQUENCY

14.1 POPULATION GENETICS

A population consisting of sexually interbreeding organisms carrying one or more particular genes, which follow the Mendelian Principles of Inheritance is called **'Mendelian Population'**.

Gene pool and gene frequencies are considered to be two important attributes of a population. A **gene pool** is the sum total of genes in reproductive gametes of a population. The nature of gene pool depends on random mating of gametes to form zygotes in the next generation. **Gene frequency** can be defined as proportions of different alleles of a gene in a population, and in a particular generation these frequencies depend on their frequencies in the previous generation and also on the proportion of various genotypes in total population.

In any population, if a character is controlled by two alleles, then the frequency of these alleles or genes can be calculated very easily by phenotypic observation of that character under homozygous and heterozygous conditions. The frequency of an allele in a population is the number of occurrence of that allele divided by the total number of alleles of that gene locus.

14.2 MEASUREMENT OF GENE FREQUENCY

In a diploid species, a population having N individuals has 2N alleles for each gene locus. If there are two alleles 'A' and 'a' of a particular gene in this population, the number of A alleles is twice the number of AA homozygotes plus the number of Aa heterozygotes, as each homozygote has two 'A' alleles, and each heterozygote has one 'A' allele. So the frequency of 'A' is the number of 'A' alleles divided by total number of alleles, i.e., 2N. If the number is denoted by 'n' then the equation can be written as:

$$n_A = 2n_{AA} + n_{Aa}$$

$$n_a = 2n_{aa} + n_{Aa}$$

If the **frequency of allele A** is denoted by 'p' then,

$$p = \frac{n_A}{2N} = \frac{2n_{AA} + n_{Aa}}{2N}$$

Similarly, if the **frequency of allele a** is denoted by 'q' then,

$$q = \frac{n_a}{2N} = \frac{2n_{aa} + n_{Aa}}{2N}$$

It must be remembered that for all alleles the total frequency will be always 1, i.e., **$p + q = 1$** and **$n_A + n_a = 2N$**.

Example 1: In human population, a sample of 100 individuals for **MN blood group** character shows 50MM, 20MN and 30 NN individuals, then the frequency of M and N allele can be calculated using the above formula.

The frequency of 'M' will be $= \frac{2 \times 50 + 20}{200} = \frac{120}{200} = 0.6$

where $2n_{MM} = 2 \times 50$ $n_{MN} = 20$, $2N = 200$.

The frequency of 'N' will be $= \frac{2 \times 30 + 20}{200} = \frac{80}{200} = 0.4$

where $2n_{NN} = 2 \times 30$, $n_{MN} = 20$, $2N = 200$.

The frequency can be calculated by another formula:

Frequency of a gene = frequency of homozygotes of that gene $+ \frac{1}{2}$ frequency of heterozygotes

In the above example,

Frequency of M = 0.5 MM + $\frac{1}{2}$ (0.2 MN) = 0.6

Frequency of N = 0.3 NN + $\frac{1}{2}$ (0.2 MN) = 0.4.

14.3 GENOTYPE FREQUENCY AND HARDY-WEINBERG EQUILIBRIUM

In a randomly mating Mendelian population the gene and genotype frequencies reach to equilibrium in a single generation. The **Hardy-Weinberg law** states that the gene and genotype frequencies in a Mendelian population remain constant generation after generation if there is no selection, mutation, migration or genetic drift.

Homozygotes (AA & aa) are produced by the union of gametes carrying similar alleles. The frequency with which a male gamete carrying allele A (frequency p) fuses with a female gamete carrying 'A' (also frequency p) will be $p \times p = p^2$. Similarly the frequency with which a male gamete carrying allele 'a' (frequency q) fuses with a female gamete also carrying 'a' will be $q \times q = q^2$.

Heterozygotes (Aa) are produced by the fusion of gametes carrying different alleles, the frequency with which a male gamete carrying allele 'A' fuses with a female gamete 'a' will be $p \times q = pq$. Similarly in opposite way the frequency with which a female gamete carrying allele 'A' fuses with a male gamete carrying 'a' will be $p \times q = pq$. So, the total frequency of heterozygotes will be 2 pq.

♂ \ ♀	A (p)	a (q)
A(p)	AA (p^2)	Aa (pq)
a (q)	Aa (pq)	aa (q^2)

Fig. 14.1: The frequencies of three genotypes produced due to random mating in a population having 'A' allele with frequency 'p' and allele 'a' with frequency 'q'

According to the theory of probability, due to random mating, the genotype frequency can easily be calculated from the following formula:

$$(p+q)\times(p+q)=p^2+2pq+q^2$$

(p + q) = Total frequency of two alleles 'A' and 'a'.

This equation is called **Hardy-Weinberg equation** which is named after British Mathematician G.H. Hardy and German Physician W. Weinberg.

As according to theory of probability the total frequency of p + q is always 1. So,

$$\mathbf{(p+q)^2=p^2+2pq+q^2=1}$$

In case of **genes showing codominance,** this equation can be used very easily to calculate the gene frequency by observing the phenotypic characters of homozygotes and heterozygotes.

Example 2: In the following example it is shown how the observed value help to calculate the frequency.

MN blood group of Human

Phenotype	Genotype	No. of Individuals	Frequency Calculation
M	L_ML_M	1787	$L_M=\frac{2\times1787+3039}{2\times6129}=0.5395$
MN	L_ML_N	3039	
N	L_NL_N	1303	$L_N=\frac{2\times1303+3039}{2\times6129}=0.4605$
	Total	6129	

If we put these two values in a frequency table we can get the expected number of phenotypes due to random mating of each allele as follows:

	L_M $p = 0.5395$	L_N $q = 0.4605$
L_M $p = 0.5395$	$p^2 = 0.2911$	$pq = 0.2484$
L_N $q = 0.4605$	$pq = 0.2484$	$q^2 = 0.2121$

So, the expected values in total population can be calculated

$$p^2 = 0.2911 \times 6129 = 1784.2$$
$$2pq = 2 \times 0.2484 \times 6129 = 3044.8$$
$$q^2 = 0.2121 \times 6129 = 1300$$

The above calculation shows that the observed number of individuals and expected number of phenotypes are very close to each other which follows the **Hardy-Weinberg Equilibrium Frequencies**.

In case of **completely dominant genes** of a population, the heterozygotes cannot be phenotypically distinguished from homozygotes for the dominant character. In this case, by observing the recessive phenotype, the recessive allele frequency can be measured very easily using Hardy-Weinberg equation. The frequency of homozygotes of recessive allele is q^2. So the frequency of recessive allele in a population is $\sqrt{q^2} = q$.

The frequency of dominant allele in that population is $(1 - q) = p$. This concept can also be applied to determine the number of individuals affected or carrying a particular disease, vis-a-vis, the frequency of the gene causing the disease. PKU (Phenylketonuria) is a serious disease of human being which is a genetically controlled metabolic disorder. If in any population 0.04% is PKU affected then the frequency of PKU allele in that population is $\sqrt{.0004} = 0.02$ and the frequency of dominant allele is $(1 - 0.02) = 0.98$.

If in a population the frequency of deleterious allele is 0.01, assuming random mating the frequency of homozygous population will be $(0.01)^2 = .0001$. Now the heterozygous frequency will be $2 \times .01 \times (1 - 0.01) = 2 \times .01 \times 0.99 = 0.0198$. So, about 1.98% population are carrier of this allele in heterozygous condition which is not expressed.

If the frequency of recessive allele is too small then it is very difficult to measure it correctly, thus it is desirable to draw a large sample which will give more error free estimation.

14.4 FACTORS AFFECTING GENE FREQUENCY

Populations change over time. The number of individuals in a population may increase or decrease depending on food resources, climate, weather and the availability of breeding areas, etc.

At the genetic level, a population may change because of mutation, migration, selection and random genetic drift. These natural processes alter allele frequencies, thereby changing the fundamental makeup of the population.

1. **Mutations** produce new alleles, though the rate of mutation is too low to be noticeable. In case of deleterious mutation, selection counteracts the effect.
2. **Migration** changes the frequencies because the immigrants from a population have different genetic make up. So, in a particular population the gene frequency changes due to genotype changes because of differential emigration from other locality.
3. **Selection** reduces the fertility or survival of certain genotypes. In many populations survival and reproductive ability are variable traits. Some individuals may die before they reproduce, whereas others may have many progeny. The differential contribution of progeny implies that the alleles that are associated with superior fitness will increase in frequency in a population.
4. **Random genetic drift** (chance difference) causes small deviations from the predicted frequencies, especially in small samples. These deviations however are expected and allowed for on the basis of statistical tests.
5. **Assortative mating** of either similar or dissimilar genotypes results in an excess of homozygotes and heterozygotes respectively which changes the gene frequency.
6. **Existence of subpopulations** which are locally mating groups in a large population, may be due to some ethnic causes or class grouping as in human, or due to low mobility of some organisms in a large area. These phenomena increases the frequency of homozygotes which is the result of inbreeding.

QUESTIONS

1. Write down the Hardy-Weinberg equation and explain its utility in measuring gene frequency.
2. The frequency of an allele in a large randomly mating population is 0.2. What is the frequency of heterozygote carriers?
3. The incidence of recessive albinism is 0.0004 in a human population. What is the frequency of the recessive allele?
4. The frequency of newborn infants homozygous for a recessive lethal allele is about 1 in 25,000. What is the expected frequency of the carriers of this allele in the population?

5. In a wild prairie grass, the genotypes TT and Tt produce plants that are 100 cm tall. In contrast tt produces only 50 cm tall. In a randomly mating population the frequency of t allele is 0.6. What is the average height of plants in this population?
6. The following data for the MN blood types were obtained from villages in North and Central America.

Group	Population size	M	MN	N
Central American	86	53	29	4
North American	278	78	61	139

   Calculate the frequencies of the L^M and L^N alleles for two groups.
7. In a population the frequency of L^M and L^N alleles were 0.78 and 0.22 respectively. What are expected frequencies of MM, MN and NN?
8. Dwarf corn is homozygous recessive for gene d which constitutes 20% of the gene pool of a population. If two tall corn plants are crossed in this population, what is the probability of a dwarf offspring being produced?
9. 80% of a population consists of tasters and 20% of non-tasters. Determine the frequency of heterozygous tasters by applying Hardy-Weinberg principle.
10. What is the frequency of heterozygotes Tt in a randomly mating population if the frequency of recessive phenotype (tt) is 0.04?

BIOSTATISTICAL SYMBOLS

x	:	Class value
$\overline{x}$	:	Sample Mean
μ	:	Population Mean
M	:	Median
M_0	:	Mode
f	:	Class frequency
n	:	Total number of observation
df	:	Degrees of freedom
Σ	:	Summation
d	:	Deviation
s^2 or σ^2	:	Variance
s or σ, SD	:	Standard Deviation
S_D	:	Standard deviation of difference
SE / $S_{\overline{x}}$	:	Standard Error
MD / δ	:	Mean Deviation
t	:	Test of significance (t-test)
F	:	Variance ratio
cv	:	Coefficient of variation
r	:	Correlation co-efficient
χ^2	:	Chi-square
SE_D/σ_d	:	Standard error of difference
p	:	Probability of success
q	:	Probability of failure
H_0	:	Null hypothesis
H_1/H_A	:	Alternative hypothesis
$\overline{D}$	:	Mean difference
u	:	Number of columns in ANOVA
ν	:	Number of rows in ANOVA
O	:	Observed value for χ^2 analysis
E	:	Expected value for χ^2 analysis

BIOSTATISTICAL FORMULAE

1. Mean $= \bar{x} = \dfrac{\sum fx}{n}$

2. Mean deviation $= \dfrac{\sum fd}{n}$

3. Degrees of freedom $= (n-1)$

4. Variance $= s^2 = \dfrac{\sum fd^2}{n-1}$

5. Coefficient of variation $=$ c.v. $= \dfrac{s}{\bar{x}} \times 100$ or $\dfrac{\sigma}{\bar{x}} \times 100$

6. Standard Deviation $= \sigma$ or $s = \sqrt{\dfrac{\sum fd^2}{n-1}}$

7. Standard Error $= SE = \dfrac{s}{\sqrt{n}}$ or $\dfrac{\sigma}{\sqrt{n}}$

8. Student's `t'-test $= t = \dfrac{\bar{x}_1 - \bar{x}_2}{SE_D}$

   (a) Unpaired `t'-test (for large sample) $= t = \dfrac{\bar{x}_1 - \bar{x}_2}{\sigma_d} = \dfrac{\bar{x}_1 - \bar{x}_2}{\sqrt{\dfrac{\sigma_1^2}{n_1} + \dfrac{\sigma_2^2}{n_2}}}$

   (b) Unpaired `t'-test (for small sample) $= t = \dfrac{\bar{x}_1 - \bar{x}_2}{\sigma_d \sqrt{\dfrac{1}{n_1} + \dfrac{1}{n_2}}}$

$$\sigma_d = \sqrt{\frac{\sum(x_1 - \bar{x}_1)^2 + \sum(x_2 - \bar{x}_2)^2}{n_1 + n_2 - 2}}$$

$$\text{or } \sqrt{\frac{\sum x_1^2 - \dfrac{(\sum x_1)^2}{n_1} + \sum x_2^2 - \dfrac{(\sum x_2)^2}{n_2}}{n_1 + n_2 - 2}}$$

(c) Paired `t'-test $= t = \dfrac{\overline{D}}{\dfrac{S_D}{\sqrt{n}}} = \dfrac{\dfrac{\Sigma D}{n}}{\dfrac{S_D}{\sqrt{n}}}$ $\left[S_D = \dfrac{\Sigma D^2 - \dfrac{(\Sigma D)^2}{n}}{n-1} \right]$

9. (a) `F'-value between treatments $= \dfrac{\dfrac{B-D}{u-1}}{\dfrac{(A-D)-\{(B-D)+(C-D)\}}{(u-1)(v-1)}}$

(b) `F'-value between replicates $= \dfrac{\dfrac{C-D}{v-1}}{\dfrac{(A-D)-\{(B-D)+(C-D)\}}{(u-1)(v-1)}}$

10. (a) Probability of either A or B $= p(A \text{ or } B) = p(A) + p(B)$

(b) Probability of A and B (independent events) $= p(A \text{ and } B) = p(A) \times p(B)$

(c) Conditional probability of A and B $= p(A \text{ and } B) = p(A) + p(B / A)$

or $p(B) \times p(A / B)$

11. Chi – square $= \chi^2 = \Sigma \dfrac{(O-E)^2}{E}$

12. Correlation co-efficient $= r = \dfrac{\Sigma\{(x-\overline{x}) \times (y-\overline{y})\}}{\sqrt{\{\Sigma(x-\overline{x})^2 \times \Sigma(y-\overline{y})^2\}}}$

13. (a) Allele frequency of `A' $= p = \dfrac{2n_{AA} + n_{Aa}}{2N}$

(b) Allele frequency of `a' $= q = \dfrac{2n_{aa} + n_{Aa}}{2N}$

BIOSTATISTICAL TABLES

Table I: *Distribution of t;* ***Table II:*** *F-value at 0.05 p;* ***Table III:*** *F-value at 0.01 p;* ***Table IV:*** *F-value at 0.001 p;* ***Table V:*** *Distribution of χ^2;* ***Table VI:*** *r value*

Table I: Distribution of t

Probability, p

Degrees of freedom (n)	.2	1	.05	.02	.01	.001
1	3.078	6.314	12.606	31.821	63.657	636.619
2	1.886	2.920	4.303	6.965	9.925	31.598
3	1.638	2.353	3.182	4.541	5.841	12.924
4	1.533	2.132	2.776	3.747	4.604	8.610
5	1.476	2.015	2.571	3.365	4.032	6.869
6	1.440	1.943	2.447	3.143	3.707	5.959
7	1.415	1.895	2.365	2.998	3.499	5.408
8	1.397	1.860	2.306	2.896	3.355	5.041
9	1.383	1.833	2.262	2.821	3.250	4.781
10	1.372	1.812	2.228	2.764	3.169	4.587
11	1.363	1.796	2.201	2.718	3.106	4.437
12	1.356	1.782	2.179	2.681	3.055	4.318
13	1.350	1.771	2.160	2.650	3.012	4.221
14	1.345	1.761	2.145	2.624	2.977	4.140
15	1.341	1.753	2.131	2.602	2.947	4.073
16	1.337	1.746	2.120	2.583	2.921	4.015
17	1.333	1.740	2.110	2.567	2.898	3.965
18	1.330	1.734	2.101	2.552	2.878	3.922
19	1.328	1.729	2.093	2.539	2.861	3.883
20	1.325	1.725	2.086	2.528	2.845	3.850
21	1.323	1.721	2.080	2.518	2.831	3.819
22	1.321	1.717	2.074	2.508	2.819	3.792
23	1.319	1.714	2.069	2.500	2.807	3.767
24	1.318	1.711	2.064	2.492	2.797	3.745
25	1.316	1.708	2.060	2.485	2.787	3.725
26	1.315	1.706	2.056	2.479	2.779	3.707
27	1.314	1.703	2.052	2.473	2.771	3.690
28	1.313	1.701	2.048	2.467	2.763	3.674
29	1.311	1.699	2.045	2.462	2.756	3.659
30	1.310	1.697	2.042	2.457	2.750	3.646
40	1.303	1.684	2.021	2.423	2.704	3.551
60	1.296	1.671	2.000	2.390	2.660	3.460
120	1.289	1.658	1.980	2.358	2.617	3.373
∞	1.282	1.645	1.960	2.326	2.576	3.291

Table I is abridged from Table III of Fisher & Yates: *Statistical Tables for Biological, Agricultural and Medical Research*, published by Oliver & Boyd, Edinburgh.

Table II: Values of F (variance ratio) at 0.05 probability

Degrees of freedom, n_2	Degrees of freedom, n_1									
	1	2	3	4	5	6	8	12	24	∞
1	161.4	199.5	215.7	224.6	230.0	234.0	238.9	243.9	249.0	254.3
2	18.5	19.0	19.2	19.3	19.3	19.3	19.4	19.4	19.5	19.5
3	10.1	9.6	9.3	9.1	9.0	8.9	8.8	8.7	8.6	8.5
4	7.7	6.9	6.6	6.4	6.3	6.2	6.0	5.9	5.8	5.6
5	6.6	5.8	5.4	5.2	5.1	5.0	4.8	4.7	4.5	4.4
6	6.0	5.1	4.8	4.5	4.4	4.3	4.2	4.0	3.8	3.7
7	5.6	4.7	4.4	4.1	4.0	3.9	3.7	3.6	3.4	3.2
8	5.3	4.5	4.1	3.8	3.7	3.6	3.4	3.3	3.1	2.9
9	5.1	4.3	3.9	3.6	3.5	3.4	3.2	3.1	2.9	2.7
10	5.0	4.1	3.7	3.5	3.3	3.2	3.1	2.9	2.7	2.5
11	4.8	4.0	3.6	3.4	3.2	3.1	3.0	2.8	2.6	2.4
12	4.8	3.9	3.5	3.3	3.1	3.0	2.9	2.7	2.5	2.3
13	4.7	3.8	3.4	3.2	3.0	2.9	2.8	2.6	2.5	2.2
14	4.6	3.7	3.3	3.1	3.0	2.9	2.7	2.5	2.4	2.1
15	4.5	3.7	3.3	3.1	2.9	2.8	2.6	2.5	2.3	2.1
16	4.5	3.6	3.2	3.0	2.9	2.7	2.6	2.4	2.2	2.0
17	4.5	3.6	3.2	3.0	2.8	2.7	2.6	2.4	2.2	2.0
18	4.4	3.6	3.2	2.9	2.8	2.7	2.5	2.3	2.2	1.9
19	4.4	3.5	3.1	2.9	2.7	2.6	2.5	2.3	2.1	1.9
20	4.4	3.5	3.1	2.9	2.7	2.6	2.5	2.3	2.1	1.8
21	4.3	3.5	3.1	2.8	2.7	2.6	2.4	2.3	2.1	1.8
22	4.3	3.4	3.1	2.8	2.7	2.6	2.4	2.2	2.0	1.8
23	4.3	3.4	3.0	2.8	2.6	2.5	2.4	2.2	2.0	1.8
24	4.3	3.4	3.0	2.8	2.6	2.5	2.4	2.2	2.0	1.7
25	4.2	3.4	3.0	2.8	2.6	2.5	2.3	2.2	2.0	1.7
26	4.2	3.4	3.0	2.7	2.6	2.5	2.3	2.2	2.0	1.7
27	4.2	3.4	3.0	2.7	2.6	2.5	2.3	2.1	1.9	1.7
28	4.2	3.3	3.0	2.7	2.6	2.4	2.3	2.1	1.9	1.7
29	4.2	3.3	2.9	2.7	2.5	2.4	2.3	2.1	1.9	1.6
30	4.2	3.3	2.9	2.7	2.5	2.4	2.3	2.1	1.9	1.6
40	4.1	3.2	2.8	2.6	2.5	2.3	2.2	2.0	1.8	1.5
60	4.0	3.2	2.8	2.5	2.4	2.3	2.1	1.9	1.7	1.4
120	3.9	3.1	2.7	2.5	2.3	2.2	2.0	1.8	1.6	1.3
∞	3.8	3.0	2.6	2.4	2.2	2.1	1.9	1.8	1.5	1.0

Table II is taken from Fisher and Yates: *Statistical Tables for Biological, Agricultural and Medical Research*, Published by Oliver & Boyd, Edinburgh.

Table III: Values of F (variance ratio) at 0.01 probability

Degrees of freedom, n_2	Degrees of freedom, n_1									
	1	2	3	4	5	6	8	12	24	∞
1	4052	4999	5403	5625	5764	5859	5982	6106	6234	6366
2	98.5	99.0	99.2	99.3	99.3	99.3	99.4	99.4	99.5	99.5
3	34.1	30.8	29.5	28.7	28.2	27.9	27.5	27.1	26.6	26.1
4	21.2	18.0	16.7	16.0	15.5	15.2	14.8	14.4	13.9	13.5
5	16.3	13.3	12.1	11.4	11.0	10.7	10.3	9.9	9.5	9.0
6	13.7	10.9	9.8	9.2	8.8	8.5	8.1	7.7	7.3	6.9
7	12.3	9.6	8.5	7.9	7.5	7.2	6.8	6.5	6.1	5.7
8	11.3	8.7	7.6	7.0	6.6	6.4	6.0	5.7	5.3	4.9
9	10.6	8.0	7.0	6.4	6.1	5.8	5.5	5.1	4.7	4.3
10	10.0	7.6	6.6	6.0	5.6	5.4	5.1	4.7	4.3	3.9
11	9.7	7.2	6.2	5.7	5.3	5.1	4.7	4.4	4.0	3.6
12	9.3	6.9	6.0	5.4	5.1	4.8	4.5	4.2	3.8	3.4
13	9.1	6.7	5.7	4.2	4.9	4.6	4.3	4.0	3.6	3.2
14	8.9	6.5	5.6	5.0	4.7	4.5	4.1	3.8	3.4	3.0
15	8.7	6.4	5.4	4.9	4.6	4.3	4.0	3.7	3.3	2.9
16	8.5	6.2	5.3	4.8	4.4	4.2	3.9	3.6	3.2	2.8
17	8.4	6.1	5.2	4.7	4.3	4.1	3.8	3.5	3.1	2.7
18	8.3	6.0	5.1	4.6	4.3	4.0	3.7	3.4	3.0	2.6
19	8.2	5.9	5.0	4.5	4.2	3.9	3.6	3.3	2.9	2.5
20	8.1	5.9	4.9	4.4	4.1	3.9	3.6	3.2	2.9	2.4
21	8.0	5.8	4.9	4.4	4.0	3.8	3.5	3.2	2.8	2.4
22	7.9	5.7	4.8	4.3	4.0	3.8	3.5	3.1	2.8	2.3
23	7.9	5.7	4.8	4.3	3.9	3.7	3.4	3.1	2.7	2.3
24	7.8	5.6	4.7	4.2	3.9	3.7	3.4	3.0	2.7	2.2
25	7.8	5.6	4.7	4.2	3.9	3.6	3.3	3.0	2.6	2.2
26	7.7	5.5	4.6	4.1	3.8	3.6	3.3	3.0	2.6	2.1
27	7.7	5.5	4.6	4.1	3.8	3.6	3.3	2.9	2.6	2.1
28	7.6	5.5	4.6	4.1	3.8	3.5	3.2	2.9	2.5	2.1
29	7.6	5.4	4.5	4.0	3.7	3.5	3.2	2.9	2.5	2.0
30	7.6	5.4	4.5	4.0	3.7	3.5	3.2	2.8	2.5	2.0
40	7.3	5.2	4.3	3.8	3.5	3.3	3.0	2.7	2.3	1.8
60	7.1	5.0	4.1	3.7	3.3	3.1	2.8	2.5	2.1	1.6
120	6.9	4.8	4.0	3.5	3.2	3.0	2.7	2.3	2.0	1.4
∞	6.6	4.6	3.8	3.3	3.0	2.8	2.5	2.2	1.8	1.0

Table III is taken from Fisher and Yates: *Statistical Tables for Biological, Agricultural and Medical Research*, Published by Oliver & Boyd, Edinburgh.

Table IV: Values of F (variance ratio) at 0.001 probability

Degrees of freedom, n_2	Degrees of freedom, n_1									
	1	2	3	4	5	6	8	12	24	∞
1	405	500	540	563	576	586	598	611	632	637
2	998.5	999.0	999.2	999.2	999.3	999.3	999.4	999.4	999.5	999.5
3	167.0	148.5	141.1	137.1	134.6	132.8	130.6	128.3	125.9	123.5
4	74.1	61.3	56.2	53.4	51.7	50.5	49.0	47.4	45.8	44.1
5	47.2	37.1	33.2	31.1	29.8	28.8	27.6	26.4	25.1	23.8
6	35.5	27.0	23.7	21.9	20.8	20.0	19.0	18.0	16.9	15.8
7	29.3	21.7	18.8	17.2	16.2	15.5	14.6	13.7	12.7	11.7
8	25.4	18.5	15.8	14.4	13.5	12.9	12.0	11.2	10.3	9.3
9	22.9	16.4	13.9	12.6	11.7	11.1	10.4	9.6	8.7	7.8
10	21.0	14.9	12.6	11.3	10.5	9.9	9.2	8.5	7.6	6.8
11	19.7	13.8	11.6	10.4	9.6	9.1	8.4	7.6	6.9	6.0
12	18.6	13.0	10.8	9.6	8.9	8.4	7.7	7.0	6.3	5.4
13	17.8	12.3	10.2	9.1	8.4	7.9	7.2	6.5	5.8	5.0
14	17.1	11.8	9.7	8.6	7.9	7.4	6.8	6.1	5.4	4.6
15	16.6	11.3	9.3	8.3	7.6	7.1	6.5	5.8	5.1	4.3
16	16.1	11.0	9.0	7.9	7.3	6.8	6.2	5.6	4.9	4.1
17	15.7	10.7	8.7	7.7	7.0	6.6	6.0	5.3	4.6	3.9
18	15.4	10.4	8.5	7.5	6.8	6.4	5.8	5.1	4.5	3.7
19	15.1	10.2	8.3	7.3	6.6	6.2	5.6	5.0	4.3	3.5
20	14.8	10.0	8.1	7.1	6.5	6.0	5.4	4.8	4.2	3.4
21	14.6	9.8	7.9	7.0	6.3	5.9	5.3	4.7	4.0	3.3
22	14.4	9.6	7.8	6.8	6.2	5.8	5.2	4.6	3.9	3.2
23	14.2	9.5	7.7	6.7	6.1	5.7	5.1	4.5	3.8	3.1
24	14.0	9.3	7.6	6.6	6.0	5.6	5.0	4.4	3.7	3.0
25	13.9	9.2	7.5	6.5	5.9	5.5	4.9	4.3	3.7	2.9
26	13.7	9.1	7.4	6.4	5.8	5.4	4.8	4.2	3.6	2.8
27	13.6	9.0	7.3	6.3	5.7	5.3	4.8	4.2	3.5	2.8
28	13.5	8.9	7.2	6.3	5.7	5.2	4.7	4.1	3.5	2.7
29	13.4	8.9	7.1	6.2	5.6	5.2	4.6	4.1	3.4	2.6
30	13.3	8.8	7.1	6.1	5.5	5.1	4.6	4.0	3.4	2.6
40	12.6	8.3	6.6	5.7	5.1	4.7	4.2	3.6	3.0	2.2
60	12.0	7.8	6.2	5.3	4.8	4.4	3.9	3.3	2.7	1.9
120	11.4	7.3	5.8	5.0	4.4	4.0	3.6	3.0	2.4	1.5
∞	10.8	6.9	5.4	4.6	4.1	3.7	3.3	2.7	2.1	1.0

Table IV is taken from Fisher and Yates: *Statistical Tables for Biological, Agricultural and Medical Research*, Published by Oliver & Boyd, Edinburgh.

Table V: Distribution of χ^2

Probability, p

Degrees of freedom (n)	.99	.98	.95	.90	.80	.50	.20	.10	.05	.02	.01	.001
1	$.0^3157$	$.0^3628$	.00393	.0158	.0642	.455	1.642	2.706	3.841	5.412	6.635	10.827
2	.0201	.0404	.103	.211	.446	1.386	3.219	4.605	5.991	7.824	9.210	13.815
3	.115	.185	.352	.584	1.005	2.366	4.642	6.251	7.815	9.837	11.345	16.266
4	.297	.429	.711	1.064	1.649	3.357	5.989	7.779	9.488	11.668	13.277	18.467
5	.554	.752	1.145	1.610	2.343	4.351	7.289	9.236	11.070	13.388	15.086	20.515
6	.872	1.134	1.635	2.204	3.070	5.348	8.558	10.645	12.592	15.033	16.812	22.457
7	1.239	1.564	2.167	2.833	3.822	6.346	9.803	12.017	14.067	16.622	18.475	24.322
8	1.646	2.032	2.733	3.490	4.594	7.344	11.030	13.362	15.507	18.168	20.090	26.125
9	2.088	2.532	3.325	4.168	5.380	8.343	12.242	14.684	16.919	19.679	21.666	27.877
10	2.558	3.059	3.940	4.865	6.179	9.342	13.442	15.987	18.307	21.161	23.209	29.588
11	3.053	3.609	4.575	5.578	6.989	10.341	14.631	17.275	19.675	22.618	24.725	31.264
12	3.571	4.178	5.226	6.304	7.807	11.340	15.812	18.549	21.026	22.054	26.217	32.909
13	4.107	4.765	5.892	7.042	8.634	12.340	16.985	19.812	22.362	25.472	27.688	34.528
14	4.660	5.368	6.571	7.790	9.467	13.339	18.151	21.064	23.685	26.873	29.141	36.123
15	5.229	5.985	7.261	8.547	10.307	14.339	19.311	22.307	24.996	28.259	30.578	37.697
16	5.812	6.614	7.962	9.312	11.152	15.338	20.465	23.542	26.296	29.633	32.000	39.252
17	6.408	7.255	8.672	10.085	12.002	16.338	21.615	24.769	27.587	30.995	33.409	40.790
18	7.015	7.906	9.390	10.865	12.857	17.338	22.760	25.983	28.869	32.346	34.805	42.312
19	7.633	8.567	10.117	11.651	13.716	18.338	23.900	27.204	30.144	33.687	36.191	43.820
20	8.260	9.237	10.851	12.443	14.578	19.337	25.038	28.412	31.410	35.020	37.566	45.315
21	8.897	9.915	11.591	13.240	15.445	20.337	26.171	29.615	32.671	36.343	38.932	46.797
22	9.542	10.600	12.338	14.041	16.314	21.337	27.301	30.813	33.224	37.659	40.289	48.268
23	10.196	11.293	13.091	14.848	17.187	22.337	28.429	32.007	35.172	38.968	41.638	49.728
24	10.856	11.992	13.848	15.569	18.062	23.337	29.553	33.196	36.415	40.270	42.980	51.179
25	11.524	12.697	14.611	16.473	18.940	24.337	30.675	34.382	37.652	41.566	44.314	52.620
26	12.198	13.409	15.379	17.292	19.820	25.336	31.795	35.563	38.885	42.856	45.642	54.052
27	12.879	14.125	16.151	18.114	20.703	26.336	32.912	36.741	40.113	44.140	46.963	55.476
28	13.565	14.847	16.928	18.939	21.588	27.336	34.027	37.916	41.337	45.419	48.278	56.893
29	14.256	15.574	17.708	19.768	22.475	28.336	35.139	39.087	42.557	46.693	49.588	58.302
30	14.953	16.306	18.493	20.599	23.364	29.336	36.250	40.256	43.773	47.962	50.892	59.703

Table V is abridged from Table IV of Fisher & Yates: *Statistical Tables for Biological, Agricultural and Medical Research*, published by Oliver & Boyd, Edinburgh.

Table VI: Value of the Correlation Coefficient (r) for different levels of significance					
Degrees of freedom (n)	.1	.05	.02	.01	.001
1	.98769	.99692	.999507	.999877	.9999988
2	.90000	.95000	.98000	.990000	.99900
3	.8054	.8783	.93433	.95873	.99116
4	.7293	.8114	.8822	.91720	.97406
5	.6694	.7545	.8329	.8745	.95074
6	.6215	.7067	.7887	.8343	.92493
7	.5822	.6664	.7498	.7977	.8982
8	.5494	.6319	.7155	.7646	.8721
9	.5214	.6021	.6851	.7348	.8471
10	.4973	.5760	.6581	.7079	.8233
11	.4762	.5529	.6339	.6835	.8010
12	.4575	.5324	.6120	.6614	.7800
13	.4409	.5139	.5923	.6411	.7603
14	.4259	.4973	.5742	.6226	.7420
15	.4124	.4821	.5577	.6055	.7246
16	.4000	.4683	.5425	.5897	.7084
17	.3887	.4555	.5285	.5751	.6932
18	.3783	.4438	.5155	.5614	.6787
19	.3687	.4329	.5034	.5487	.6652
20	.3589	.4227	.4921	.5368	.6542
25	.3233	.3809	.4451	.4869	.5974
30	.2960	.3494	.4093	.4487	.5541
35	.2746	.3246	.3810	.4182	.5189
40	.2573	.3044	.3578	.3932	.4896
45	.2428	.2875	.3384	.3721	.4648
50	.2306	.2732	.3218	.3541	.4433
60	.2108	.2500	.2948	.3248	.4078
70	.1954	.2319	.2737	.3017	.3799
80	.1829	.2172	.2565	.2830	.3568
90	.1726	.2050	.2422	.2673	.3375
100	.1638	.1946	.2301	.2540	.3211

Table VI is abridged from Table VII of Fisher & Yates: *Statistical Tables for Biological, Agricultural and Medical Research,* published by Oliver & Boyd, Edinburgh.

PLANT BIOTECHNOLOGY

15 PLANT BIOTECHNOLOGY: AN INTRODUCTION

The term '**Biotechnology**' was coined in 1917 by a Hungarian Engineer, Karl Ereky, to describe a process for large scale production of pigs. The word biotechnology indicates the interaction between the science of biology and technology.

Biotechnology has been broadly defined as, '**the development and utilisation of biological processes, forms and systems for obtaining maximum benefits to human and other forms of life'.**

According to the definition adopted by the European Federation of Biotechnology, created in 1978, 'Biotechnology makes it possible, through an integrated application of knowledge and techniques of biochemistry, microbiology, genetics and chemical engineering, to draw benefit, at the technological level, from the properties and capacities of microorganisms and cell cultures. Biotechnology offers the possibility of producing, from widely available renewable resources, substances and compounds essential to life'.

Modern biotechnology enables an organism to produce a totally new product which the organism does not or can not produce normally through the incorporation of the technology of **'Genetic engineering'**.

The knowledge of totipotency of plant cell, different aspects of plant tissue culture techniques, cell fusion techniques alongwith the use of rDNA technology has developed a new area of **'Plant biotechnology'**.

15.1 AREAS AND BRANCHES

Many areas of biotechnology have developed through the interaction between various branches of biology and engineering; biochemistry, biophysics, cell biology, chemistry, embryology, ecology, microbiology, genetics, molecular biology and immunology—all these disciplines of biology and on the engineering side chemical and biochemical engineering have helped to develop the different areas and branches of modern biotechnology (Table 15.1).

Table 15.1: Areas and Branches of Biotechnology along with their products

Areas/Branches	Products
1. Recombinant DNA Techniques (Genetic Engineering)	Fine chemicals, Enzymes, Vaccines, Hormones, Antibodies, Antibiotics
2. Plant Cell Culture/Animal Cell Culture Techniques	Fine chemicals (alkaloids, steroids, essential oils, dyes), Somatic embryos, Monoclonal antibodies, Diagnostic and Therapeutic kits
3. Process Engineering	Harvesting, Pretreatment, Filtration, Effluent treatment, Water recycling, Product extraction, Novel reactors
4. Waste Product Treatment and Utilization	Herbicide/Insecticide detoxification, Sewage treatment and utilization, Byproduct utilization
5. Traditional Fermentation	Alcohol, Organic acids, Amino acids, Polysaccharides, Antibiotics
6. Biofuels	Methane (biogas from wastes), Hydrogen, Alcohol (from waste)
7. Enzymes and Biocatalysts	Food processing, Fine chemicals, Diagnostic kits, Chemotherapy, Biosensors
8. Nitrogen Fixation	Microbial or Algal biofertilizer.

Recombinant DNA Techniques and Genetic Engineering: The knowledge of molecular biology has provided the ability to extract a gene coding for a desired product and transfer it to another organism which has helped in production of useful proteins, hormones, vaccines, enzymes, etc. by some microorganisms, plant cells or animal cell cultures. The use of this technique is essential for four reasons: First, large scale production of desired product; Second, supply of natural source may be limited; Third, the purity of naturally available material is doubtful; Fourth, is the cost of the natural product may be higher.

Plant Cell Culture Techniques: The ability to culture plant cells on a large scale, either for the production of biomass or in order to extract the desired product from cell cultures (high yielding), has become a highly desirable technology. Plant cells are important and rich source of high value drugs and many secondary metabolites. With the advancement of this technology the production becomes more economic to produce low cost and high volume materials.

Process Engineering: Laboratory scientists can clone a novel gene, discover a new antibiotic or enzyme catalysed process but it requires to transfer the knowledge to the scale of operation required to make a useful product in significant quantity. Harvesting, pretreatment, filtration of the raw material, reactor design and their control, recovery/reuse of biocatalyst or organism, product extraction and analysis—all these require the help of chemical or biochemical engineers.

Waste Treatment and Utilisation: The treatment of sewage and waste water from domestic and industrial sources is the largest biotechnological industry. Wastes and byproducts of activities in agriculture, forestry and food industry can be used for various purposes, particularly to produce energy. Many types of wastes are suitable for biological treatment and some may even yield useful products.

Except anaerobic or aerobic waste treatment, biodegradation is the process by which waste materials such as oil spill, herbicides, pesticides, etc. are degraded by the action of microbial systems.

Traditional Fermentation: Traditionally, fermentation has meant the production of potable alcohol from carbohydrates. However, fermentation—the application of microbial metabolism to transform simple raw materials into valuable products—can produce useful substances e.g. citric acid, antibiotics, biopolymer and single cell proteins. The potential of this process is endless and varied as microorganisms themselves.

Biofuels: Biotechnology can provide two solutions towards the problems of fuels in our world today—one is new kind of fuels, another is alternative carbon feedstock. Sugarcane waste can be used as source for alcoholic fuels. Animal sludge can be used for biogas production, some of aquatic herbs (*Eichornea*) are also used for biogas production—which is alternative source of fuels. The most aesthetically pleasing biotechnological fuel would be hydrogen derived from the biophotolysis of water, which is based on the combination of the photosystems of plants with bacterial-derived hydrogenase enzymes and light.

Enzymes and Biocatalysis: Recently great interest has been aroused in extending the use of enzymes in food processing, chemical production, analytical and diagnostic systems, and in the treatment of disease. This modern interest is largely a result of the greater understanding of the function and nature of enzymes and also the new technologies involved in the handling of biocatalysts, particularly their immobilisation so that they may be fixed in a reactor or on a sensor probe.

Biofertilizer: Living plants used as biofertilizers have become a means of supplementing the availability of the scarce source of chemical fertilizers and also enriching the soil. Leguminous plants enrich the fertility of soil due to presence of *Rhizobium*; other biofertilizers are *Azotobacter*, *Azospirillum* and *Azolla*; other phosphate solubilizing microorganisms and mycorrhizal fungi have also found a place in biofertilizer technology.

15.2 SCOPE AND PROSPECTS

Each and every organism performs its function within its optimum limits but the excitement of modern biotechnology lies in the fact that the application of different scientific methods (like genetic engineering) can enhance the natural capabilities of production much more as well as miracle can happen i.e., bacteria can produce mammalian hormone, and also plant cell can produce plastic granules.

Biotechnology in Agriculture: Biotechnology is making a new ground in the food or agriculture area. Different plant cell, tissue and organ culture techniques can help in micropropagation of different kinds of plants for different purposes, phytomass or biomass production to get some alkaloids or secondary metabolites, anther cultures are used for haploid production, gametic and somatic cell/tissue cultures are used for gametoclonal or somaclonal variation or for production of artificial seeds.

Genetic engineering and transformation techniques have helped in production of transgenic plants carrying desirable traits like disease resistance, insect resistance, herbicide resistance and also may be used for increasing photosynthetic efficiency, nitrogen fixing ability, improved storage proteins, hybrid crops, etc.

Protoplast culture and fusion have helped in production of somatic hybrids between sexually incompatible species, permitting transfer of desirable traits from wild or unrelated crop species to our cultivated crop plants.

Biotechnology in Industry and Medicine: Industrial microbiology is another area receiving major attention of biotechnologists; many pharmaceutical drugs and chemicals are now being produced in microbial cells in larger quantity utilizing techniques of genetic engineering. In the field of medicine, insulin and interferon synthesized by bacteria are already in the market. A large number of vaccines for immunization against deadly diseases, DNA probes and monoclonal antibodies for diagnosis of various diseases, human growth hormone and other pharmaceutical drugs for treatment of diseases are being produced through microbes. Gene therapy for some lethal diseases and DNA fingerprinting or autoantibody fingerprinting techniques are great boon in medical and forensic science.

Biotechnology and Environment: Biotechnology is also being used for dealing with environmental problems like pollution control, depletion of natural resources for non-renewable energy, restoration of degraded lands, biodiversity conservation. Biopesticides, biofertilisers and biosensors (for detection of metals), cleaning of metals and spilled oils, biomonitoring in different industries are all the result of biotechnological developments.

Biotechnology in Protein and Metabolic Engineering: Protein engineering, metabolic engineering are also the recently developed major areas of biotechnology. Protein engineering helps the production of superior enzymes and storage proteins with the help of computer aided protein modelling. Biotechnology research has helped in the production of metabolites at the industrial scale by the use of computer aided introduction of different genes and regulatory elements called metabolic engineering.

In the area of bioinformatics recently databases are available through information network througout the world which help the development of biotechnologial research in many areas.

QUESTIONS

1. Define biotechnology. State different areas and branches of biotechnology.
2. Write down the scope and prospects of biotechnology.

16 PLANT CELL AND TISSUE CULTURE: PRINCIPLES AND APPLICATIONS

A plant consists of different organs and each organ is composed of different tissues, which in turn is made up of different individual cells. German botanist Gottlieb Haberlandt (1902) developed the concept of *in vitro* cell culture. He was the first to culture isolated, fully differentiated cells in a nutrient medium containing glucose, peptone and Knop's salt solution. Using different types of cells from different plant parts he could culture the cells under aseptic condition, but failed to induce division.

16.1 CELLULAR TOTIPOTENCY

Though Haberlandt was able to culture isolated cells but failed to induce division *in vitro*. The culture of embryogenic tissue by Hannig (1904) and root-tip culture by Kotte and Robbins (1922) helped in solving the problems of cell divisions in *in vitro* conditions. During the period of 1934-1954, the main focus was the development of callus from different kind of explants.

The two main aspects of plant tissue culture are:

1. Multiplication of a single cell *in vitro*.
2. Development of a whole plant from proliferated tissue of this cell.

Various experimental studies on single cell culture and cell suspension culture; development of cell mass and then induction of somatic embryogenesis from carrot suspension culture (Steward et.al. 1966) contributed to the establishment of the concept of totipotency of somatic cells under experimental conditions.

The technique of cloning isolated single cells *in vitro* has demonstrated the fact that somatic cells, under appropriate conditions, can differentiate to a whole plant. This potential of a cell to grow and develop a multicellular or multiorganed higher organism is termed as **cellular totipotency.**

In plant tissue culture, more often an explant (an excised piece of differentiated tissue or organ) is used to initiate their growth in culture. The non-dividing differentiated quiescent cells of the explant, when grown on a medium, first undergo changes to achieve the meristematic state. The phenomenon of mature cells reverting to a meristematic state and forming undifferentiated callus tissue is termed as **dedifferentiation.** The ability of the component cells of the callus to differentiate into a whole plant or a plant organ is termed as **redifferentiation.** These two phenomena of dedifferentiation and redifferentiation are the inherent property of plant cell i.e., the potential lies within the genes which remain inactive in differentiated tissues or organs and are able to express only under adequate cultural conditions (Fig. 16.1).

In animals, the differentiation is irreversible. This is in contrast to plants where even highly mature or differentiated cells have the ability to regress to a meristematic state as long as they are viable. Tissue culture techniques offer an excellent opportunity to study

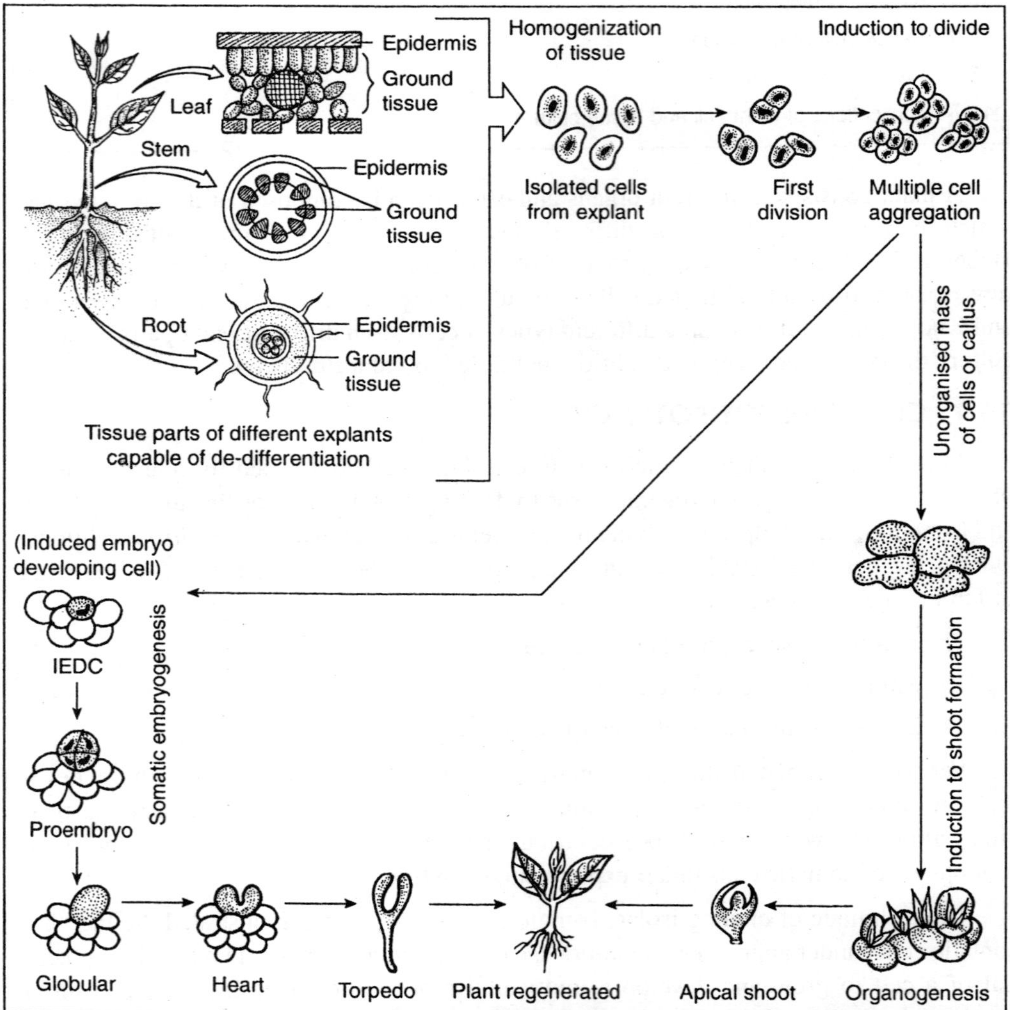

Fig. 16.1: Cellular Totipotency used in tissue culture

factors responsible for differentiation of cells. These factors control **cellular totipotency** through cytological, histological and organogenic differentiation.

16.2 LABORATORY REQUIREMENTS

'Plant tissue culture' or *in vitro* cultivation of plant parts need some basic requirements

(a) Cultivation should be done under aseptic conditions.

(b) The isolated plant part should get an appropriate environment which will help to divide the cell and to get an expression of internal potential.

Basic facilities for plant tissue culture operations involving any type of *in vitro* procedures must include certain essential elements:

(a) Washing and storage facilities;

(b) Media preparation, sterilisation and storage room;

(c) Transfer area for aseptic manipulations;

(d) Culture rooms or incubators for maintenance of cultures under controlled conditions of temperature, light and humidity;

(e) Observation or data collection area;

(f) Transplantation area.

Washing and Storage Facilities

An area with large sink (lead lined to resist acids and alkalis) and draining area is necessary with provision for running water, draining-boards or racks and ready access to a deionised, distilled and double-distilled apparatus. Space should also be available to set up drying ovens, washing machines, plastic or steel buckets for soaking labware, acid or detergent baths, pipette washers, driers and cleaning brushes. For storage of washed and dried labware, the laboratory should be provided with dustproof cupboards or storage cabinets.

Media Preparation Room or Space

This part is the central section of the laboratory where most of the activities are performed i.e., media preparation and sterilisation of media and glasswares needed for culture. There should be sufficient working bench as well as storage space.

The following items are essential in the room (Fig. 16.2A–D):

(i) Different types of glasswares

(ii) Different kinds of balances

(iii) Required chemicals

(iv) Hot plates and Stirrer

(v) Water bath

(vi) pH meter

(vii) Autoclave and Hot air oven

(viii) Microwave oven

(ix) Vortex, Shaker

(x) Centrifuge

(xi) Refrigerator and Freezer

(xii) Storage cabinet (Dustfree)

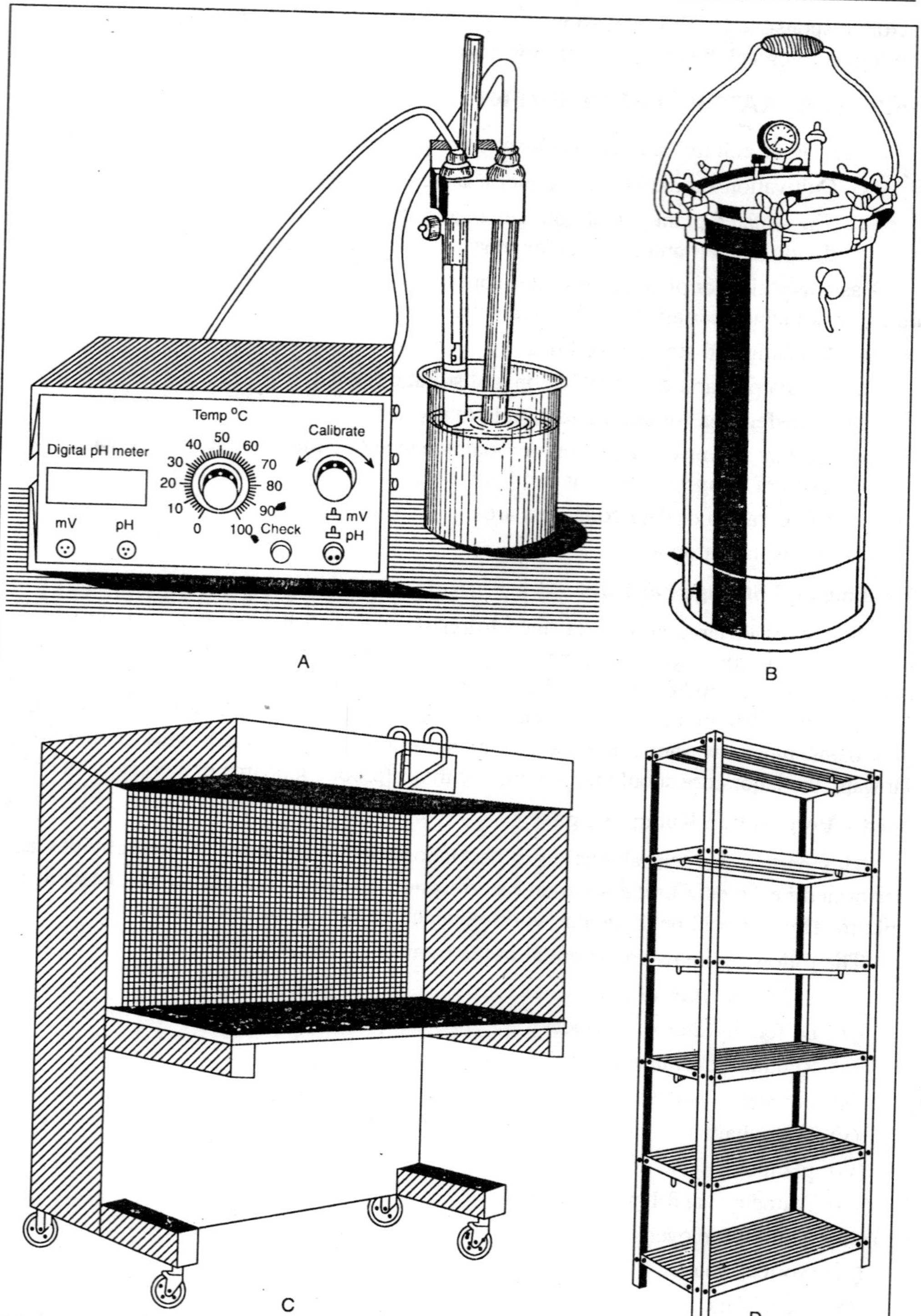

Fig. 16.2: A. A digital pH meter, B. A simple portable autoclave, C. A laminar air flow, D. Design for a skeleton rack for keeping culture vessels and incubation of culture

Transfer Area

Tissue culture techniques can only be successfully carried out in a very clean laboratory having dry atmosphere with protection against air-borne microorganisms. For this purpose a sterile dust-free room/cabinet is needed for routine transfer and manipulation work.

The '**laminar air flow cabinet**' (Fig. 16.2C) is the most common accessory used for aseptic manipulations now-a-days. The cabinet may be designed with horizontal air flow or vertical air flow where the air is forced into the cabinet through a bacterial HEPA (High Efficiency Particulate Air) filter. The air flows over the working bench at a constant rate which prevents the particles (microorganisms) from settling on the bench. Before operation in the laminar air flow cabinet, the interior of the cabinet is sterilised with the ultraviolet (UV) germicidal light and wiping the floor of cabinet with 70% alcohol. **Inoculation chamber**, a specially designed air tight glass chamber fitted with UV light, may also be used as transfer area.

Culture Room

Plant tissue cultures should be incubated under conditions of well-controlled temperature, illumination, photoperiod, humidity and air circulation. Incubation culture rooms, commercially available incubator cabinets, large plant growth chambers and walk-in-environmental rooms satisfy these requirements. Culture rooms are constructed with proper air-conditioning, perforated shelves (Fig. 16.2D) to support the culture vessels, fitted with fluorescent tubes having a timing device to maintain the photoperiod, black curtains may be used to maintain total darkness. For the suspension cultures, gyratory shakers are used. Air conditioners and heaters are used to maintain the temperature around 25 ± 2°C and humidity is maintained by uniform forced air-ventilation. The lighting is also done in a measured amount i.e., 40-200 fc (foot-candle).

Data Collection Area

The growth and development of tissues cultured *in vitro* are generally monitored by observing cultures at regular intervals in the culture room or incubators where they have been maintained under controlled environmental conditions. Arrangement should be there where the observations can be done under aseptic conditions using microscope. Special facilities are required for germplasm conservation i.e., cryopreservation accessories should be there.

Transplantation Area

Plants regenerated from *in vitro* tissue culture are transplanted to soil in pots. The potted plants are ultimately transferred to greenhouse but prior to transfer the tissue culture grown plants are allowed for acclimatisation under well humid condition and controlled temperature and under controlled entry of sunlight.

16.3 CULTURE MEDIA

Growth and morphogenesis of plant tissues *in vitro* are largely governed by the composition of the culture media. Although the basic requirements for culturing plant tissue is same but in practice there are some specific nutritional requirements for promoting optimal growth from different kinds of explants in case of different plant species.

Components of Media

The principal components of most plant tissue culture media are inorganic nutrients (macronutrients and micronutrients), carbon sources, organic supplements, growth regulators, vitamins, amino acids and gelling agent for solid or semisolid media. The various culture media formulated for plant tissue culture are Murashige and Skoog's medium, Gamborg's medium, White's medium, etc. which differ (Table 16.1) mainly in quantity of the components.

Table 16.1: Composition (mg/l) of three basal media for plant tissue cultures

Ingredient	Murashige and Skoog (MS) medium	Gamborg et al. (B5) medium	White's medium
NH_4NO_3	1650	–	–
KNO_3	1900	2500	80
$Ca(NO_3)_2.4H_2O$	–	–	288
$CaCl_2.2H_2O$	440	150	737
$MgSO_4.7H_2O$	370	250	–
KH_2PO_4	170	–	–
$Na_2SO_4.10H_2O$	–	–	460
$(NH_4)_2SO_4$	–	134	–
$NaH_2PO_4.H_2O$	–	150	19
KCl	–	–	65
KI	0.83	0.75	–
H_3BO_3	6.2	3.0	–
$MnSO_4.4H_2O$	22.3	–	1.5
$MnSO_4H_2O$	–	10.0	0.75
$ZnSO_4.7H_2O$	8.6	2.0	–
$Na_2MoO_4.2H_2O$	0.25	0.25	–
$CuSO_4.5H_2O$	0.025	0.025	0.001
$CoCl_2.6H_2O$	0.025	0.025	–
Na_2EDTA	37.3	37.3	2.67
$FeSO_4.7H_2O$	27.8	27.8	27.8
Sucrose	30,000	20,000	20,000
Inositol	100	100	–
Nicotinic acid	0.5	1.0	0.5
Pyridoxine-HCl	0.5	1.0	0.1
Thiamine-HCl	0.1	10.0	0.1
Glycine	2.0	–	3.0

(i) **Inorganic Nutrients:** A variety of mineral elements (salts) supply the macro- and micronutrients required in the life of a plant. Elements required in concentration greater than 0.5 m mol/l referred to as macronutrients and those less than 0.05 m mol/l as micronutrients. The macronutrients include six major elements like Nitrogen (N), Phosphorus (P), Potassium (K), Calcium (Ca), Magnesium (Mg) and Sulphur (S) which are present as salts in various media. The micronutrients are iron (Fe), Manganese (Mn), Zinc (Zn), Boron (B), Copper (Cu), Molybdenum (Mo). Among these iron is used in the medium as chelated form with EDTA (ethylene-diamino-tetraacetic acid).

(ii) **Carbon and Energy Source:** Plant cells and tissues in the culture medium lack autotrophic ability and therefore need external carbon for energy. The most preferred carbon source in tissue culture media is sucrose; glucose supports equally for good growth while fructose is less efficient.

(iii) **Vitamins:** The plant cells and tissues are capable of synthesizing different vitamins but in *in vitro* condition they are being produced at sub-optimal level. Hence it is necessary to supplement the media with required vitamins such as Thiamine (B_1), Nicotinic acid (B_3), Pyridoxine (B_6), Pantothenic acid (B_5) and myoinositol. Except these other vitamins like Biotin, Folic acid, Ascorbic acid are also used in some media.

(iv) **Amino Acids:** *In vitro* grown plant cells or tissues are capable of synthesizing amino acids but the addition of amino acids to media is important for stimulating cell growth and protoplast culture and for establishment of cell lines. Glutamine, asparagine, glycine, arginine, cysteine are the commonly used amino acids.

(v) **Growth Regulators:** Three broad classes of growth regulators mainly auxins, cytokinins and gibberellins are used in tissue culture. The growth, differentiation and organogenesis of tissues become feasible only on the addition of one or more of these classes of hormones to a medium.

Various kinds of auxins like IAA (Indole-3-Acetic Acid), NAA (Napthalene Acetic Acid), 2, 4-D (2, 4-dichlorophenoxy acetic acid) are used to induce cell division, elongation of stem, internodes and rooting.

Cytokinins like BAP (Benzyl amino purine), Kinetin (Furfuryl amino purine), Zeatin or 2iP (Isopentynyl adenine) are used which are mainly concerned with cell division, modification of apical dominance and shoot differentiation in tissue culture.

The ratio of auxin and cytokinin is important with respect to morphogenesis in the culture system. For embryogenesis, callus initiation and root initiation the requisite ratio of auxin to cytokinin is high, while the reverse leads to organogenesis and shoot proliferation.

Gibberellins (GA_3) are occasionally used growth regulators to induce plantlet formation from adventive embryos developed in culture.

(vi) **Other Organic Supplements:** Culture media are often supplemented with variety of organic extracts which have the constituents of an undefined nature e.g. casein hydrolysate, coconut milk, malt extracts and fruit juice.

(vii) **Activated Charcoal:** The addition of activated charcoal to culture media is reported to stimulate growth and differentiation by reducing toxicity by removing toxic substances (e.g. Phenols).

(viii) **Antibiotics:** For transformation experiments addition of antibiotics to culture media is required to prevent the growth of *Agrobacterium* which retards the cell or tissue growth.

(ix) **Gelling Agent:** Gelling or solidifying agents are commonly used for preparing semisolid or solid tissue culture media to provide solid surface area for growth. Agar (a polysaccharide obtained from seaweeds), Gelatin (commercially available as Phytagel, Alginate or Gelrite) are commonly used solidifying agents at a concentration of 0.8–1.0% (which are not any kind of nutrient), depending on the type of tissue and the species of plant and culture methods.

Media Preparation

For media preparation it is convenient to use stock solutions of major salts (20X), minor salts (200X), organic supplements (200X), growth regulator (1 m mol or 10 m mol). For final media preparation the stock solutions are mixed together in appropriate quantities. All the stock solutions are stored in proper plastic or glass containers at low temperature. After preparation of media the pH is adjusted at 5.5–6.0 and agar is mixed whenever it is required. The media is poured into the desired type of **culture vessels** (culture tubes, conical flasks, petriplates, etc.), properly plugged with non-absorbent cotton and autoclaved.

16.4 ASEPTIC MANIPULATION

Microorganisms are omnipresent in the environment and in culture media they grow more faster than the plant cell or tissue. Furthermore, they produce metabolic wastes or toxic substances in the media which inhibit the growth of plant cell or tissue. So it is very much essential to carry out all the operations under total aseptic condition. For this purpose there are several steps to be followed:

A. Sterilisation of the culture vessels and necessary instruments

B. Sterilisation of nutrient media

C. Sterilisation of culture rooms and transfer-area

D. Sterilisation of explant

E. Aseptic transfer of explant/subculturing.

A. Sterilisation of the Culture Vessels and Instruments

For sterilisation of culture vessels, glasswares, instruments, etc. it is better to use steam sterilisation techniques i.e., by *autoclaving* at high pressure (15 pound/inch^2) and high temperature (121°C) for 15-20 mins. All the items should be properly covered with aluminium foil before sterilisation. Glasswares, metal instruments can also be sterilised by exposure to dry hot air oven (160°–180°C) for 2–4 hrs. Plastic labware of special types, cotton plugs, gauze, filters can be sterilised using steam sterilisation technique. The autoclaved instruments like forceps, scalpels, needles, spatula are further sterilised by flame sterilisation technique before use by dipping in 95% alcohol, and followed by flaming and cooling.

B. Sterilisation of Nutrient Media

The nutrient media used in tissue culture are commonly sterilised by autoclaving and filter sterilisation. Macro-, micronutrients and other stable compound mixtures are autoclaved, whereas the thermolabile compounds are filter-sterilised separately and mixed with the media whenever necessary. Culture media containing high conc. of sugars when autoclaved the carbohydrate get decomposed under high pressure. Some vitamins, amino acids, plant extracts and hormones are thermolabile, they require filter sterilisation. The solutions are passed through a bacterial membrane filter under pressure (pore size 0.2 μm), the sterilised solution is then mixed separately with autoclaved media.

C. Sterilisation of Culture Rooms and Transfer-area

The floor and walls of culture rooms are cleaned by gently washing with detergent then by wiping them with 2% sodium hypochlorite solution and then using ethanol. Commercially available disinfectants like Lysol can be used for this purpose. The process of surface sterilisation of culture rooms should be done at regular intervals.

The transfer-area is also sterilised once or twice a month by washing with common brand of antifungal agent. Transfer rooms or areas are further sterilised by UV light. UV radiation is harmful for eyes, so UV radiation should be done before operation. Where laminar air flow cabinet is used for transfer, the surface is cleaned by ethanol and chamber is sterilised by UV light before work in progress. Then the sterile air is flowed through HEPA filter during works.

D. Sterilisation of Explant

All different kinds of plant materials or explants should be surface sterilised by a variety of chemicals before using in tissue culture (Fig. 16.3). It is the eradication of surface microorganisms with the help of different chemicals. The type and concentration of different chemical sterilant to be used for sterilisation of different types of explants and exposure time must be decided experimentally. Sometimes the sterilisation procedure may lead to lethality to plant tissue, so the use of disinfectants should be tested. Here are few disinfectants which are used commonly for surface sterilisation of different explants (Table 16.2).

(a) 1% solution of Sodium hypochlorite (NaClO), commercial bleach having 5% active chlorine can be used.

(b) 4%–10% solution of Calcium hypochlorite [$Ca(ClO)_2$] can be used, it enters within the plant tissue slower than sodium hypochlorite.

(c) 1% solution of Bromine-water.

(d) 0.01–0.1% solution of Mercuric Chloride ($HgCl_2$) which is an extremely toxic substance for plant tissue, repeated rinsing with water is very much essential.

(e) 70% Ethyl alcohol is used for sterilisation of plant material dipping them for 30 sec–2 mins.

(f) 10% Hydrogen peroxide (H_2O_2) solution is effective for end surface sterilisation.

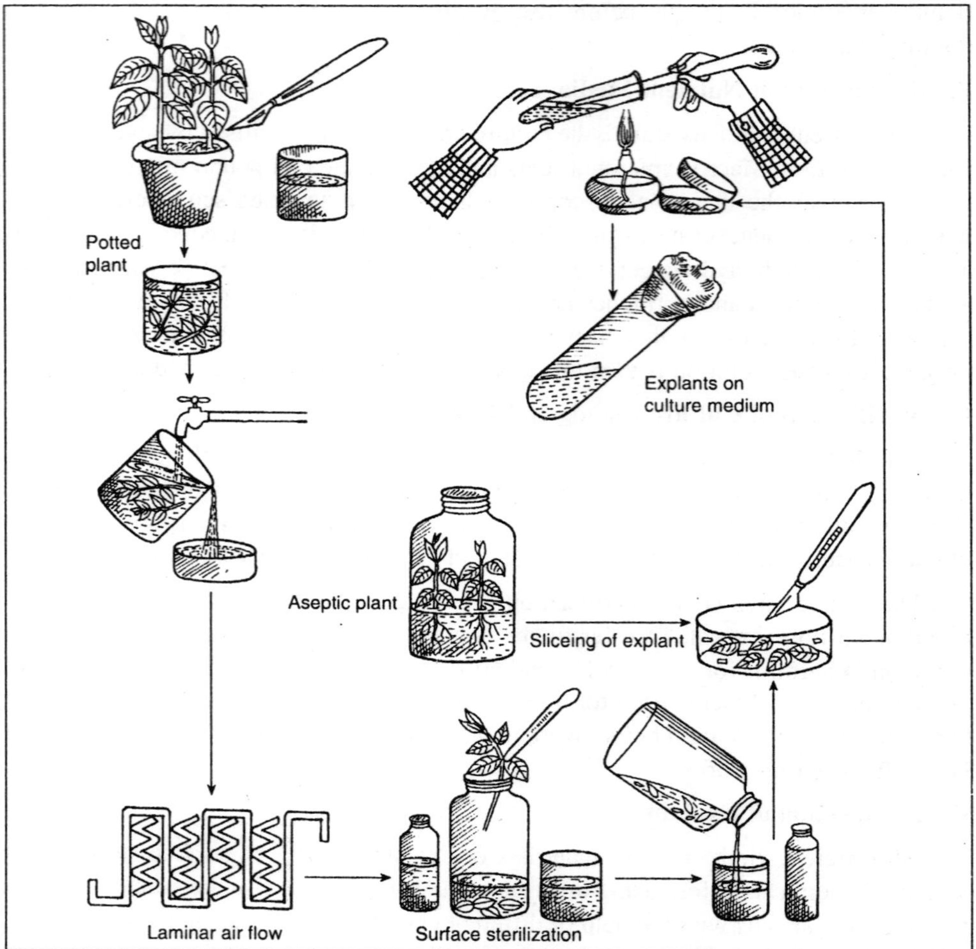

Fig. 16.3: Flow diagram illustrating the procedure for surface sterilization of plant material and inoculation of explant for culture

(g) 1% Silver nitrate solution.

(h) 25–200 mg/l antibiotic solution.

Table 16.2: A number of commonly used disinfectants

Name of disinfectants	Conc. used	Duration of treatment (min)
Sodium hypochlorite	0.5–1%	5–30 min.
Calcium hypochlorite	5–10%	5–30 min.
Hydrogen peroxide	10–12%	5–15 min.
Bromine water	1–2%	2–10 min.
Silver nitrate	1%	5–30 min.
Mercuric chloride	0.01–1%	2–10 min.
Ethyl alcohol	70–90%	few sec.–2 min.
Antibiotics	25–200 mg/l	30–60 min.

All these sterilants should be washed out properly before using the explant as the retention of these chemical substances may affect the establishment of successful tissue culture. But in most of the cases it becomes difficult to determine the optimal conditions for each kind of tissue. So to avoid this problem the explants can be taken from aseptically grown plants developed from the surface sterilised seeds as these seeds are more resistant to chemicals due to presence of seed coat. For this purpose, the seeds are surface sterilised and then cultured aseptically in basal nutrient media. These give rise to aseptic seedlings from which the different explants can be used. Explants from such seedlings need no further sterilisation (Fig. 16.4).

But for anther culture and shoot tip culture the explants are collected from outside grown plant. For these kinds of explants addition of few drops of surfactant (Trito-X or Tween-20) to the solution or treating the plant material in a solution of Cetavlon for 2 min. before exposing to sterilant may enhance sterilisation efficiency.

E. Aseptic transfer of Explant/Subculturing

Control of contamination largely depends upon the operator's technique while transferring the sterilised explant/subculturing into the sterile culture vial containing nutrient

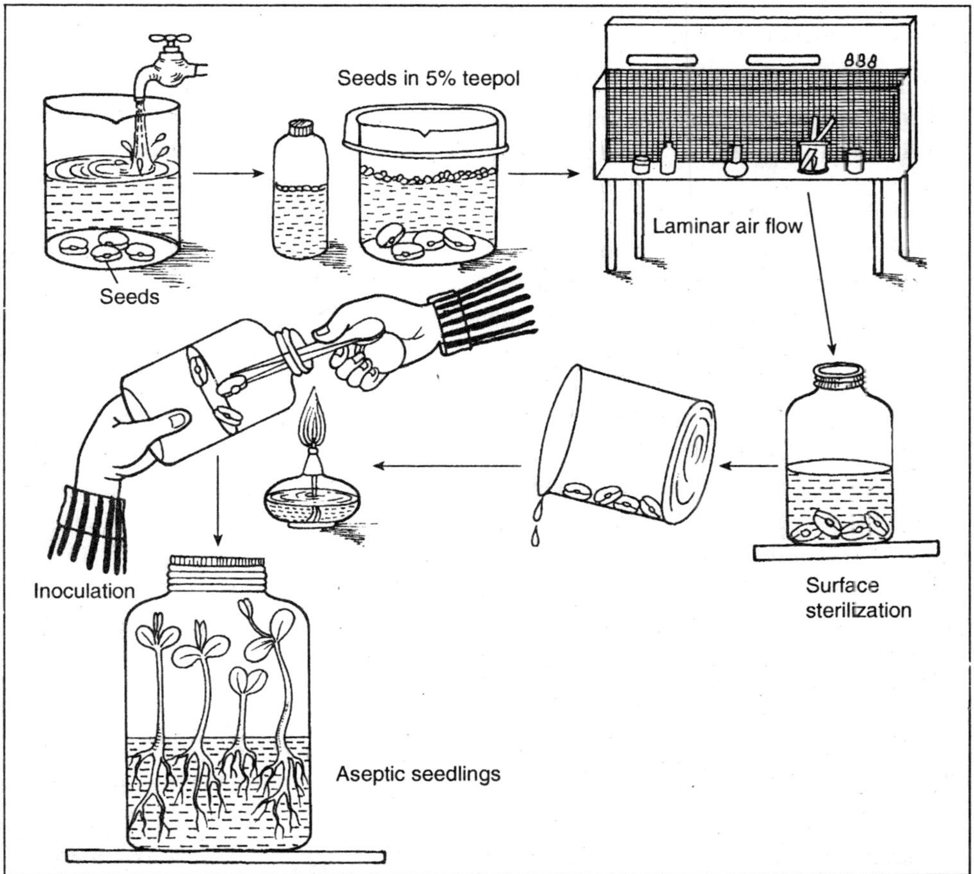

Fig. 16.4: Flow diagram illustrating the preparation of aseptic plants from seeds

media under aseptic condition. Dust from the surface, hair, hands and clothes are the potential sources of contamination. Before starting the transfer procedure the surface of transfer area and hands should wiped with 95% ethanol; sterile clothes (aprons) should be used. All the metalic equipments used for transfer (innoculating needle, forcep, scalpel) should be dipped into 95% ethanol and then flammed and cooled. The tissue material should not touch the edge of culture vessel during transfer.

16.5 TISSUE CULTURE TECHNIQUES

For *in vitro* plant cell, tissue and organ culture a proper organised method should be followed under totally aseptic condition. There are some general steps which are essentially followed for all kind of experiments, whereas some specific steps are taken for different types of culture. The general steps can be categorised under following heads:

(i) Preparation of nutrient media,

(ii) Sterilisation of instruments, culture vessels and media,

(iii) Sterilisation of transfer area and culture room,

(iv) Sterilisation of explants to be used,

(v) Inoculation of explant into the media,

(vi) Incubation of culture,

(vii) Subculturing when required,

(viii) Transplantation of the regenerated plantlets.

(i) to (iv) steps have been discussed under 16.3 and 16.4.

(v) **Inoculation of Explant**

Successful control of contamination largely depends upon the precautions taken to prevent the entry of microorganisms at the time of transferring the sterilised explants on the nutrient medium. Dust, hair, hands and clothes are the potential sources of contamination. The inoculating chamber should be dust free, the operator should wear sterile headgear and clothes (aprons) before entering the culture area. The hands should be wiped with 95% alcohol and the transfer area also should be cleaned and wiped with 95% alcohol before starting the transfer process. Talking or sneezing should be avoided during transfer of explant into the media. The neck or mouth of culture container should be flamed, the transferring instruments also to be flamed and dipped in alcohol. Care should be taken so that the explant should not touch the edge of culture vessel, and after transferring the mouth should be closed by cap or by cotton plug and petridishes to be sealed by 'Parafilm'. During transfer it is also to ensure that the plant tissue should be exposed to the media properly.

(vi) **Incubation of Culture**

After inoculation, the cultures are incubated in culture room or in a BOD incubator at 25±2°C temp. For certain plant or for some particular culture type below or above 25°C is needed.

Some tissues grow well under low light condition (approx. 1000 lux), for regeneration light and dark periods are needed, and for regenerated plantlet well lighted (approx.

3000 lux) condition and 16h light with 8h dark period is needed. The illumination in the culture room is provided by cool white fluorescent light placed approx. 18″ above the culture racks.

Low humidity causes the quick dessication of culture medium and high humidity is favourable for contamination of culture medium. Specific relative humidity (20–98%) is to be maintained in the culture room and uniform air circulation is to be done properly.

For cell suspension culture agitation and aeration is secured by the use of shaker systems; open platform orbital shakers or the orbital incubators fitted with fluorescent lights to provide different day/night regimes.

(vii) **Subculturing**

The growth and development of tissues cultured *in vitro* are generally monitored by observing the cultures at regular intervals in the culture room or incubators. Based on the observations either with hand-lens or with the aid of simple microscope under aseptic conditions, the explants may be required to transfer to new media (freshly prepared) or with new ingradients or hormone composition depending on the state of growth of cell or tissue. The same precautions and full aseptic conditions are maintained during the transfer process also. The delaying of this process may lead to inhibition of proper development of tissues and also delaying the regeneration of plantlets. In case of suspension culture the change of media or fresh inoculation at quick intervals is needed and also for callus culture the subculturing of the callus tissue is needed to get the callus tissue in dividing conditions.

(viii) **Transplantation of the Regenerated Plant**

Plants regenerated from *in vitro* tissue culture are transplanted to soil in pots. Prior to transfer to pots the acclimatisation of these regenerated plants are needed. The plants at this time develop adequate root systems and cuticular leaf surface structure so that it can withstand the field environmental condition. The process of acclimatisation needs the humid chamber and a slow process to make the plantlet habituated from high humid condition to normal atmospheric humidity. The greenhouse or the growth chamber should have artificial light system also which includes a mixture of fluoroscent and incandescent lamps designed to provide balanced wavelengths of light for plant growth and photosynthesis. The greenhouse facilities are needed for winter crops and summer crops differently for maintenance of proper temp. required, air circulation and the relative humidity. The potted plants are grown in field for further observation, flowering and normal seed setting to get the next progeny.

16.6 TYPES OF TISSUE CULTURE

Plant consists of different types of tissues and organs which have different types of individual cells. Plant tissue culture refers to the *in vitro* cultivation of plants, seeds, plant parts (tissues, organs, embryos, anthers, pollen, single cells, protoplasts) on nutrient media under aseptic conditions (Fig. 16.5).

(a) **Seed Culture**

Seeds may be cultured *in vitro* to generate seedlings or plants. It is the best method for raising the sterile seedling. The seed culture is done to get the different kinds of explants from aseptically grown plants which help in better maintenance of aseptic tissue

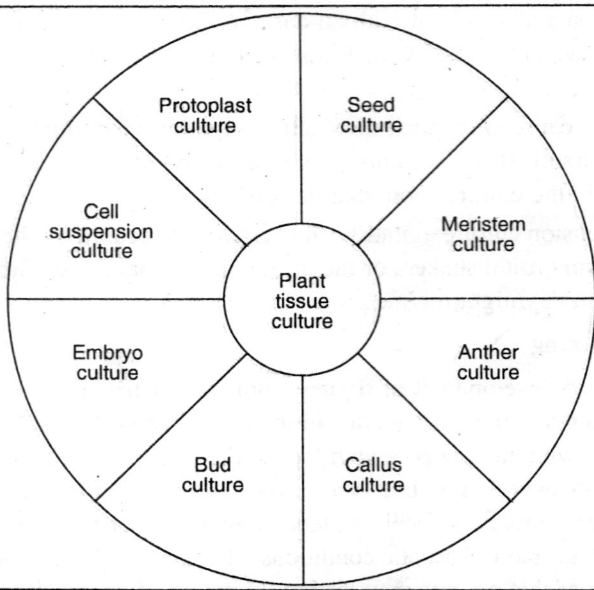

Fig. 16.5: Types of tissue culture

culture. In case of Orchid, the seed culture technique is mostly used because the cotyledons, root primordia or endosperm, all are poorly developed. So they need special nutrients for their germination and seedling development.

(b) **Embryo Culture**

Embryo culture is the sterile isolation and growth of an immature or mature embryo *in vitro* with the goal of obtaining a viable plant. In some plants seed dormancy may be due to chemical inhibitors or mechanical resistance structures covering the embryo. Excision of embryos and culturing them in nutrient media help in developing viable seedlings. Emrbyo developed from wide hybridisation between two different species may not mature fully due to embryo-endosperm incompatibility. So, the isolation and culture of hybrid embryos prior to abortion, help in overcoming the post-zygotic barrier and production of interspecific or intergeneric hybrids.

(c) **Meristem Culture**

The apical meristem of shoots of angiosperms and gymnosperms can be cultured to get the disease free plants. Meristem tips, between 0.2–0.5 mm, most frequently produce virus-free plants and this method is referred to as meristem-tip culture. This method is more successful in case of herbaceous plants than woody plants. In case of woody plants, the success is obtained when the explant is taken after the dormancy period is over. After the shoot tip proliferation, the rooting is done and then the rooted plantlet is potted.

(d) **Bud Culture**

Buds contain quiescent or active meristems in the leaf axils, which are capable of growing into a shoot. Single node culture, where each node of the stem is cut and allowed to grow on a nutrient media to develop the shoot tip from the axil which ultimately develops into new plantlet. In axillary bud method, where the axillary buds are isolated from the leaf axils and develop into shoot tip under little high cytokinin concentration.

(e) **Callus Culture**

Callus is basically more or less unorganised dedifferentiated mass of cells arising from any kind of explant under *in vitro* cultural conditions. The cells in callus are parenchymatous in nature, but may or may not be homogenous mass of cells. They are meristematic tissue, under special circumstances they may be again organised into shoot primordia or may develop into somatic embryos. The callus tissue from different plant species may be different in structure and growth habit. The callus growth is also dependent on factors like the type of explant and the growth conditions. After callus induction it can be subcultured regularly with appropriate new medium for growth and maintenance.

(f) **Cell Suspension Culture**

The growing of individual cells that have been obtained from any kind of explant tissue or callus referred to as cell suspension culture. These are initiated by transferring pieces of tissue explant/callus into liquid medium (without agar) and then placed them on a gyratory shaker to provide both aeration and dispersion of cells. Like callus culture, the cells are also subcultured into new medium. Cell suspension cultures may be done in batch culture or continuous culture system. In the later system, the culture is continuously supplied with nutrients by the inflow of fresh medium with subsequent draining out of used medium but the culture volume is constant. This culture method is mainly used for the synthesis of specific metabolite or for biomass production.

(g) **Anther Culture**

An important aspect of plant tissue culture is the haploid production by anther culture or pollen culture which was first established by Guha and Maheswari (1964, 1966) in *Datura*. During the last few decades, much progress has been made in different crops like rice, wheat, maize, mustard, pepper and others. The anthers bearing the uni-nucleate microspores are selected and allowed to grow in medim to produce callus from the pollen mass. Then the triggering of these androgenic calli is directed to produce the embryos and haploid plants are developed from these androgenic embryos. The anther culture can be done with the isolated anthers on solid medium where anther wall will break open and the androgenic calli will be formed from the pollen. In pollen culture, microspores of uninucleate stage are collected in liquid media and can be grown in suspension culture. In suspension, the uninucleate pollens may give rise to calli mass or the globular mass from which the plants can be raised either through embryogenic or organogenic pathway.

(h) **Protoplast Culture**

It is the culture of plant protoplasts i.e., culture of cells devoid of cell wall. Isolated protoplasts are usually cultured in either liquid or semisolid agar media plates. Protoplasts are isolated from soft parenchymatous tissue by enzymatic method and then viable protoplasts are purified and cultured. The protoplast culture is aimed mainly to develop genetically transformed plant where the transgene is put successfully within the plant protoplast and the transgenic plant is regenerated from that transformed protoplast. Another aspect of protoplast culture is somatic hybridization of two plant species through protoplast fusion.

16.7 APPLICATIONS OF PLANT CELL AND TISSUE CULTURE

In recent years, the plant cell, tissue and organ culture have become an important tool for crop improvement. According to M. S. Swaminathan the significance of biotechnological approach of plant cell and tissue culture can be divided into following categories:

(i) Tissue culture applications in order to capitalise upon the totipotency of cells.

(ii) Cell and protoplast cultures coupled with DNA vectors to overcome problems caused by barriers to gene transfer through sexual means.

(iii) Culture of plant cells for the production of useful compounds.

For last few years, the research in this field have experienced the emergence of various techniques to be applied in plant improvement (Fig. 16.6). Various aspects of plant improvement through tissue culture technology are as follows:

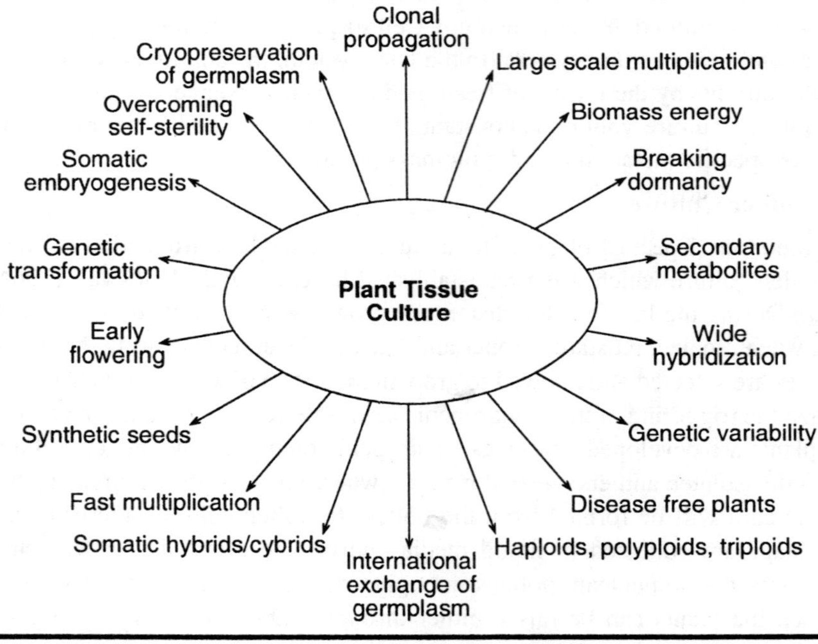

Fig. 16.6: Plant improvement through tissue culture technology

1. Clonal Propagation and Micropropagation

Plant population derived from a single donor plant is called a clone and the multiplication of genetically identical copies of that cultivar is called clonal propagation which may be an useful tool to get a large population of plant species having desirable traits. Micropropagation is achieved through multiplication of shoot tips or axillary buds cultured *in vitro*. This technique is very much used in horticulture and silviculture—in the plants which have long seed dormancy, tree species, orchids and many fruit plants. This micropropagation technique is also helpful for supplying the plant material throughout the year involving large scale multiplication i.e., grower and breeder gets a large number plant stocks irrespective of seasonal variation. In tissue culture from a callus mass large numbers of shoot meristems can be regenerated within a very short time and space. As a result a large number of plantlets can be produced from such callus tissue. The most obvious advantage of this technique is the large scale production of plants of same genetic stock.

2. Biomass Energy

In recent years, the interest has aroused in commercializing the *in vitro* propagation of forest trees. Micropropagation has been successfully done in many trees of economic importance like *Acacia nilotica, Albizia lebbeck, Azadirachta indica, Butea monosperma, Dendrocalamus strictus, Shorea robusta, Tectona grandis* and *Cedrus deodara, Cryptomeria japonia, Picea smithiana, Pinus sylvestris.* All these plant species are useful in forestry for biomass energy production. Development of automated procedure, plant delivery systems using somatic embryos and artificial seeds are also in progress.

3. Secondary Metabolites

Production of many useful compounds like alkaloids (Codeine, Vincristine, Quinine, etc.), Steroids (Diosgenin), Glycosidic compounds (Digoxin) and many other essential oils (Jasmine), flavouring and colouring agents (saffron) can be done by plant cell culture. This aim can be achieved by selection of specific cells producing high amount of desired compounds and development of a suitable medium. In general, secondary metabolites produced by plant cell cultures are rather small in amount but by clonal selection the particular high yielding clone of cells can be isolated. Sometimes the plant cell culture may provide the helpful way for more production of secondary metabolite by feeding the culture with inexpensive product precursors (**biotransformation**) or by manipulating their biosynthetic control mechanisms.

4. Genetic Variability

The variability generated by the use of a tissue culture cycle has been termed as **somaclonal variation** by Larkin and Scowcroft. This genetic variability is due to cells of various ploidy level and genetic constitution of the initial explant or also may be developed due to different cultural conditions. The chromosomal instability in the cultured cells play an important role in polyploidization of cells and genetically variable plants can be raised. Such kind of variations may show some useful characters such as resistance to a particular disease, herbicide resistance, stress tolerance, etc. and also some agronomical traits like tiller number, panicle size, flowering time, plant height, lodging resistance, yield, nutrient content and different kinds of morphological variations in leaf.

5. Somatic Embryogenesis and Synthetic Seed

Direct or indirect somatic embryogenesis may be achieved from pro-embryonic cell of the direct explant or the embryoids developed within the callus tissue from induced embryogenic cells. The potential application of this technique is the mass production of adventitious embryos which ultimately develop into complete plantlet in maturing media.

These somatic embryos can be encapsulated with suitable nutrient containing alginate medium which are called **artificial seeds** or **synthetic seeds.** As the somatic embryos are derived from a single cell, this method is very much useful for production of disease free propagule. This artificial seed production is also desirable in case of asexually propagating plants.

6. Breaking Dormancy

Using embryo (zygotic) culture technique the seed dormancy period can be reduced or eliminated and the breeding cycle can be shortened in many of the plants like *Malus* sp, *Ilex* sp. and *Telia americana* etc. The life cycle of *Iris* was reduced from 2–3 years to less than one year. It was possible to obtain two generations of flowering in *Rosa* sp. Embryo abortion in unsuccessful crosses may be recovered by culture of immature embryo of different hybrids.

7. Haploid Plants

Haploid plants can be obtained through anther or pollen culture (**androgenesis**) or through ovaries or ovule culture (**gynogenesis**). The anther culture and haploid plant production has been attempted in many of the crop plants, where these haploids are of immense importance for production of homozygous diploid or polyploid lines by colchicine treatment within a very short period specially in case of fruit trees. These androgenic haploids can also be used for production of different kinds of aneuploids like monosomic, nullisomic, trisomic, etc. and also for the induction of mutagenesis and doubling of those mutated lines. Many of the recessive traits can be made expressed in double haploids such as low glucosinolate content in *Brassica,* salt tolerance and disease resistance in rice, etc. Generation of exclusively Y chromosome containing plant is possible also through haploid production as in case of *Asparagus*. The triploid or polyploid can also be produced by using protoplast fusion technique of this kind of androgenic haploids which may be used for different breeding programmes.

8. Somatic Hybrids

Isolation and regeneration of plant from the protoplasts *in vitro* has opened up a new avenue in various fields of plant breeding and in plant biotechnology. **Somatic hybridisation**, i.e., the asexual hybridisation using isolated somatic protoplasts is a new tool to make the wide hybridisation successful. Products of fusion between two protoplasts (heterokaryon) could be cultured to regenerate a new somatic hybrid plant of desired genotype. This technique has been mainly used for introgressing many useful criteria from the wild genotype to cultivated crop variety. Success has been achieved obtaining somatic hybrid plants between sexually compatible and incompatible plants. Production of **cybrid**, i.e., the fusion between two protoplasts—one partner with nucleus and another partner with cytoplasm, is also of immense importance in the plant breeding programme, mainly for production of male sterile line with the help of extranuclear genome.

9. Transgenic Plants

The genetically modified (**GM**) plants, in which a functional foreign gene has been incorporated by biotechnological method, are called transgenic plants. A number of transgenic plants has been produced carrying genes for different traits like insect resistance, herbicide tolerance, delayed ripening, increased amino acid and vitamin content, improved oil quality, etc. The different methods of introduction of foreign genes, direct (electroporation, microinjection or particle bombardment) or indirect (*Agrobacterium* mediated), have been applied either in plant tissue culture method such as embryogenic or organogenic plant development from different plant parts or in protoplast culture system. The direct DNA uptake through protoplast is the most ideal method for production of transgenic plants. Any gene of interest that may be of eukaryotic or prokaryotic origin can be used for this purpose but should be expressed.

10. Germplasm Conservation

Many of the important crop species produce recalcitrant seeds with early embryo degeneration. Also many of the plants are vulnerable to insects, pathogens and various climatic hazards. Maintenance of these plants are very difficult. Mainly the plant species which are endangered, rare and threatened with extinction are needed to be conserved by *ex-situ* method of germplasm conservation. Plant tissue culture may be applied for this purpose. *In vitro* germplasm storage collection provides a cost effective alternative to growing plants under field conditions, nurseries or greenhouses. Furthermore, the cryopreservation of cells and tissue, revival of these tissue and regeneration of plants from tissue through tissue culture technique really effective in **conservation biotechnology.** **Cryopreservation** involves storage of cells, tissues, etc. at a very low temperature using liquid nitrogen.

QUESTIONS

1. Define cellular totipotency. Discuss the different types of tissue culture.
2. State the laboratory requirements and the composition of culture media in tissue culture.
3. Discuss aseptic manipulations required in the tissue culture laboratory.
4. Write the applications of plant cell and tissue culture technology.

17 Callus Culture and Cell Suspension Culture

Differentiated organ/tissue in plants may be subjected to dedifferentiation in *in vitro* condition. The path of dedifferentiation leads to the establishment of callus culture which when subjected to grow on liquid medium under agitation will result in suspension culture (Fig. 17.1).

17.1 CALLUS CULTURE: INITIATION AND MAINTENANCE

Callus culture represents clumps of unorganised parenchyma tissue formed through vigorous proliferation by cell division from any kind of explants under cultural condition, showing no polarity. Any kind of permanent but living tissue such as parenchyma, collenchyma, cortical tissue, pith cells, phloem tissue, cambium cells or meristematic tissue may take part in callus formation. The permanent tissue cells become rejuvenated and the influence of endogenous or exogenously applied growth substances leads to the stimulation of cell division.

Initiation

Induction of cell division in the permanent tissue is highly dependent on the high auxin content (e.g., 2, 4-D) in the medium. The hormone requirement for callus induction may be auxin alone, cytokinin alone or auxin and cytokinin in different ratios. The type of growth regulator requirement and its concentration in the medium depends strongly on the genotype and endogenous hormone content of an explant. Transformation of any kind of tissue explant into proliferating callus mass depends highly on the plant genotype, the source of origin of the explant and the physiological state of the tissue at culture. For the initiation of callus cultures (Fig. 17.2A), tissues from young seedling or from juvenile part of the mature plant either grown *in vivo* or *in vitro* are generally taken, either sterilised (when grown *in vivo*) or cut aseptically (when grown *in vitro*) and inoculated aseptically on a nutrient medium provided with different combinations of exogenous growth

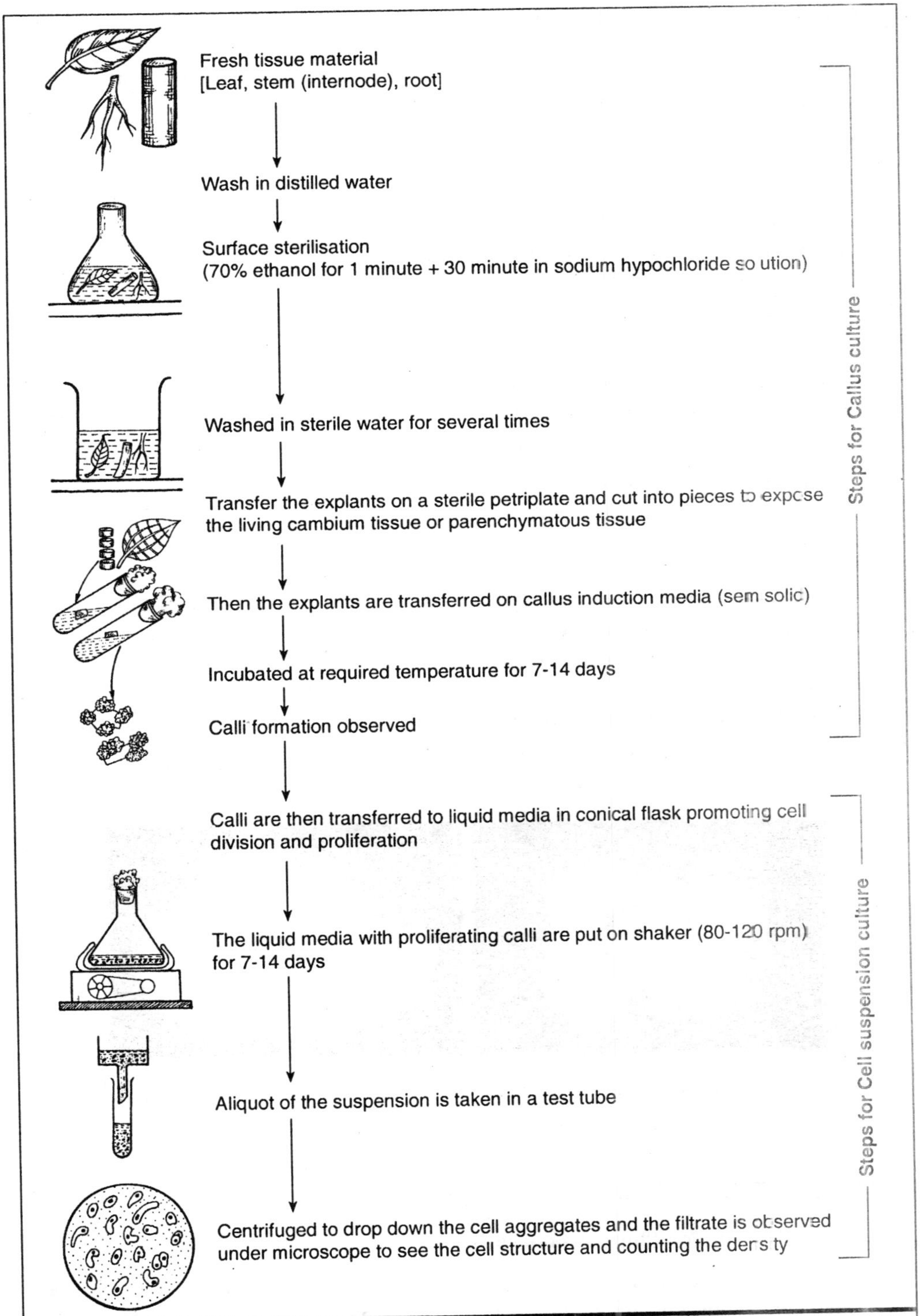

Fig. 17.1: Schematic flowchart for callus culture and cell suspension culture

substances. If the callusing is difficult to induce or if juveline tissue is needed then immature embryos or seedlings or parts of these are used. Endogenous hormone level depends on the type of explants and its original position in the plant which has an important influence in processes of cell division.

Maintenance

After callus induction, the callus is grown further on a new medium, which is referred to as subculturing. During subculturing the callus tissue shows sigmoid growth pattern on a particular media. The time interval for subculturing is calculated according to the growth pattern of the particular tissue type and genotype of the plant. When the callus growth reaches to the stationary phase of growth then it needs to be cultured on to a fresh medium.

Callus tissue from different explants of different plant species may be different in structure and growth habit: white or coloured, soft (watery) or hard, friable (easy to separate the cells) or compact. Callus growth can be monitored by—

(i) Fresh weight measurement (most convenient)

(ii) Dry weight measurement (more accurate)

(iii) Mitotic index measurement (for cell divisional rate).

Generally, maintenance of callus tissue growth needs growth hormones in order to grow as long as they are maintained through serial subcultures (Fig. 17.2B). But it has been observed also that the callus tissue in some plant species may become habituated after prolonged culture. This means that the callus tissue is able to grow on a standard maintenance medium or basal medium which is devoid of growth hormone. This property of callus tissue is called as **habituation.**

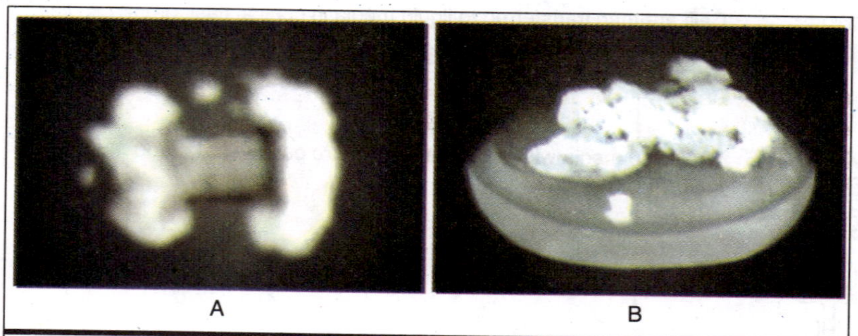

Fig. 17.2: Initiation of callus and its growth

17.2 SIGNIFICANCE OF CALLUS CULTURE

1. Source of Tissue for Plant Regeneration

Callus tissue is a good source of experimental tissue material from which whole plant can be regenerated by changing the nutrient and hormonal constituents in the culture medium. This kind of plant regeneration is known as **organogenesis.** Again by manipulation of hormonal constituents in the medium the cells of calli can be transformed into embryonal mass and the whole plant can be regenerated from those somatic embryos, the process known as somatic **embryogenesis**.

2. Source of Chromosomal Variation (Somaclonal Variation)

Chromosomal variation may occur genetically or epigenetically in the cells of callus tissue. If the explant contains the meristematic tissue then ordinary diploid cell mass develops. But endoreduplication is a natural phenomenon in the cells of permanent tissue as they remain in mitotically blocked condition, as a result when these cells are stimulated to divide the callus tissue may have such genetically variable cells. By serial subculturing the number of diploid or polyploid cells may increase or decrease in number, and after prolonged subculturing only one karyotype get established. If the polyploid cells are mostly found in the callus then the polyploid plant can be obtained through the process of regeneration. Again if callus tissue is derived from the haploid cells (such as microspore) then the haploid plants can also be obtained.

3. Source of Secondary Metabolites

If any of the medicinally important secondary metabolite is available from certain type of tissue material of a particular plant then that tissue material may be used as explant to get the callus tissue. The particular secondary metabolite can be obtained by extraction of that particular callus tissue. This process can reduce the time and expenditure as it does not need to have the whole plant cultivation and sacrificing it from extraction.

4. Source of Tissue for Cell Suspension Culture

When the callus tissue is of friable and soft type then the cell suspension culture can be initiated by culturing the callus tissue in liquid medium, continuously agitated by moving in rotatory shaker. More friable callus tissue is an ideal material for the dispersion of cells. As cell division occurs in the callus tissue the cells on the surface layer shed off and disperse in the medium. Here also the produced secondary metabolite may leach out in the medium and can be obtained from the medium by distillation. Cell suspension culture is useful in the **cell plating** technique to develop single cell culture which is important in the production of mutant cell lines and exploitation of somaclonal variation.

17.3 SUSPENSION CULTURE: INITIATION AND MAINTENANCE

A cell **suspension culture** consists of cell aggregates dispersed and growing in moving liquid media. It is normally initiated by transferring pieces of undifferentiated and friable calli to a liquid medium agitated by a suitable device (Fig. 17.3A–C).

Agitation in the medium helps in two ways, it exerts a mild pressure on cell aggregates, breaking them into smaller clumps and single cells; secondly agitation maintains uniform distribution of cell and cell clumps in the medium. It also helps in good gaseous exchange between culture medium and air.

Suspension culture can also be achieved by using **mechanical method** i.e., sterile explants like stem, leaf, root or any kind of soft tissue can be gently grinded by glass homogenizer and then the homogenate containing intact cells or small tissue masses can be used for initiation of suspension culture.

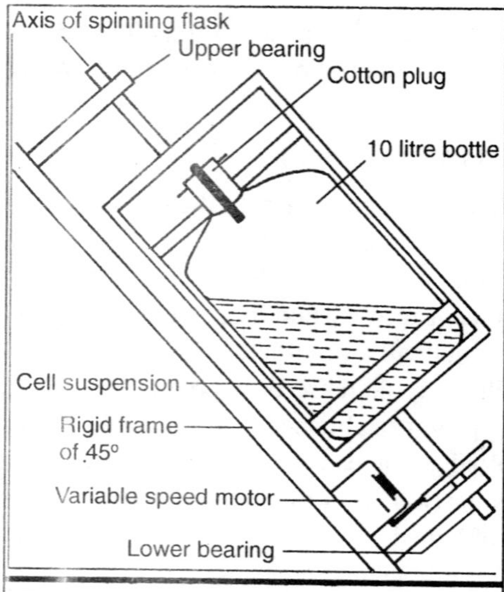

Fig.17.3A: 10 litre spinning culture apparatus

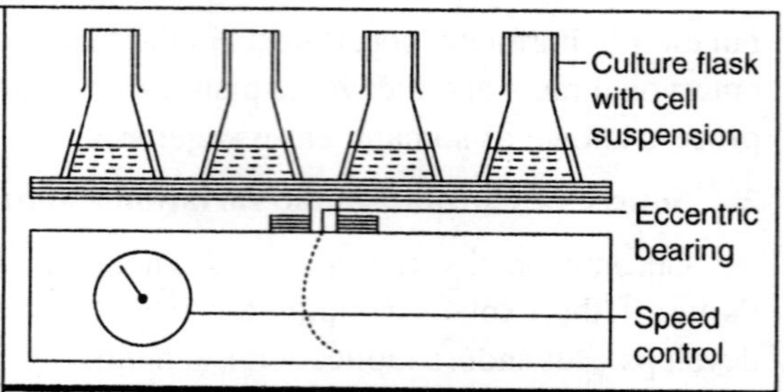

Fig. 17.3B: Side view of a platform shaker loaded with suspension cultures contained in conical flasks

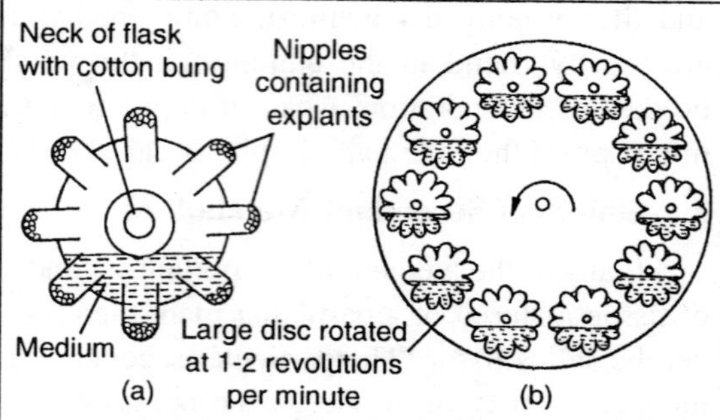

Fig.17.3C: (a) Detail of a nipple flask, (b) Large disc loaded with 10 nipple flasks used for growing cell suspension cultures

Another method is **enzymatic method** which may also be applied for isolation of single cells by the use of pectinases to digest the pectin wall. Orbital shakers are widely used for initiation and serial propagation of plant suspension culture.

The time period required for establishment of suspension culture from its callus tissue is called the initiation phase. During this phase the callus tissue breaks up, the cells grow and divide until the depletion of some nutrient in the medium. To know the time period for subculturing of a particular species the growth measurement in suspension culture is very much needed. The growth measurement can be done by counting the cell number under simple microscope after staining and macerating the small aggregates. The growth can also be measured by collecting the cell mass and by taking the fresh weight or dry weight of the cell mass i.e., the packed cell volume (PCV) measurement.

During this period, five phases of growth can be observed:

(i) **Lag phase:** Where the cells prepare to divide.

(ii) **Exponential phase:** Where the rate of cell division is highest.

(iii) **Linear phase:** Where the cell divisional rate slows down but the cell expansion takes place.

(iv) **Deceleration phase:** Where the rate of cell division and cell elongation both decreases.

(v) **Stationary phase:** Where the number of cells and their size remain constant.

After the initiation phase, the suspension culture can be passed through a nylon mesh to remove the larger clump and allowing the single cells or smaller cell aggregates to transfer into new medium for further culture. In the subsequent passages, the cell suspension is subcultured by taking a small aliquot and transferring into new medium containing flasks. The concentration of growth hormones like auxin and cytokinin both play a critical role for the growth of cell suspension. The cells in suspension culture may vary in shapes and sizes. Those may be oval, round, elongated, etc. and mainly of thin walled.

Frequent subculturing in a suitable medium ensures to achieve rapid growth rate, uniform and viable cells. For subculturing the initial inoculum density is very critical for starting of the cell division. Very low density of cells are unable to start growing and also very high density of cells is inhibitory, i.e., after attaining certain density the subculturing is needed.

17.4 TYPES OF SUSPENSION CULTURE

- Suspension culture
  - Batch culture
    - Slowly rotating culture
    - Shake culture
    - Spinning culture
    - Stirred culture
  - Continuous culture
    - Chemostats
    - Turbidostats

Batch Culture

A batch culture is a cell suspension culture grown in a fixed volume of nutrient culture medium, cell suspension increases in biomass by cell division and cell growth until a factor in the cultural environment becomes limiting and the growth ceases. These cultures are maintained continuously by propagating a small aliquot of the inoculum in the moving medium and transferring it into fresh medium at regular intervals. The biomass growth in batch cultures follows a fixed pattern, i.e, sigmoid pattern of growth curve. Batch cultures are not ideal system for studying the various aspects of cellular behaviour, as there is no steady state period in which relative concentration of metabolites or enzymes are constant. This type of culture is generally used where the secondary metabolite is leached out in medium at the stationary phase or the recovery of drug is done by collecting the whole cell mass from the medium at a time.

Continuous Culture

The large scale cultures are grown under steady-state condition for long periods by inflow of fresh medium and drawing out of the used medium keeping the culture volume constant. Continuous culture may be of two types: **closed** or **open** types.

In the **open type** the addition of fresh medium is balanced by the outflow of old medium including the harvest of cells, which maintains indefinitely a constant submaximal growth, i.e., majority of cells are in the similar metabolic state. This open system may also be of two different types:

(i) **Chemostats** in which growth rate and cell density are held constant by a fixed rate of input of a growth limiting nutrient medium. The growth limiting substance is so adjusted that its increase or decrease is reflected by increase or decrease in the growth rate of cells.

(ii) **Turbidostats** in which fresh medium flows in response to increase in the turbidity so as to maintain culture at a fixed optical density of suspension.

The **closed type** of continuous suspension culture is that where the cells are retained and inflow of fresh medium is balanced by outflow of corresponding volumes. The cells from the outflowing medium are separated mechanically and the same medium is added back to the culture. As a result in this type of culture cell biomass continues to increase as the growth proceeds.

Continuous Synchronous Culture

In batch and continuous culture the cells divide randomly. In batch culture the doubling time of the successive generations is likely to be changing, but in the 'open type' continuous culture use of chemostat/turbidostat helps to bring all the cells averagely at the same stage of cell cycle. Thus instead of this average condition, in the **continuous synchronous culture** there is a pattern of change which coincides with the cell cycle and repeats itself with each successive doubling of the cell population. The cell thereby amplified by the size of the synchronous populations may be examined at any stage of the cell cycle and at any desired growth rate. This has the advantage that enzymes or metabolites occurring only at certain stages of cell cycle can be obtained at maximum yields, a most useful criteria for consideration in a possible industrial production of such compounds.

Synchronization of Suspension Culture Cells

A **synchronous culture** is defined as a culture in which the cell cycles of majority of cells or proportion of cells are synchronous, i.e., majority of cells proceed through each cell cycle phase simultaneously. In general the cells in suspension culture remain in a synchronous condition, hence it is essential to manipulate the growth conditions in such a way to achieve a high degree of synchronisation. It is expressed as percentage synchrony of cells in suspension. The effective procedures of synchronization differs from species to species.

The methods employed to achieve synchronization of suspension cultures may be grouped under two categories: **Physical** and **Chemical.**

Physical methods helpful in achieving synchrony are:

(i) **Selection by volume:** Synchronization may be achieved by selecting a particular size of cell aggregates by using the cell fractionation technique.

(ii) **Temperature shock:** Low temperature shocks combined with nutrient starvation are reported to induce synchronization of suspension cultures.

Chemical methods used for achieving synchrony are:

In this method the cells are starved of an essential nutrient or the cell cycle is stopped by a biochemical inhibitor and then allowed to undergo simultaneous divisions either by supplementing the nutrient or withdrawing the inhibitor chemical.

(i) **Starvation:** Cultures starved of nitrogen, phosphorus, or carbonate, result in the arrest of cell growth during the G_1 or G_2 phase of the cell cycle. After a period of starvation when these growth limiting compounds are supplied to the medium, the stationary cell enters in divisional phase synchronously. Growth hormone starvation also can bring the synchronization.

(ii) **Inhibition:** Inhibitors of DNA synthesis (5-amino-uracil, 5-fluorodeoxypurine, hydroxyurea, etc.) in cell culture help to accumulate the cells at the G_1/S boundary. Removal of the inhibitor is followed by synchronous division of cells.

(iii) **Mitotic arrest:** Suspension cultures in exponential growth are supplied with 0.02% (w/v) colchicine for 4-8 hr inhibit the spindle formation. Then removal of colchicine will help to induce the synchrony among the cells in suspension culture.

17.5 UTILITY OF SUSPENSION CULTURE

Plant cell suspension culture is potentially valuable for studying the biosynthesis of secondary metabolites. If plant cells are cultured under conditions that allow the expression of a certain degree of differentiation, they have the potential to produce, either by *de novo* or by *biotransformation* of specific precursors, a wide range of secondary products. Suspension culture is also useful in producing somatic embryos and isolating different cell lines including mutant lines.

1. Source of Secondary Metabolites

Cell suspension cultures have been shown to produce a number of valuable secondary metabolites, which includes alkaloids, glycosides (steroides and phenolics), terpenoids, variety of volatile oils, pharmaceutical products, etc.

Many useful secondary metabolites could be produced under controlled condition of consistent product quality by the use of particular cell line. Temperature, photoperiod as well as nutrition and changes in the level of hormones may modify or induce or repress the biosynthetic activities of cultured cell. With the help of controlled metabolic processes the improvement in yield can be done with less labour and costs.

2. Biotransformation

This involves the ability of plant cell cultures to catalyse the biotransformation of a readily available or inexpensive precursor into a more valuable final product. The conversion of a small part of molecule into industrially important chemicals by means of biological systems is termed as **biotransformation.** The use of plant cell culture in this process requires the selection of cell lines which have the capabilities of enzymatic activity for specific reaction of interest.

3. Source of Somatic Embryos and Cell Line

Embryonic cell suspension culture offers the possibility for large scale clonal propagation. Somatic embryos from suspension cultures can be useful for long term storage in germplasm banks. This embryogenic culture serves as an important tool for selection of cell line under different biotic and abiotic stress. Selection of specific cell lines producing high amounts of a useful metabolite can be done in suspension culture either through single cell clone or by cell aggregate cloning.

4. Source of Protein

Apart from secondary metabolites, plant cell cultures have proved to be a source of a variety of proteins which in future may become of economic importance as well as enzymes, peptides. Plant cell cultures have infinite possibilities for producing useful natural substances in industry.

5. Mutant Production and its Use

Cell suspension cultures are of great use for mutagenesis studies. The mutagens can be added to the medium and the mutagenic cell lines can be obtained which may be further used for raising of mutant plants or new mutant cell lines. These cell lines may have the ability to produce new or novel products of secondary metabolites, or the increased enzymatic activities for the biosynthesis of a particular product, or a defect in enzymes located at a branching point of a metabolic pathway resulting in increased product formation. The another culture derived haploid cell line can also be used successfully to derive mutant cell line.

17.6 SINGLE CELL CULTURE AND CELL PLATING

The single cells can be isolated from a variety of tissue and organ of intact plant or from callus tissue or from cell suspension. Single cells can be isolated from intact plant tissue or callus tissue by using either mechanical method or chemical method.

Mechanical isolation means chopping of the tissue material by fine scalpel which releases few intact single cells or by glass homogenizer the tissue may be crushed where the homogenate containing only few intact cells can be cultured. In **chemical method** the macerozyme or pectinase can be used to dissolve the middle lamella and releasing single cells from intact tissue. In most cases the single cells are isolated from friable callus tissue or cell suspension culture. The cell aggregates or clumps are removed from the suspension by using mesh only and then the isolated single cells in the filtrate can be cultured either in liquid or in solid media.

Methods of Single Cell Culture

There are various methods of single cell culture (Fig. 17.4A–C):

(a) Paper raft nurse culture technique
(b) Microchamber technique
(c) Microdroplet technique
(d) Nurse culture technique
(e) Cell plating technique.

(a) **Paper raft nurse culture technique:** The isolated single cells are placed aseptically on nutrient medium soaked filter paper and placed on a actively growing callus tissue. After small aggregate formation those are transferred to fresh media.

(b) **Microchamber technique:** With the help of paraffin oil and cover glass a microchamber is formed on a glass slide and droplet containing single cells in medium is placed inside this microchamber and incubated for division.

(c) **Microdroplet technique:** In this technique the single cells are cultured in special kind of apparatus named Cuprak dishes which have two kinds of chambers, small outer chamber filled with water and large inner chamber carrying numerous wells each filled with microdroplet of medium containing single cells.

(d) **Nurse culture technique:** In this technique the growth or division of single cell is induced by nurse callus. It has been observed that the single cells plated near the callus tissue divide quickly to give rise to calus tissue. This happening must be due to some leaching effect on the division of single cells by the growing callus mass.

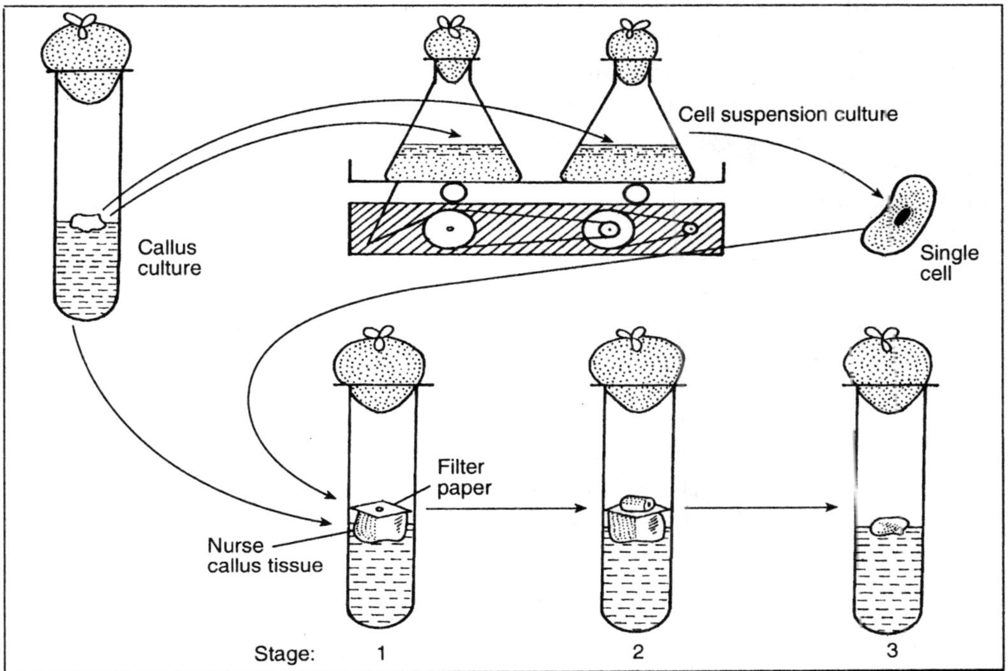

Fig. 17.4A: Growth of single cells using a 'nurse' technique. Stage 1: a single cell taken from a friable callus is placed on upper surface of filter paper which is in contact with nurse callus. Stage 2: the single cell divides and daughter cells proliferate to form colony. Stage 3: when colony reaches a suitable size. It is transferred to fresh medium where it gives rise to a single cell clone

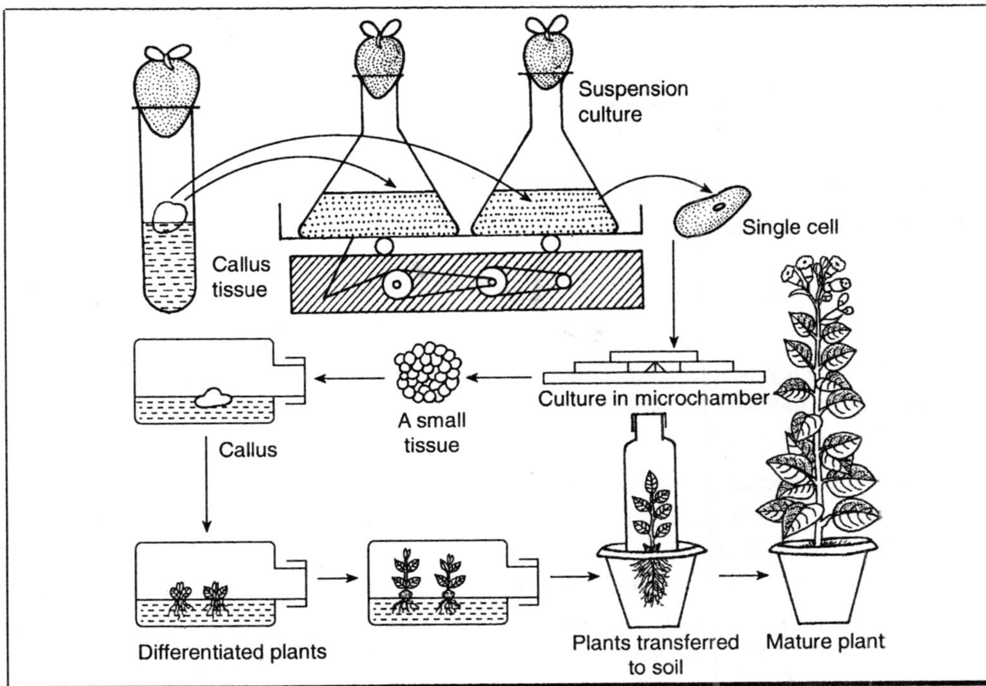

Fig. 17.4B: Development of plant from a single cell cultured in microchamber

(e) **Cell plating technique:** The basic technique of plating is to first count the cell number without maceration stage, this will enable a known number of cell units to be established per unit volume of plating media. Both the cell suspension and nutrient medium containing agar are prepared in double concentration separately. The equal volumes of suspension and the agar medium cooled at 35°C are mixed and then dispersed rapidly into petriplate in such a manner that cells are evenly distributed in a thin layer (~1 mm thick). The dishes are then sealed and incubated for cell division, which will give rise to callus (Fig. 17.4C).

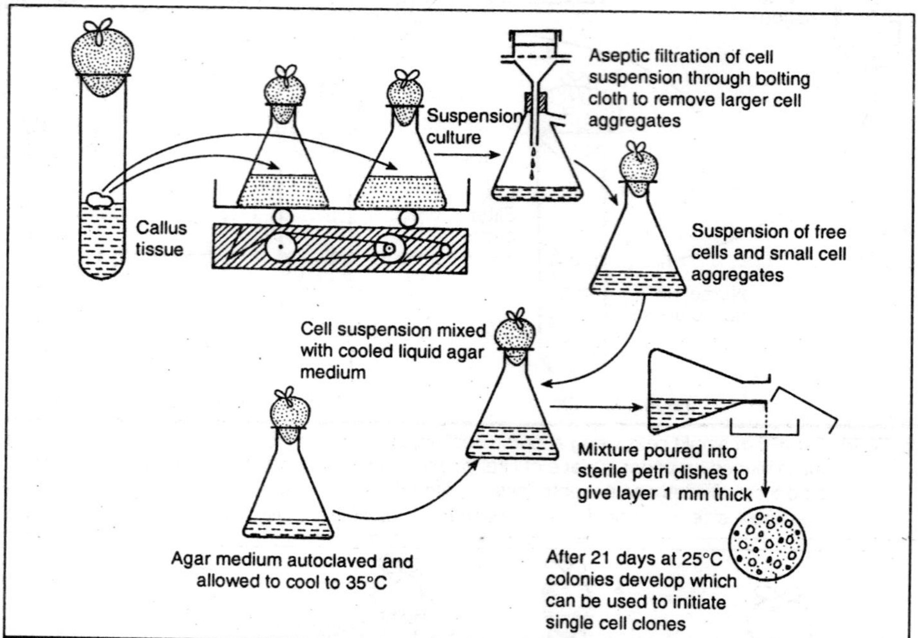

Fig. 17.4C: Procedure for obtaining single cell clones using a petri dish plating technique

Bergmann (1960) first introduced this most popular technique of plating of cell suspension and this technique is very much useful to calculate the plating efficiency.

$$\text{Plating Efficiency (PE)} = \frac{\text{No. of colonies/Plate at the ends of experiment}}{\text{No. of cellular units initially plated/plate}} \times 100$$

The plates may be observed under an inverted microscope and single cells develop into callus, this method ensures the isolation of pure single-cell clones. Usually, plating at cell densities of 10^3–10^5 cells/ml or more yields a high plating efficiency.

Efforts have been made to develop a synthetic medium for cells plated at low density. Cells plated in a culture medium synthesize necessary metabolites and threshold conc. of those helpful for cell division to start. At initial high cell density the equilibrium is reached much earlier than at a low cell density. It has been observed that below a critical cell density the cells fail to divide. This problem can be overcome by supplementing the minimal medium with some undefined factors like coconut milk, casein hydrolysate or yeast extract and following some different culture techniques.

QUESTIONS

1. Define callus. How is callus culture initiated and maintained? Mention the significance of callus culture.
2. What is suspension culture? Discuss the different types of suspension culture. State the utility of suspension culture.
3. What is single cell culture? State the different methods of single cell culture with special reference to cell plating technique.

18 PLANT REGENERATION AND MICROPROPAGATION

Totipotency, the inherent property of cell, may be exploited in plant system for regeneration of plants. The regeneration path may follow either **organogenesis** or **embryogenesis** leading to the formation of complete plantlets. This has created a new vistas of *in vitro* technique of non-sexual way of rapid propagation in plants called **micropropagation**.

18.1 ORGANOGENESIS: DIRECT AND INDIRECT

In plant tissue culture, **organogenesis** is a process of differentiation by which plant organs like roots, shoots, buds etc. are formed from the unusual points of origin of an organized explant where a preformed meristem is lacking. Plant development through organogenesis is the formation of organs either *de novo* or adventitious in origin and plant regeneration via organogenesis is a monopolar structure.

Plant production through organogenesis can be achieved by two modes: (i) Emergence of adventitious organs directly from the explant. (ii) Organogenesis through callus formation with *de novo* origin (Fig. 18.1).

Direct Adventitious Organ Formation

Every cell of plant is derived from the original zygote through mitotic divisions containing the complete genome. The formation of adventitious buds depends on the reactivation of genes concerned with embryonic phase of development. The addition of growth regulators like auxin and cytokinin in the medium is required to initiate shoot formation from different kinds of tissue explant. Adventitious *in vitro* regeneration may give a much higher rate of shoot production than is possible by proliferating axillary shoots. This technique is much used for multiplication in micropropagation system (Fig. 18.2A).

In suitable medium supplemented with growth hormones the somatic tissues of higher plants are capable of regenerating adventitious buds/shoots. These buds are formed

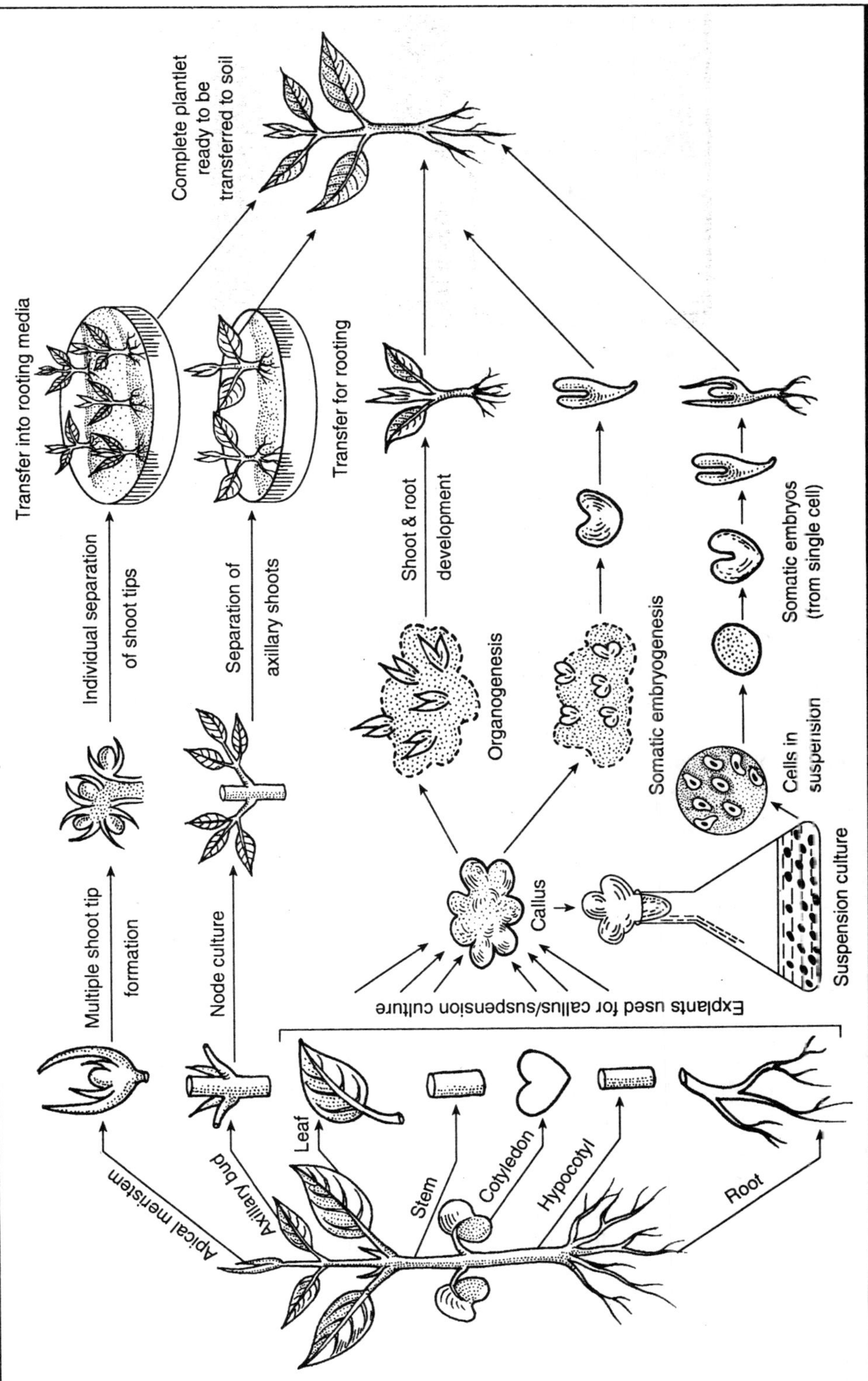

Fig. 18.1: Schematic diagram showing the different explants used and the pathways for Apical shoot tip culture, Node culture, Callus culture and Suspension culture

Fig. 18.2: A. Direct organogenesis, B. Organogenesis through callus formation

directly from a plant organ or any piece of tissue without forming any callus structure. This type of organogenesis is mostly found in herbaceous plants. Promotion of bud formation by cytokinin occurs in several plant species, though the requirement of exogenous cytokinin and auxin in the process varies with the tissue system of different species. In some plant species, the adventitious buds are produced during vegetative reproduction.

Organogenesis through Callus Formation

Sometimes plant regeneration from cultured explants (cotyledons, hypocotyl, stem, leaf, shoot apex, shoot, root, young inflorescence, embryos, etc.) involves the initiation of basal callus formation and then shoot bud differentiation (Fig. 18.2B). For different species different kinds of explants may be necessary for successful plant regeneration. Though the explants with mitotically active cells i.e., meristems, shoot tips, axillary buds, immature leaf, immature embryos are good for callus initiation and also can be used successfully to initiate plantlets through organogenesis.

Two modes of cell culture are generally used for organogenic path:

(i) The cultivation of cell clusters on a solid medium;

(ii) The cultivation of cell suspensions in liquid medium.

Medium and Growth Regulators Requirement

The application of growth regulators are critical for morphogenesis, which varies greatly depending on the tissue type. Callus tissue is comprised of wide range of cell types, the meristematic cells are interspersed within the vacuolated cells. During subculturing the meristematic cells may be favoured by the medium composition and organized growth of meristematic tissue may lead to formation of meristemoids. In these structure the vascularisation starts due to appearance of tracheidal cells in the callus which ultimately initiates the formation of shoots and roots. Lowering the auxin and increasing the cytokinin concentration is traditionally performed to induce shoot organogenesis from callus (Fig. 18.3A). Auxin alone or in combination with low concentration of cytokinin are important in the induction of root primordia (Fig. 18.3B). Organogenesis can be induced in either cell suspension or callus culture cells; transfer of cells from callus forming medium to regeneration medium and then continued subculturing helps in organ formation.

Fig. 18.3A: Shoot regeneration from callus

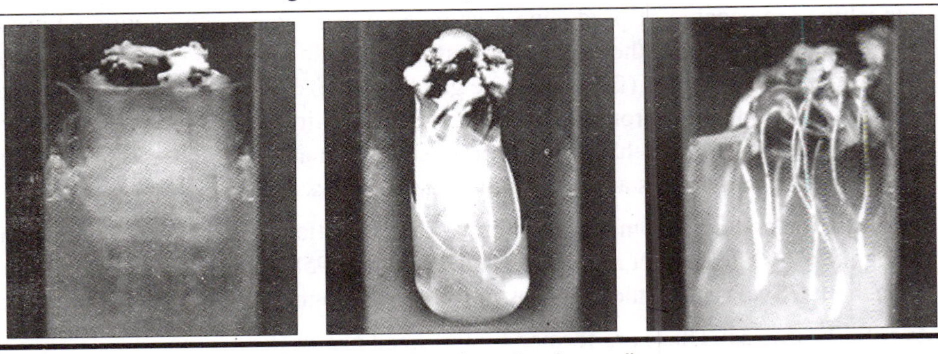

Fig. 18.3B: Root formation from callus

Fig. 18.3C: Different stages of organogenesis and whole plant formation and its transfer to soil and greenhouse

Organogenic differentiation is an outcome of the process of dedifferentiation followed by redifferentiation of cells. Dedifferentiation favours unorganized cell growth and the resultant developed callus has meristems randomly divided. Most of these meristems, if provided appropriate "*in vitro*" conditions, would redifferentiate shoot buds and roots. The whole plant regeneration from cultured cells may occur either through shoot-bud differentiation or somatic embryogenesis. All these events establishes the **totipotency** of somatic cells.

18.2 SOMATIC EMBRYOGENESIS

Somatic embryogenesis is the process in which a single cell or a small group of cells follow a developmental pathway that leads to reproducible regeneration of non-zygotic embryos which are capable of producing a complete plant. These non-zygotic embryos may originate directly from other organs or parthenogenetic embryos (without fertilization) or androgenetic embryos (from the male gametophyte). In general somatic embryos are those which are formed from the somatic tissue in cultural i.e., *in vitro* condition. Embryos formed in cultures have been variously designated as accessory embryos, adventive embryos, embryoids and supernumerary embryos.

The initiation and development of embryos from somatic tissues in plant culture was first recognised by Steward et.al. (1958) and Reinert (1958–1959) in *Daucus carota*. Kohlenbach (1978) classified the embryos in following manner:

I. **Zygotic Embryos:** These are formed by fertilized egg or the zygote.

II. **Non-zygotic Embryos:** These are formed by cells other than zygote.

(a) **Somatic Embryos:** These are formed by sporophytic cells (except zygote), directly arising from other embryos or organs which are termed as **adventive embryos**.

(b) **Parthenogenetic Embryos:** These are formed by unfertilised egg.

(c) **Androgenetic Embryos:** These are formed by the male gametophyte i.e., microspore or pollen grains.

Somatic embryos should closely resemble their bipolar structure as in the case of zygotic embryos. There should be appropriate root, shoot and cotyledonary development.

Sharp et. al. (1980) described mainly two routes for somatic embryogenesis:

1. **Direct Embryogenesis:** The embryos initiate directly from the explant without callus formation and here some cells which are called as '**Pre-embryonic determined cells**' (PEDC) initiates embryonic development, only those cells need to be released. Such cells are found mostly in embryonic tissues, certain tissues of young *in vitro* grown plants, hypocotyl, nucellus, embryo-sac, etc.

2. **Indirect Embryogenesis:** Here, the embryos are developed through cell proliferation i.e., callus formation. The cells from which embryos arise are called as '**Induced embryogenic determined cells**' (IEDC). Here growth regulators with specific cultural conditions are required for initiation of callus and then redetermination of those cells into the embryo development. Somatic embryos arise from single cells located within clusters of meristematic cells either in callus mass or in suspension. Such cells develop into proembryos with polarity following a pattern that tends to mimic the

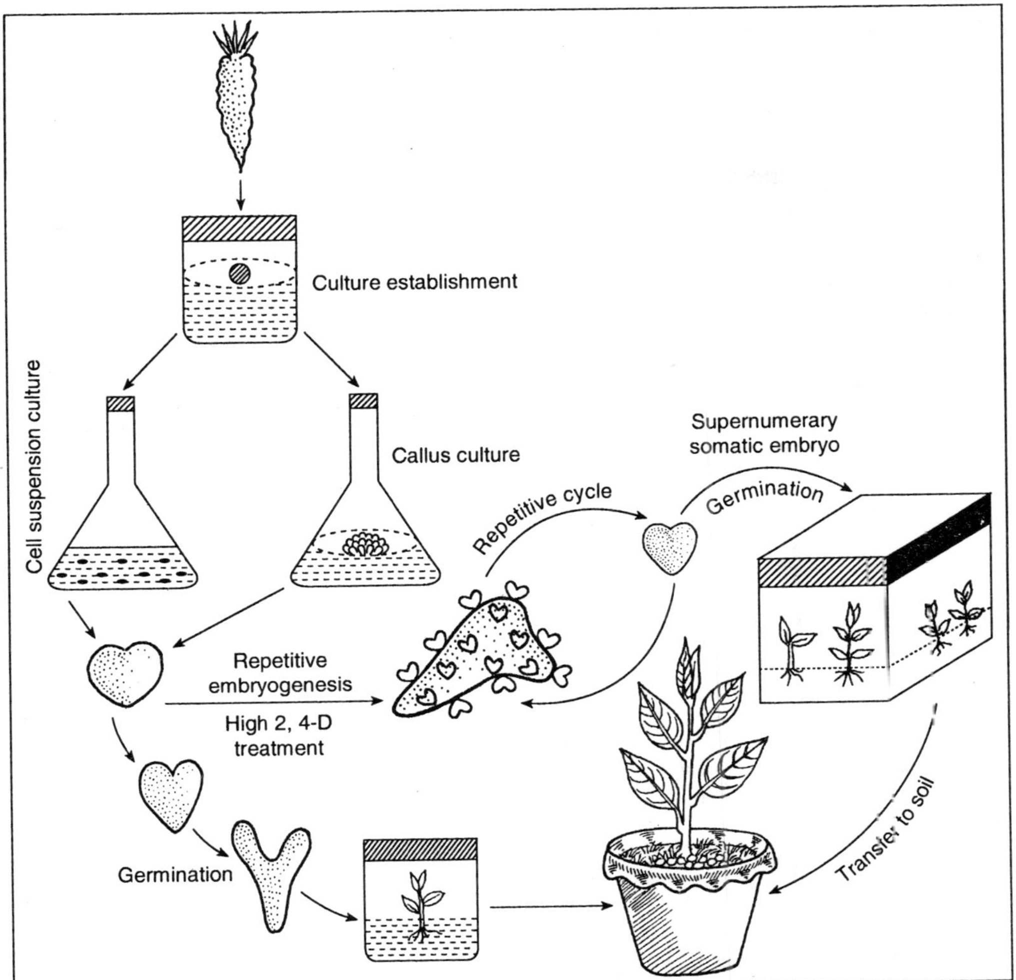

Fig. 18.4: Mass propagation of plants through somatic and secondary somatic embryogenesis

general pattern associated with the development of *in vivo* embryos in the ovule. Proembryo initials may be single cells or multicellular groups. When the conditions are suitable these embryos germinate to produce plantlets (Fig. 18.4).

Medium and Growth Regulators Requirement

Somatic embryogenesis encompasses various stages from callus initiation to embryo development, maturation and subsequent plantlet formation. For many species different types of media are required for the whole process. The presence of auxin in the media is generally essential for embryo initiation, but lowering the auxin content further in the medium is helpful for embryogenic clump formation which ultimately develops into mature embryo. In general it has been observed that the auxin or auxin in combination with cytokinin in different ratios appear essential for the onset of growth of somatic embryos. Sometimes the medium lacking auxin but addition of low levels of cytokinin in combination with ABA may prove beneficial for somatic embryogenesis. Sometimes the addition of charcoal into the medium has been proved to be useful for somatic embryo development. The germination of somatic embryos does not require any growth regulators, application of GA_3 or Zeatin may be helpful for shoot and root development.

Difference between Zygotic Embryogenesis and Somatic Embryogenesis

In angiosperms, the ovule contains a haploid egg cell or ovum which is a female gamete gets fertilised by the male gamete resulting in the formation of unicellular zygote. This zygote gives rise to multicellular embryo which is known as zygotic embryo and the process must be called as **zygotic embryogenesis**. Embryos may be formed from the unfertilised egg or any other cell of female gametophyte or the sporophytic tissue then these are called as **non-zygotic embryos.** In *in vivo* there is no embryo development from any somatic cell of the plant, i.e., the somatic cell does not have the embryogenic potential in general. But due to the totipotency nature of plant cell the somatic cell having the complete set of genome may be induced in cultures to form the organised bipolar struc-

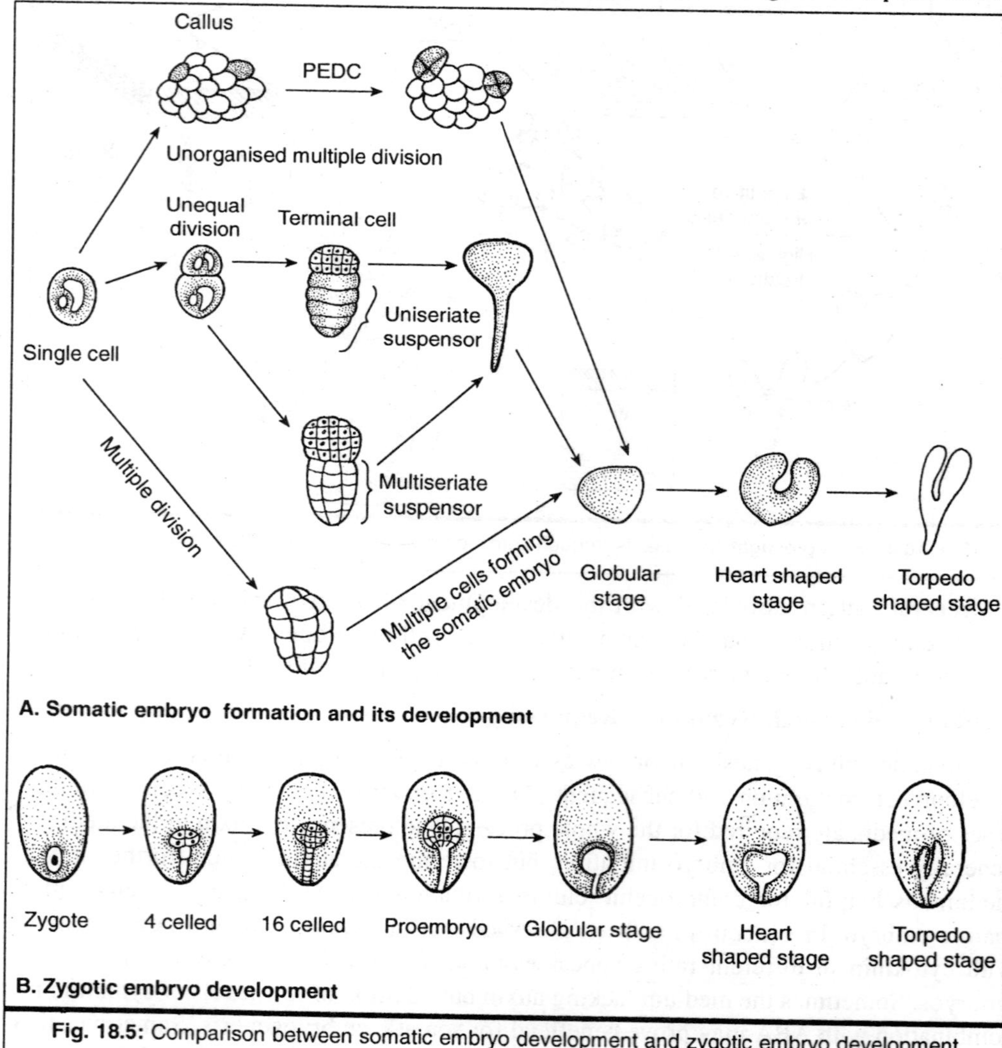

Fig. 18.5: Comparison between somatic embryo development and zygotic embryo development

tures bearing cotyledons resembling the zygotic embryo. This phenomenon is called *in vitro* **somatic embryogenesis.** The term '**embryoids**' are used to distinguish these somatic embryos from the zygotic embryos. Any sporophytic cell or gametophytic cell may undergo somatic embryogenesis due to some changes occurring at molecular level i.e., due to induction of some genes, protein production and high metabolic activity.

The stages of developmental process of both somatic embryo an zygotic embryo resemble each other morphologically. In both the cases the single somatic cell (sporophytic or gametophytic) or the zygotic cell undergoes divisional phases to produce the **globular stage,** then the **heart** or **torpedo stage** at the end of which the cotyledons or first leaves begin to develop. In dicots two cotyledons and in monocot only one cotyledon emerges. After this stage the developmental process differs in zygotic and somatic embryo (Fig.18.5).

The zygotic embryos undergo a prolonged maturation phase and many storage proteins are being synthesized which is primarily regulated by abscisic acid. Then a dormancy period is required for further germination. But the somatic embryos can grow and differentiate to produce the root and shoot apices with leafy development without any dormancy period. Another important morphological or developmental difference is that the somatic embryos are devoid of suspensor and endosperm.

Unlike zygotic embryos or seed embryos, the somatic or adventive embryos are devoid of an endosperm as a food reserve. Fluid drilling devices are harnessed to sow the naked embryos directly into the soil, or the embryos are encapsulated in plastic strips or pellets together with a little nutrient, aimed at using embryos for large scale planting.

Applications of Somatic Embryogenesis

(i) **Large scale propagation compared to zygotic embryos:** Induction of somatic embryogenesis forms the ultimate goal in free cell suspension cultures relying on the totipotency of the cell and could reasonably be exploited for micropropagation. Each cell of the suspension cultures can be induced to produce somatic embryos which can be maintained in an arrested state of development by cold storage or using mitotic inhibitors until the time of sowing. Somatic embryogenesis is highly desirable and holds out promise for rapid multiplication in a shorter time, with a shoot-root axis.

(ii) **More useful than organogenesis:** The mass production of adventitious embryos in cell culture is still regarded by many as the ideal propagation system. The adventitious embryo is a bipolar structure that develops directly into a complete plantlet and there is no need for a separate rooting phase as with shoot culture.

(iii) **Useful for mutagenic studies and mutant production:** The somatic embryos generally arise from single cells, so it may be advantageous for mutagenic studies. Also the plantlets arising from such somatic embryos are more homogeneous in nature, so the mutant gene expression can be studied well.

(iv) **Useful for genetic manipulation technique:** In plant biotechnological application, during foreign gene transfer if the transformed cell gives rise to plantlet via somatic embryogenesis then there is least possibility of chimera formation. So for transgenic plant production this method of multiplication system is very much useful.

(v) **Useful for pathogen-free plant production:** Plants derived from this kind of somatic embryos may be free from viral or other pathogens. So it may be an alternative approach of disease free plant production.

(vi) **A good source of protoplast culture:** Embryogenic cultures are specially valuable in providing a source of regenerable protoplasts in the graminaceous and coniferous plants. Protoplasts from these cultures were induced to divide to form a cell mass from which the embryoids, even plantlets are regenerated on a suitable nutrient medium.

(vii) **Conservation of genetic resources:** Somatic embryos which originate from single cells and subsequently regenerate mostly genetically uniform plants are good materials for genetic resource conservation. Embryogenic cultures as well as somatic embryos remain viable upon storage at ambient temperature, cold storage or cryostorage.

18.3 SYNTHETIC SEED (ARTIFICIAL SEED)

Somatic embryos are not enclosed by seed coats and due to microbial contamination and desiccation they are not able to survive if these are sown directly into field soil. The aim of somatic embryos encapsulation is to produce an analog to true seeds. The procedure is to mix the somatic embryos in a substance to make it well protected in that matrix and it is rigid enough to allow for rough handling. So, the **artificial** or **synthetic seeds** are the seed like structures where the somatic embryos derived from tissue culture are encapsulated by **hydrogels** and can be used directly to be sown in soil as a substitute of natural seeds (Fig. 18.6). These will germinate normally into plants. This covering helps to give physical protection, avoid desiccation and encapsulation should carry nutrients, growth regulators to help in germination. Antibiotics may be added to avoid contamination and also it should be durable and non-toxic.

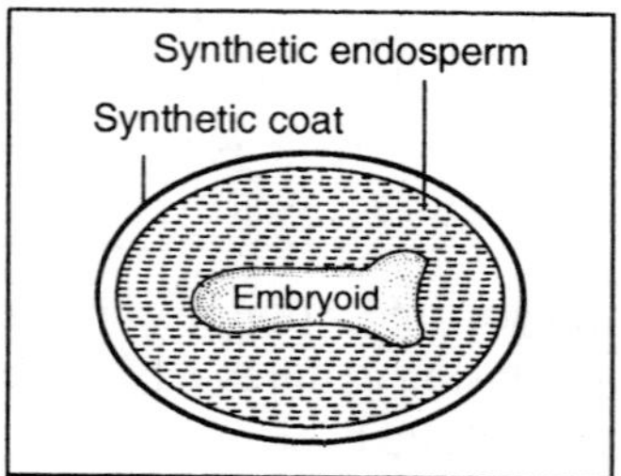

Fig. 18.6: Diagram of an encapsulated seed of plant

The difference between the seed with zygotic embryo and artificial seed with somatic embryo are:

Natural seeds	Synthetic seeds
1. Hard seed coat present.	No seed coat, only encapsulated.
2. Embryos are much protected within cotyledons or endosperm.	Embryos are not protected within any kind of maternal tissue.
3. Embryos undergo controlled desiccation by the maternal tissue and have a natural dormancy period.	Embryos do not pass through any kind of desiccation and they the do not have any dormancy period.
4. The natural seeds have their own storage reserves like endosperm or cotyledons to provide food during germination.	The artificial seeds do not have their own storage tissue, the nutrients or growth regulators can be supplied within the encapsulating material.

Methods of Artificial Seed Production

For making synthetic seeds the somatic embryos should be coated or encapsulated in protective covering and handled in such a condition so that, these can be delivered in field condition with good germination rate (Fig. 18.7).

The synthetic seeds can be produced using the following procedures:

(i) Hydrated condition, using sodium-alginate or calcium alginate hydrogels as the encapsulation material;

(ii) Using desiccated embryos in coated condition;

(iii) Using polymer, self breaking capsules releasing nutrient;

(iv) Using fluid drilling method.

Radenbergh et. al. (1986) developed **hydrated artificial seeds** by mixing somatic embryos with sodium-alginate, followed by dropping into a solution of $CaCl_2$ to form calcium-alginate beads. About 30–55% embryo had the germination capacity producing seedlings.

Kin and Janick (1989) developed the **desiccated artificial seeds** applying synthetic seed coats. They mixed equal volume of embryo suspension and 5% solution of poly-ethylene oxide, a water soluble resin, which subsequently dried to form strip containing multiple embryos. The embryos used here are desiccated by hardening treatments with 12% sucrose and using 10^{-6} M ABA followed by chilling.

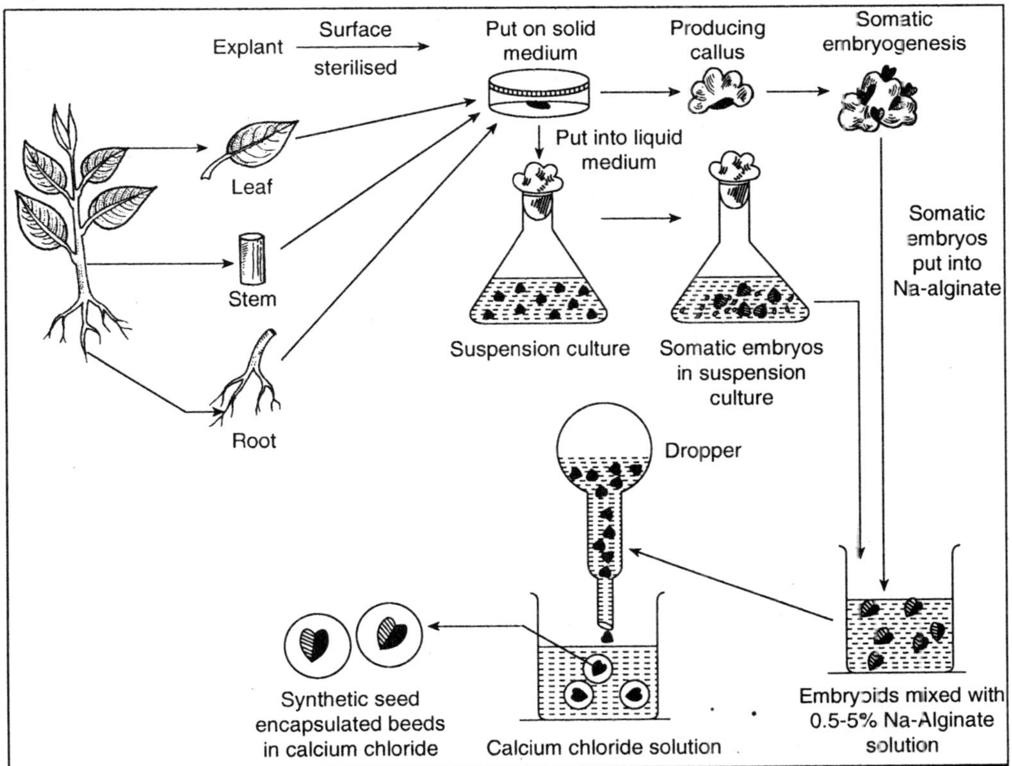

Fig. 18.7: Steps for Synthetic or Artificial seed production

Onishi et. al. (1994) and Sakamoto et. al. (1995) developed the synthetic seeds using **automated encapsulation process** by using coating material Elvax 4260 (ethylene vinyl acetate acrylic acid tetrapolymer). It helped to remove the stickyness of coating material and maintain uniform structure of beads.

Many other techniques have been developed recently which help to release the nutrient within the bead, **self breaking beads**, pharmaceutical type of capsules. Cellulose acetate mini-plugs have recently been developed to promote good sowing quality of synthetic seeds i.e. good germination rate in field condition.

Gray (1987) used the technique of desiccation of somatic embryos and then storage at normal temperature. When rehydrated again germinated to produce viable plantlet but not in good amount.

Senratna et. al. (1990) used ABA at the last stage of embryo development which promoted their tolerance to desiccation. Such desiccated embryos had good germination ability either *in vitro* or in soil condition.

Thorpe and Stasolla (2001) concluded that heatshock treatments, osmotic stress, nutrient depletion also can induce a degree of desiccation tolerance comparable to ABA treatment and have no detrimental effect on germination of embryos.

Potential Uses of Synthetic Seeds

(i) In some plants it requires a long time to reach at reproductive phase and seed production. In these cases the synthetic seeds can be helpful to get the propagules in a short period.

(ii) Generally the reproduction phase in a plant is season dependent, in these cases the somatic embryos and the synthetic seeds can be produced at any time as required.

(iii) In case of natural seeds there is a particular dormancy period, but in artificial seeds there is no dormancy period, thus it is more helpful for propagation. It can reduce the life cycle period of a plant.

(iv) Artficial seeds can also be helpful where there is no successful seed production after sexual hyrbidization. The somatic embryos can be obtained from somatic hybrids obtained through protoplast fusion.

(v) Artificial seeds are helpful in case of meiotically unstable genotypes, where the normal seed set is of low frequency.

(vi) In cases where the embryo germination is difficult, the artificial seed can provide the beneficial adjuvants i.e., growth promoting seed substances, plant nutrients, etc. through the artificial coats.

(vii) In genetic manipulation of crop plants the production of artificial seed may be useful.

(viii) For somatic hybrid plant production through protoplast culture and fusion the artificial seed technology is helpful.

(ix) In the production of genetically modified (GM) crops through transgenesis, the synthetic seed production may be applied.

(x) In general, synthetic seed technology is useful for rapid easiest way of non-conventional method of propagation in crop plants.

18.4 MICROPROPAGATION

Multiplication of genetically identical copies of a cultivar by asexual reproduction is called **clonal propagation.** The *in vivo* clonal propagation is often difficult, expensive and even unsuccessful. Tissue culture method offers an alternative way of clonal propagation which is popularly known as **micropropagation.** Here in this method a multiple number of miniature of vegetative shoots are produced from a clone within a short time and space (Fig. 18.8). Use of tissue culture for micropropagation was initiated by G. Morel (1960) who found this as the only commercially viable approach for orchid propagation.

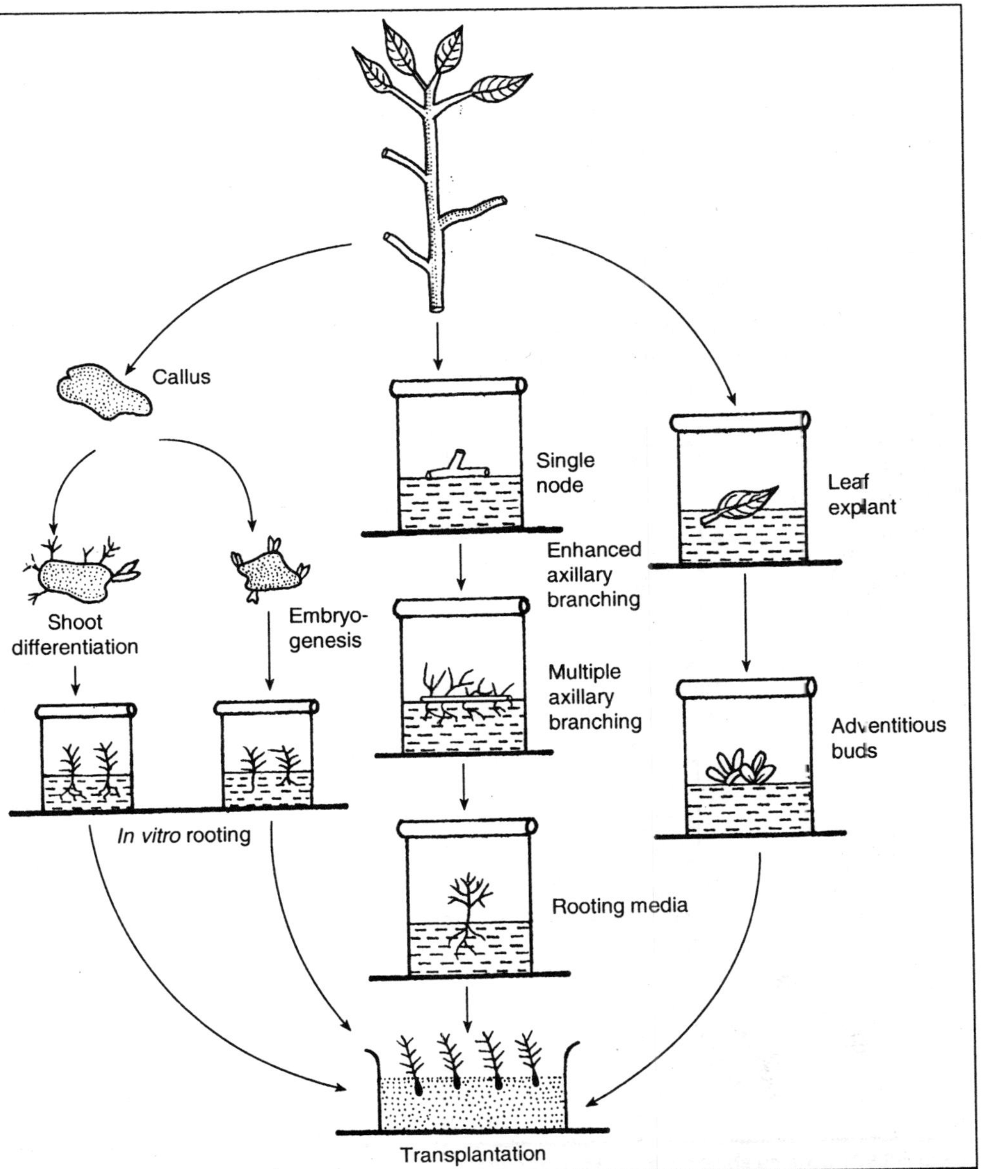

Fig. 18.8: Overview of micropropagation

The different techniques of single cell and protoplast culture also enable thousand plants to be derived within a short space and time. The products of these rapid vegetative propagation can also be regarded as clones when it is established that the cells they comprise are genetically identical.

Methods of Micropropagation

(i) **Multiplication by meristematic tissue of axillary and apical shoots:** Axillary and apical shoots contain quiescent or active meristems depending on the physiological state of the plant. When these shoot tips are cultured on a basal medium containing no growth regulators, these typically develop into single seedling like shoots with strong apical dominance (Fig. 18.9A-B).

On the contrary, when the shoots of the same explant material are grown on culture media containing cytokinin, axillary shoots develop precociously which proliferate to form clusters of secondary and tertiary shoots. These clusters when subdivided and transferred onto fresh medium again these will form similar clusters. This subdivision process may be continued indefinitely when provided with basic nutrients. About 5–10 multiplication rates on a 4–8 weeks of micropropagation cycle may ultimately lead to extremely impressive clonal propagation range of $0.1–3.0 \times 10^6$ within a year.

(ii) **Multiplication by adventitious shoots:** Axillary buds are "preformed meristems" present at the leaf axils, whereas adventitious buds may arise from any plant structures

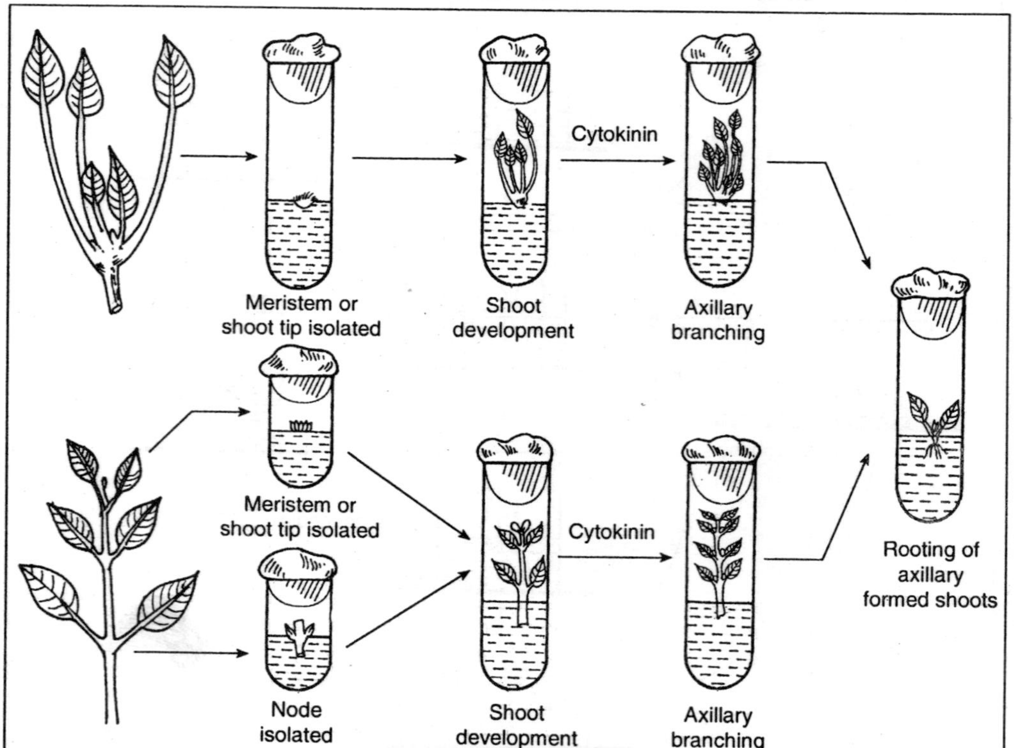

Fig. 18.9A: Scheme showing the *in vitro* micropropagation of plants by the axillary bud method. Upper row: as applied to rosette plants; bottom row: applied to plants which elongate

Fig. 18.9B: *In vitro* micropropagation in *Asparagus*

and this regeneration is often dependent on the presence of organised plant tissue. The explants may be stems, internodes, leaf blades, cotyledons, root elongation zone, bulb, corms, tubers, rhizomes, etc. If these explants are induced by using appropriate level of growth regulators in the medium they will form the meristematic zone which regenerate multiple shoots on a suitable culture medium. High number of adventitious shoots may arise only from a single epidermal cell.

Continuous propagation by adventitious shoot proliferation from bulbs and corms can be achieved by cultivating two vertically split piece of shoot bases. Clusters of shoots may develop from the abaxial surfaces of developing leave and scales.

Trimming of the shoot apices is a good approach which has been found to ensure continuously productive cultures of different hybrid plants for indefinite period.

(iii) **Multiplication by adventitious embryo formation:** It is another useful approach followed for many important plant species. Adventitious embryos can arise directly from a group of cells within the original explants or from primary embryoids. There are many plant species which develop embryos *in vivo* from diverse kind of explants e.g. the orchid leaf tips produce large number of embryoids whereas *Citrus* and *Mangifera* have shown to develop polyembryos from nucellar tissue. These adventitious embryoids are diploid in nature and can be used as clonal material for micropropagation. In *in vitro* similarly the adventitious embryos originated from different explants can be good materials for clonal propagation.

(iv) **Multiplication through callus culture:** Direct plantlet formation from the explants in culture are more desirable in case of micropropagation or clonal propagation. But shoot formation may occur through organogenesis or embryogenesis from the callus produced from the explant. The limitation of this system is that the callus cells are not genetically stable so it can not be called as a single clone and this process is more time consuming. Also the plant regeneration capacity may decline due to the genetic unstable condition.

In case of some plant species genetically stable calli have also been derived, in these types of calli slow growing meristematic zones are formed from the peripheral layer. The meristematic layers invariably comprise of diploid cells expressing totipotency. These types of calli can be subdivided into smaller parts and which may produce multiple shoots from each.

Stages of Micropropagation

The process of micropropagation or clonal propagatioin is a complicated process which can be subdivided into four prominent stages of operation.

Stage 0: This is the initial step of micropropagation in which stock plants has to be grown under controlled condition before using for culture initiation.

Stage I: The preparation of explants from stock plants is followed by its establishment in a suitable culture medium. The steps involved in this stage are (a) explant isolation, (b) surface sterilisation, (c) washing, (d) establishment of explant on appropriate culture medium.

Stage II: This stage involves the multiplication of shoots or rapid somatic embryo formation using a defined culture medium. Various approaches followed for micropropagation include:

(i) Multiplication through the growth and proliferations of meristems excised from apical and axillary shoot of the parent plant.

(ii) Induction and multiplication of adventitious meristems through processes of organogenesis or somatic embryogenesis directly on explants.

(iii) Multiplication of calli derived from any kind of explant and subsequently shoot development either through organogenesis or embryogenesis. The harvest cycle generally takes 4–8 weeks. Either the shoots are marketed directly or carried over to the next stage for further development.

Stage III: Shoots obtained from stage II are transferred to the next rooting or storage medium. These shoots are directly established in soil as micro-cuttings to develop roots. The shoot handling at this stage differs from species to species. When the shoots are directly transferred to soil then the following procedures are to be maintained—

(i) each shoot should be rooted individually,

(ii) hardening of the shoots to increase their resistance,

(iii) allowing the plants capable of autotrophic development rather than the heterotrophic development in culture,

(iv) fulfilling the requirements for breaking dormancy.

Stage IV: Transfer of plantlets to sterilized soil for hardening under greenhouse environment is achieved. This step is to ensure the successful transfer of the plantlets of stage III or unrooted shoot apices of stage II into the suitable compost mixture or soil in pots under controlled condition of light, temperature, humidity. For marketting sometimes these platlets are established in an artificial growing medium such as soil-less mixes, rockwood plugs or the sponges. It takes 4–16 weeks for marketting of the finished products.

Advantages of Micropropagation

1. The technique of micropropagation is an alternative approach to conventional methods of vegetative propagation, which has the enhanced rate of multiplication.

2. A million of shoot tips can be obtained from a small, microscopic piece of plant tissue within a short period of time and space.
3. The advantage in this type of propagation is that as shoot multiplication usually has a short cycle (2–6 weeks) and each cycle results in logarithmic increase in the number of shoots.
4. This method of propagation is more advantageous in case of bulb or corm producing plants, because minitubers or minicorms for plant multiplication are available throughout the year irrespective of season.
5. Smaller size of propagules are advantageous for storing and transporting as it takes lesser space.
6. The propagules can be maintained in soil-free environment which facilitates their storage on a large scale.
7. Stocks of germplasm can be maintained for many years using this method of propagation.
8. This method is more applicable where disease free propagules are wanted. This *in vitro* technique helps to raise pathogen free plant and to maintain them.
9. Micropropagation is very useful in case of dioecious plants, because there the seed progeny yield is 50% male and 50% female, but this technique helps to get the progeny according to the desired sex.
10. A major advantage of micropropagation happens to be the minimum growing space required in commercial nurseries. Thousands to millions of plantlets can be maintained within the culture vials. This is specially useful for maintaining the horticultural species.
11. This method is more helpful in case of slow growing plants where the seeds are produced after a long term and the seeds are the only propagule. This method can overcome the difficulty to obtain the propagules.
12. Through seed production genetically uniform progeny is not possible always. But the micropropagation method will help to maintain the genetic uniformity in the propagules.
13. Micropropagation is one of the finest ways of plant multiplication by *in vitro* technique of plant tissue culture. The newer tissue material obtained through r DNA technology or haploid culture or somatic hybridization can be the source of tissue material for micropropagation, as it is the easiest method for obtaining the multiple propagules.

Commercial Uses of Micropropagation

Micropropagation of orchids demonstrated profit all over the world, besides orchid about 600 species of other ornamental plants have been successfully cloned and some of them are commercially exploited (*Chrysanthemum*, Carnation, *Gerbera* and *Anthurium*), list of names is increasing day by day, the advantage is rapid cloning of selected colour individuals. Except the horticultural flowering plants, a large number of forest trees, fruit trees, oil producing plants (*Eucalyptus*, Oil-palm), vegetables and many edible crops like peach, apple, pear, cherry, apricot and palm have been cloned and put into use commercially (Fig. 18.10).

Fig. 18.10: Micropropagation in different plants

Breeding and selection by sexual hybridization is a very slow procedure because of long generation time and they are difficult to propagate vegetatively. So the *in vitro* cloning of more than 100 woody species over a wide range of families has been successfully achieved.

Forest trees are important sources of our biodiversity and also they provide food, fuel, construction and industrial products for us. These resources are diminishing at an unprecedented rate. Large scale, cost-effective and nursery-friendly *in vitro* technique like micropropagation protocols have been developed for selective forest trees; about 70 forest tree species of Angiosperm and about 30 forest tree species of Gymnosperm.

Table 18.1: Relative cost component associated with micropropagation of a typical crop

Cost component	*Percent*
Laboratory, Direct labour	15
Utilities	9
Depreciation	7
Supervision (Laboratory and Greenhouse)	23
Planting, Direct labour	9
Other production cost	37

The most important aspect of commercial micropropagation is the economics involved and the unit cost of a plantlet. Research achievements in the field of micropropagation is not always economically viable and not accepted by commercial houses. The investment in commercial tissue culture business will depend to a large extent on cost of the laboratory set up, type of plant to be propagated and the skill involved.

The commercial nursery man should start with those crop species for which published methods are available and also it is essential that the grower must have some training in tissue culture and plant husbandry. Another approach followed in micropropagation is to automate it at its various stages. In this connection the **bioreactors** are being used for large scale multiplication of somatic embryos, shoots and bulbs. **Automatisation** have also been used for subculture of shoots during stages III and IV of micropropagation. This reduces the cost of labour component in micropropagation.

QUESTIONS

1. What do you mean by organogenesis? Discuss direct and indirect organogenesis for plant regeneration.
2. Define embryogenesis. How does it differ from organogenesis. State the different routes of somatic embryogenesis.
3. Write the differences between zygotic embryo and somatic embryo. Mention the applications of somatic embryogenesis.
4. Define synthetic/artificial seed? How does it differ from natural seed? Write the methods of artificial seed production. State the potential uses of synthetic seeds.
5. Define micropropagation. Write the sources of tissue and the methods of micropropagation. State the different stages in micropropagation technique.
6. Discuss the advantages of micropropagation. How far the technique is subjected to commercial use?

19 ZYGOTIC EMBRYO CULTURE AND EMBRYO RESCUE

Zygotic embryo culture is the aseptic isolation and growth of sexually produced embryos *in vitro* with the objective to give rise to normal viable seedling. After sexual hybridization programme when the embryo development is not perfect then to overcome the low seed viability, inability of germination or self sterility of seeds, the zygotic embryo culture is needed. This technique is also useful to explore the nutritional or physical requirements for embryonic development avoiding the seed dormancy and accelerating the seed germination rate which is helpful to shorten the life cycle of that plant.

19.1 TYPES OF EMBRYO CULTURE

Culture of mature and intact seed embryo: Mature embryos are excised from ripe seeds and cultured mainly to avoid inhibition of seed germination due to seed dormancy. It also helps to know the various parameters of embryonic growth and the metabolic and biochemical aspects of dormancy and germination.

Culture of immature embryos or proembryos: Proembryos means the early developmental stages of embryos that precede cotyledon initiation. Proembryos or immature embryos originating from normal plant species which fail to germinate are excised from the immature seeds and cultured with complex nutrient medium required to raise the plant.

Culture of adventive embryos from polyembryonic seeds: In case of *Citrus*, polyembryos are developed from the nucellar tissue, which are adventive in nature. These embryos can be exploited in culture to give rise to plantlets.

Culture of intact seed containing undifferentiated embryo: In case of orchid, seed contains spherical mass of embryonic tissue which is not differentiated into radicle and plumule and the seed does not contain the storage tissue. This type of seed containing undifferentiated embryo is cultured and treated as embryo culture.

Culture of non-viable or abortive embryo: In many interspecific or intergeneric breeding experiments, the embryos produced are non-viable or abortive which do not germinate normally. Culturing of these inviable embryos supplemented with various nutrients *in vitro* may give rise to plants.

19.2 PROCEDURE OF EMBRYO CULTURE

Source Material

Physiologically uniform embryos of the same size and same growth phase should be used for culture, which can be achieved if the plants are raised and maintained under controlled condition. For supply of materials regularly the selected plants should bear the flowers at regular intervals; to obtain the embryos of a specific age, artificial pollination of freshly opened flowers is necessary and it may be desirable to prepare a chart showing the different stages of embryo development with days after pollination. Where the embryos become abortive after development of seeds, the embryos should be excised before the onset of abortion.

Surface Sterilization

As the embryo grows within the seed which is within a totally sterile environment, so surface sterilization of the embryo is not needed unless there is an injury on seed-coat, or any kind of systemic infection is present. Otherwise, mature seeds, entire ovules or fruits are surface sterilized and the embryos are removed aseptically from the surrounding tissue.

In case of orchid seeds the entire seed contains the morphologically undifferentiated spherical embryo. As there is no nutritive tissue and the seed coat is very thin, the whole seed is cultured in place of embryo.

... of Embryo

...ryo-excision operation is performed aseptically in a laminar air flow cabinet ... of a stereomicroscope, and using the dissecting tools like forceps, needles, ... blades, etc. If the seed coat is hard then seeds are soaked in water and ...plitted open and embryos are excised. But in case of immature embryos ...mbedded in endosperm, the incision is made at the micropylar end and ... to isolate the young embryo.

...irements in Culture Medium

...inly two phases of embryo development, heterotrophic phase—where ...endent on the nutritional tissue or maternal tissue, another is the ...-where the embryo is metabolically capable of synthesizing sub... ...owth, i.e., becoming independent for nutrition.

...n for embryo culture differs for young or immature embryos ...yos. When immature or heterotrophic embryos are cultured, ...comes progressively complex to permit the expression of the ...al of an embryo. This is dependent upon the progressive acti... ...ystems or biochemical pathways concerned with the synthesis of substances necessary for the growth of embryos at a specific age. But the embryos of mature stage are completely autotrophic and grow on a simple medium comprising of mineral salts with carbohydrate energy source.

Mineral salts are used according to MS, B5 or White's media with some modifications. Among the carbohydrates, sucrose is the main source, addition of maltose, lactose, raffinose or mannitol may be required for some species of embryo culture. The carbohydrate concentration may also vary according to the maturity of embryo. Presence of ammonium salt has been found essential for proper growth and differentiation of embryos. Various amino acids or their amides are sometimes essential for embryo growth. Casein hydrolysate (CH), an amino acid complex has been widely used as an additive to embryo culture media. Coconut milk helps the young excised embryos to increase in length of various species like sugarcane, barley, tomato, carrot, or some interspecific hybrids. Among the growth regulators, GA is used at low levels which effectively stimulates excised heart shaped embryos.

The incubation temperature for embryonic growth and germination may vary with the genotype and species. Initial dark incubation for 4 days of embryo culture is essential, following which these can be grown to maturity under light condition according to requirement.

The protocol of embryo culture has been depicted in Fig. 19.1.

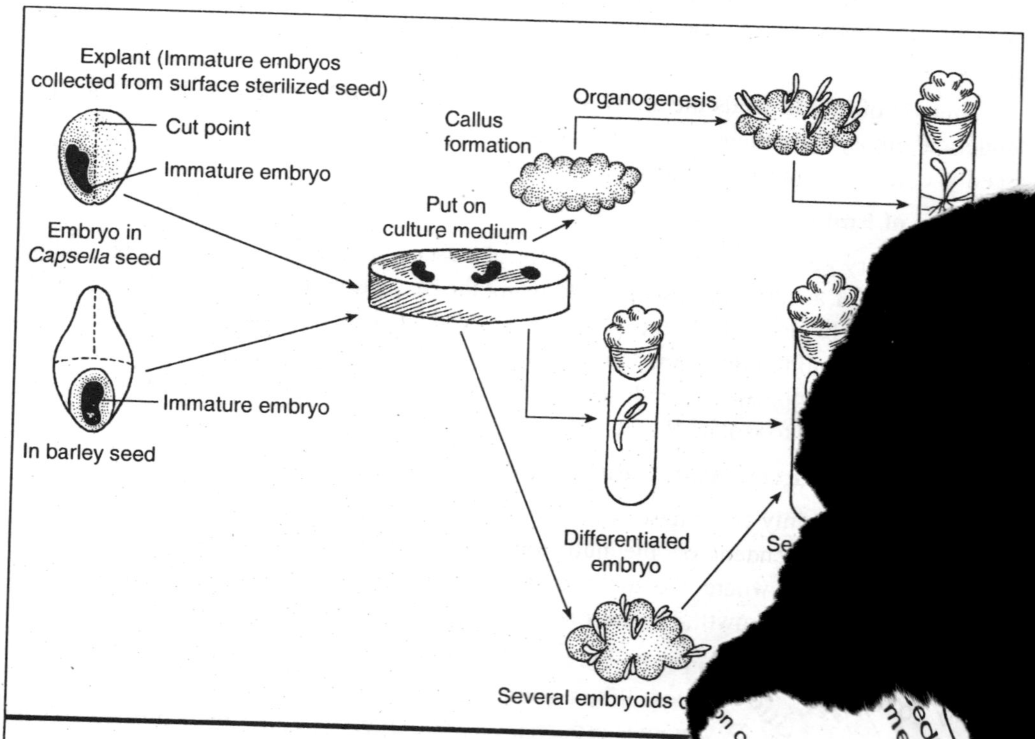

Fig. 19.1: Procedure of Embryo Culture

Morphogenesis in Partially Differentiated Embryo

Embryo development is a process from zygote formation to embryo germination, the stages can be classified into five stages:

Stages	Characteristics
I. Cleavage and differentiation	Cell division with little growth; differentiation of all major tissues.
II. Growth	Rapid cell expansion and division.
III. Maturation	Little or no cell division or expansion, synthesis and storage of reserve materials.
IV. Dormancy	Developmental arrest.
V. Germination	Renewed cell expansion and division; embryo growth.

Detailed cultural studies have revealed the role of suspensor in embryo development. Suspensor is an ephemeral structure which is found at the radicular end of the proembryo that attains its maximum development by the time the embryo reaches the globular stage. In some cases the presence of suspensor with the immature embryo is essential for maturity. The requirement of the suspensor may be substituted by the addition of GA or ABA to the culture medium.

After full development of the embryo it enters into a phase of dormancy characterised by metabolic quiescence and development arrest of cells. Dormancy may last for different time periods for different kinds of crop species.

Excised immature plant embryos on a nutrient medium tend to bypass the stage of dormancy and cease to undergo the linear embryogenic mode of development. The phenomenon of seedling formation without completing normal embryogenic development is called **precocious germination**. Casein hydrolysate, high sucrose level, O_2-tension, elevated temperature, high light intensity and also ABA like substances prevent precocious germination of embryo.

In many plant species fully developed seeds contain embryos that lack differentiation into radicle, plumule and cotyledons. In these cases the globular stages are attained and the embryonal end proximal to the micropyle regarded as the 'radicular pole' and distal to the micropyle is regarded as the 'plumular pole'. In case of orchid, the plumular pole enlarges to form a spherical structure which turns green and after attainment of certain size differentiates into roots and shoots.

19.3 APPLICATIONS OF EMBRYO CULTURE

1. Rescuing Embryos from Incompatible Crosses

In case of interspecific or intergeneric crosses there may be normal fertilization but the embryo abortion results in the formation of non-viable seeds. Of several cases studied, poor or abnormal development of endosperm caused embryo starvation and eventual abortion. In these cases the hybrid embryos can be isolated before abortion and can be cultured into a cellular endosperm of anyone of the parents or using the endosperm of any closely related third species. This technique is called **embryo-endosperm and transplant technique** (Fig. 19.2).

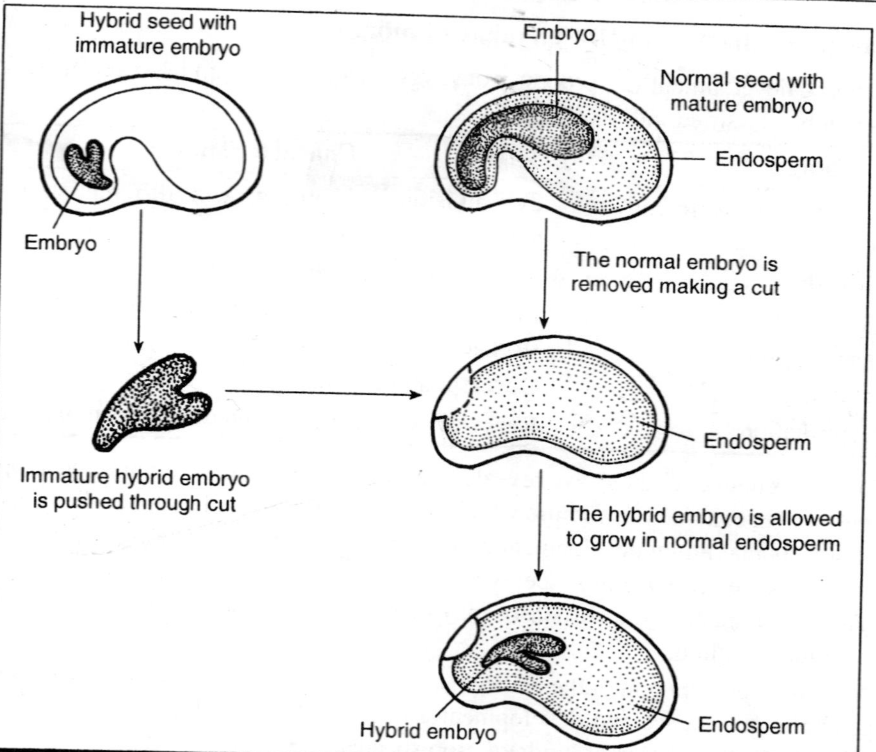

Fig. 19.2: Method of Embryo Rescue by embryo-endosperm transplantation

Isolation of hybrid embryos at the immature stage i.e., before abortion and their culture *in vitro* may circumvent strong post-zygotic barriers. To raise the rare hybrids by rescuing embryos from incompatible crosses is the most popular application of embryo culture. With the improvement of embryo culture technique a large number of inter-specific and intergeneric hybrid plants have been raised.

2. Overcoming Dormancy and Shortening Breeding Cycle

Long periods of seed dormancy may lead to delay in breeding work in horticultural and crop plants. Using embryo culture technique the seed dormancy can be overcome. Seed dormancy may be caused by some metabolic inhibitors produced within the seed or due to presence of hard seed coat. In these cases the mature embryo culture is the best way to overcome seed dormancy. But when the cause of seed dormancy is not well understood, the immature embryos are excised and cultured to by-pass the seed-dormancy.

3. Overcoming Seed Sterility

In early ripening fruit cultivars, seeds do not germinate because their embryo remains still immature. Using embryo culture method it is possible to raise seedlings from sterile seeds of early ripening stone fruits.

4. Embryo Callus has Organogenic Potential

An embryo callus is reported to possess a high regenerative capacity compared to those derived from mature organs such as the leaf, stem and root. The age of embryo influences the regenerative ability of callus. Generally no differentiation occurs in case of callus obtained from mature embryo but the immature embryos of many monocots can be of potential use for production of callus which may be used as good explants to initiate the plant regeneration through organogenesis. To obtain a callus with morphogenic potential excised embryos are placed on an agar medium with the scutellum facing up in the presence of 2, 4-D alone or in combination with cytokinin.

5. Production of Monoploids and Diploids

Embryo culture technique has been used to raise the monoploids in barley, where the interspecific fertilization between *Hordeum vulgare* and *H. bulbosum* takes place but during further development one chromosome set gets eliminated, resulting in the formation of monoploid of barley which can be rescued by embryo culture. When a tetraploid and a diploid species are crossed to get the seedless triploid plant (as in case of *Citrus*) the embryo culture through rescuing of immature embryos is important.

6. Biotechnological Application

Plasma membranes of desiccated zygotic embryos are generally with large pores. The mature embryos in excised condition can be soaked or imbibed in a solution containing the vector DNA with desirable clone, which can enter within the cell and get transformed within the embryonic cell and can be perpetuated within the developing embryo. Another method is application of biolistic method i.e., by using particle gun the DNA clone can be pushed within the embryonal tissue which can be transformed easily and can produce the plant with transformed clones.

QUESTIONS

1. What is the purpose of embryo culture? Mention the different types of embryo culture.
2. Write down the procedure of embryo culture.
3. What do you mean by embryo rescue? Discuss the applications of embryo culture.

20 PROTOPLAST CULTURE AND SOMATIC HYBRIDIZATION

Any single cell of plant tissue can be cultured and used to regenerate into full grown plantlets depending on the totipotency nature of plant cell. But cell has a barrier of cell wall that poses problems to many experimental manipulations which can be overcome by removing the cell wall. The need for protoplast culture has arisen thus. The technique has also helped a lot to get the somatic hybrids as well as to do genetic manipulations by uptake of DNA and plant organelles.

20.1 PROTOPLAST ISOLATION

Each plant cell has a definite cellulose wall which are attached with each other by middle lamella made up of pectin. Protoplast is the content which is surrounded only by plasmalemma. So to get the isolated protoplast it is essential to remove the pectin material to obtain the single cells and then to remove cell wall by enzymatic method.

A. Sources of Explant for Protoplast Isolation

Protoplasts can be isolated directly from the different parts of whole plant which bears the soft parenchymatous tissue (e.g., young fully expanded soft leaves) or indirectly from the *in vitro* grown plant tissue (e.g., callus tissue). Protoplast yield and viability depends greatly on the type of tissue material from which it is isolated and the method used. Cell suspension cultures may also provide a very good source of protoplast.

Before isolation of protoplast, the soure material if it is from *in vivo* grown plant then it should be properly surface sterilized using the proper method of sterilization. Then any of the methods either mechanical or enzymatic can be used to isolate the protoplast.

B. Methods of Protoplast Isolation

The essential step of protoplast isolation is the proper use of **osmoticum.** The plant protoplast is an osmotic system where the plasma membrane acts as the semipermeable membrane which equilibriates the outward and inward flow of water. If the inward pressure is more then due to heavy inflow of water the protoplast will burst. So during protoplast isolation the enzymes applied to isolate the protoplast and the tissue material should be placed in proper **osmolyticum** or **plasmolyticum.** Generally mannitol, a sugar alcohol at high concentration, an easily transportable component provides a good stable osmotic environment which prevents bursting of protoplast (Fig. 20.1).

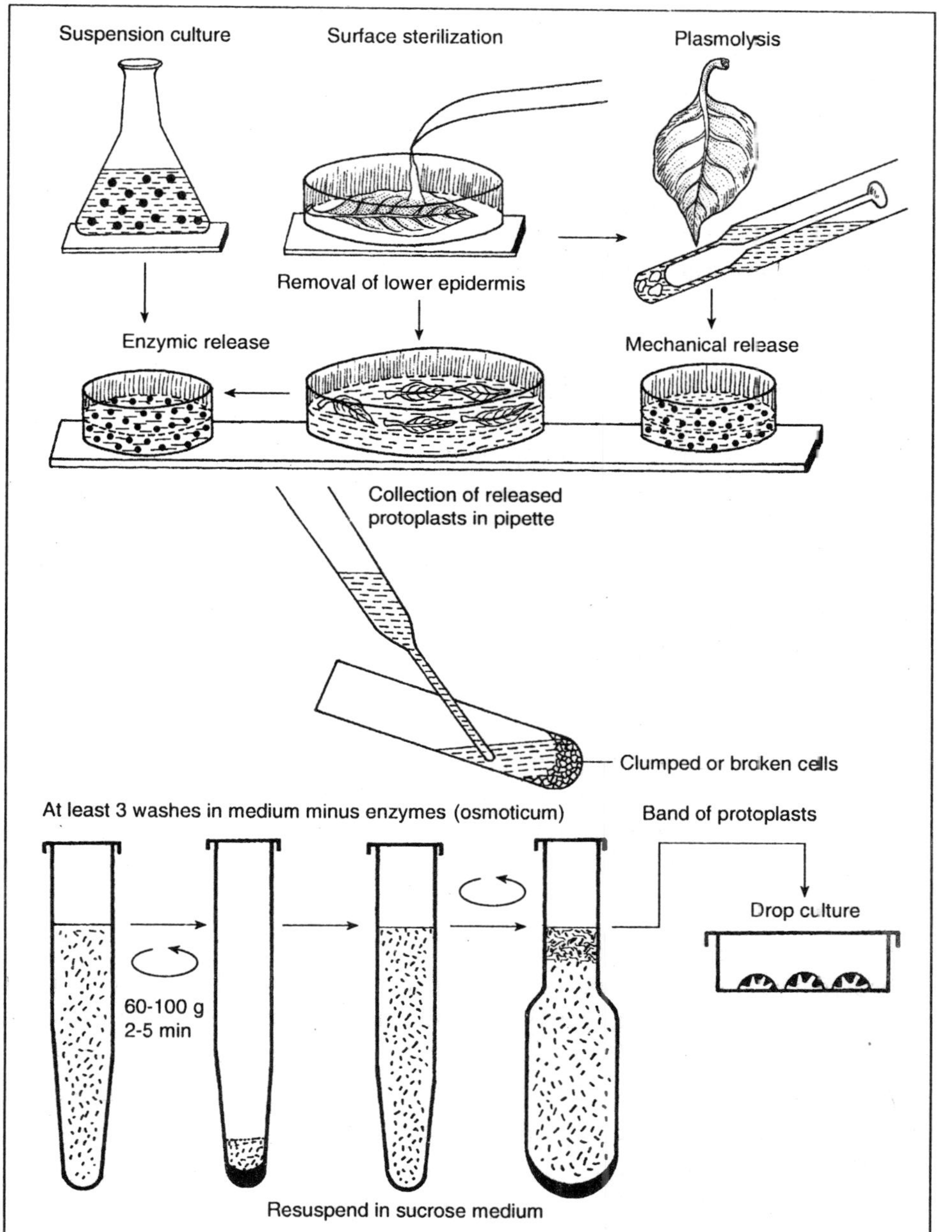

Fig. 20.1: Isolation of Protoplasts

Protoplasts may be isolated by any one of the two following ways:

(a) Mechanical method (Non-enzymatic method); (b) Enzymatic method.

(a) **Mechanical method:** Any soft parenchymatous tissue is kept in a plasmolyticum. The plasmolysed tissue is then finely chopped into pieces and the intact cells (plasmolysed) are released into the medium from the cut surface. The suspension is then allowed for deplasmolysis and the released protoplasts attain their original size.

The yield of protoplast in this method is very low, for large scale of protoplast yield the enzymatic method is followed.

(b) **Enzymatic method:** Young fully expanded soft leaves, or *in vitro* grown callus tissue or cell suspension culture grown cells can be used as the source material. The tissues or cells are incubated in plasmolyticum for 1 hr before enzymatic treatment. The intact tissue materials cut into smaller pieces to increase the surface area of enzymatic activity. The enzymes can be used either sequentially in **two step method** or in a single step by **mixed enzymatic method**.

The **enzymes used** are of three main **categories:**

(i)	(ii)	(iii)
Cellulases	Hemicellulase	Pectinase
Cellulysin	Helicase	Macerozyme-R-10
Driselase	Rhozyme	Pectinol

The concentration of enzymes used and the time period of incubation varies greatly depending on the tissue type. In two step method, the pectinase and hemicellulases are applied first and then cellulase is applied for complete removal of cell wall.

After release of protoplast into the suspension, for removal of enzymes the protoplasts are collected in centrifuge tube as pellet and washed several times with the osmoticum.

C. Purification of Protoplasts

For purification, the protoplasts suspended in osmoticum is centrifuged using sucrose (20%) solution. The viable protoplasts float on the top surface of sucrose solution forming a band. These protoplasts are then collected, resuspended in osmoticum and washed several times. Finally the protoplasts suspended in a measured volume of protoplast culture medium after counting the number with the help of **hemocytometer.** The viability of protoplast is checked with the help of Fluorescin diacetate staining or phenosafranine or calcofluore white.

D. Isolation of Subprotoplasts

Subprotoplasts do not contain the entire content of plant cells, these can be of different types such as cytoplasts, karyoplasts or miniprotoplasts or microprotoplasts or microplasts.

(i) **Cytoplasts** lack the nucleus and contains the entire cytoplasm of a cell or part of it. These are often used for cybrid production.

(ii) **Karyoplasts** or **Miniprotoplasts** contain a nucleus surrounded by some cytoplasm and the original outer plasma membrane.

(iii) **Microprotoplasts** contain fraction of both nuclear and cytoplasmic material.

(iv) **Microplasts** lack the nuclear material and contain only fraction of cytoplasm and outer membrane.

E. Protoplast Viability Test

The most frequently used staining methods for checking protoplast viability are:

1. **Fluorescein diacetate (FDA)** dissolved in acetone is used at a conc. of 0.01% and intact viable protoplasts only fluoresce when observed under UV.
2. **Phenosafranine** is also used at a conc. of 0.01%, which is specific for dead protoplasts that shows red in colour.

20.2 PROTOPLAST CULTURE

Isolated protoplasts are cultured either in a liquid medium or semisolid agar medium in a thin layer or as small drops of nutrient medium in petridish. The medium for protoplast culture requires the same component as required for callus or suspension culture. Increasing the calcium concentration only helps to maintain the integrity of the membrane. Generally the media require more amount of sugar. The vitamins and growth substances are used as per requirement for cell division, callus formation and then differentiation.

Plating density is another important criteria for protoplast culture, a density of 1×10^4 to 1×10^5 protoplast per ml is optimal, such high densities are helpful for earlier division of plant protoplasts whereas the density is reduced subsequently during progress of culture.

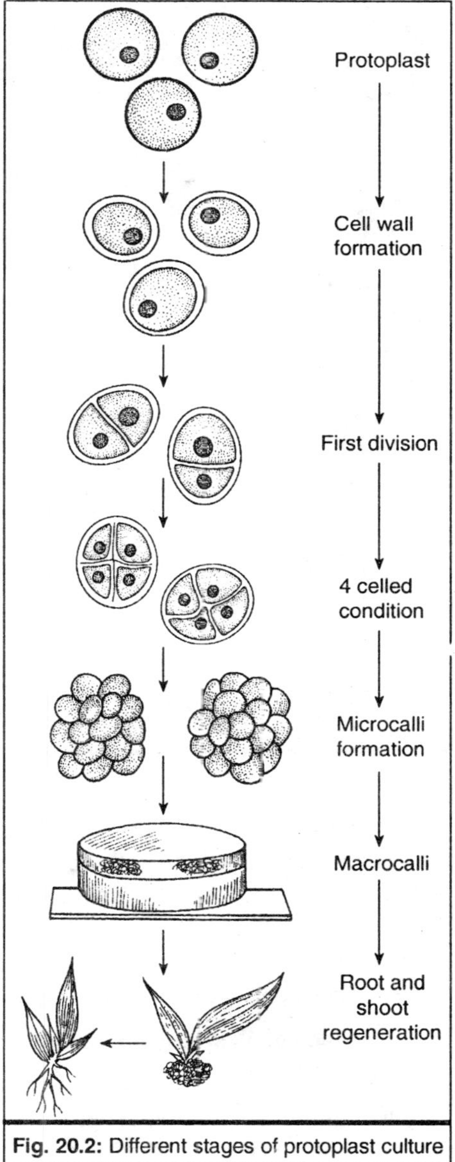

Fig. 20.2: Different stages of protoplast culture

A. Stages of Protoplast Culture

Protoplast culture has mainly four stages of development. The viable protoplast in culture regenerates its own wall around them and then it is prepared for **cell division.** Sequential cell division leads to **callus formation.** From this callus by organogenetic or embryogenetic **differentiation** the **plantlets** or **embryos** may develop (Fig. 20.2). But as these are all derived from a single protoplast so all regenerated are of same kind of genetic constituent.

Cell wall regeneration is the first prerequisite for cell division, after wall formation the walled cells expand and divide into two cells equally which looks like '8'. After the first division each

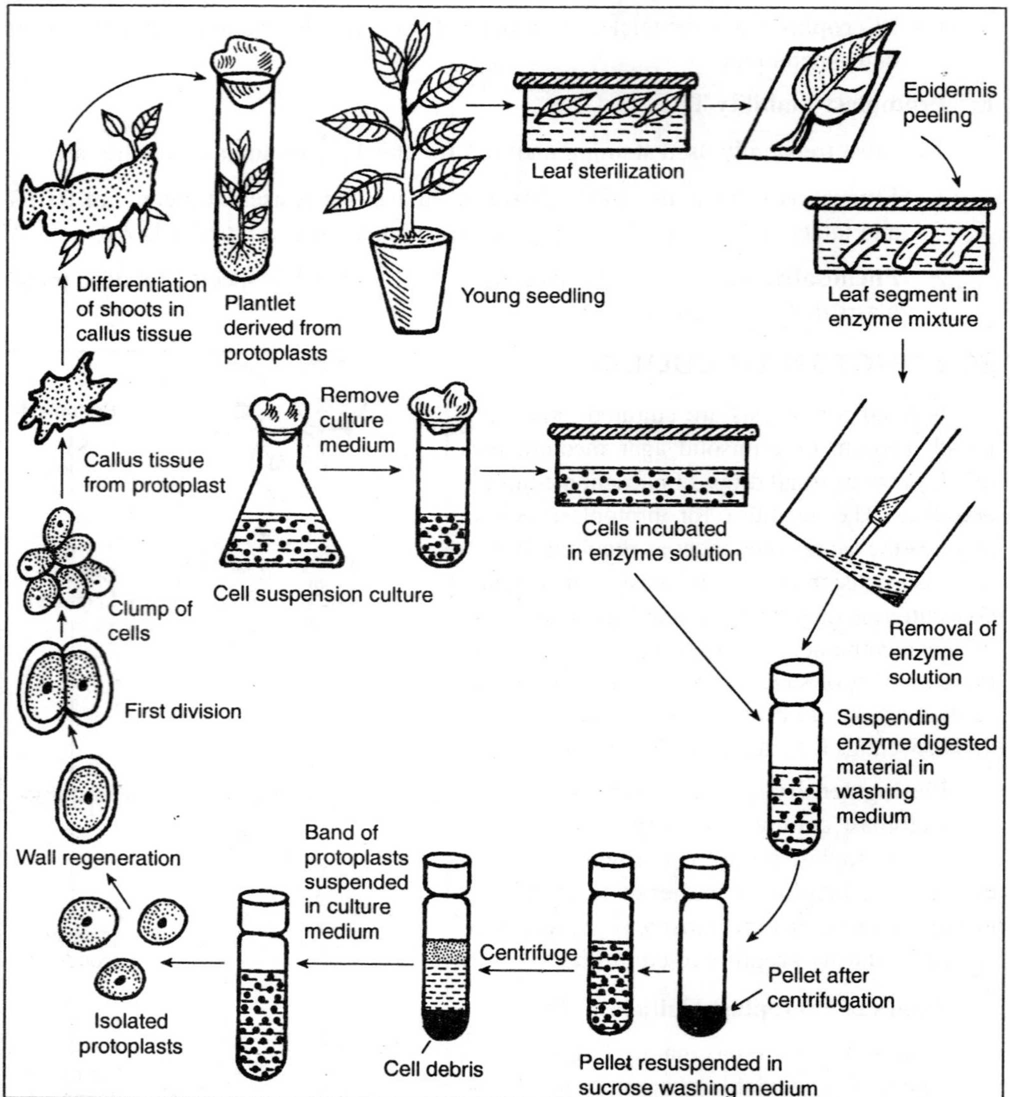

Fig. 20.3: Steps involved in protoplast isolation, culture and regeneration

daughter cell divides into two cells. Repeated division results in the formation of cell clump or cell aggregates. All the cells derived from protoplasts do not undergo division. Several factors such as genotype of donor plant, culture medium, hormones as well as physical factors are important factors for division of protoplasts and callus formation. After callus formation these are subcultured at regular intervals for further differentiation by using different hormone combinations which is called plant regeneration medium (Fig. 20.3).

B. Methods of Protoplast Culture

There are different methods of protoplast culture such as liquid culture, agar culture, droplet culture, co-culture, hanging droplet culture, immobilised/bead culture and feeder layer technique (Fig. 20.4A-D).

1. **Liquid culture:** This method is generally preferred in most cases during early developmental stages of protoplasts, because it allows easy dilution and transfer, protoplasts easily get divide in liquid media, osmotic pressure of the medium can be regulated and can be effectively reduced during further growth of protoplasts. The disadvantage of this method is that it does not permit the isolation of single colonies derived from one parent cell.
2. **Agar culture:** Agarose is most frequently used to solidify protoplast culture media. Protoplast suspension is taken at double density and mixed with melted agar medium at 45°C and mixed well and plated in small petridish. Here the protoplasts remain in same position and immobilised, proper plating efficiency can be obtained but the medium change can be done only after visible calli formation.

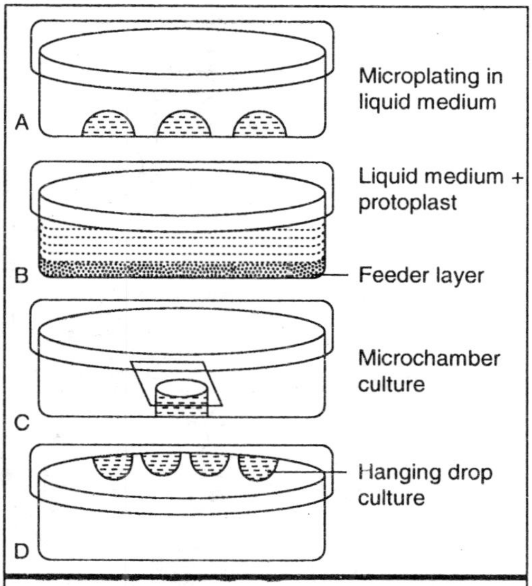

Fig. 20.4A-D: Different cultural techniques for protoplast culture

3. **Droplet culture:** Suspending protoplasts in liquid culture media are placed on petridishes in the form of droplet, the cultured protoplasts clump together at the centre of droplets. The liquid medium can be changed at regular interval.
4. **Co-culture:** Sometimes to induce division the newly isolated protoplast suspension is mixed with a reliable fast growing protoplast suspension and mixed protoplasts are plated. Some growth factors help to induce the proper growth and development of the isolated protoplasts.
5. **Hanging droplet technique:** Culture of protoplasts can be done in an inverted droplet on the inner surface of the lid of petridish, a very small number of protoplasts can be cultured in this way. A thin layer of liquid medium is kept in the petridish to keep the environment inside the petridish humid.
6. **Bead culture:** The protoplasts suspension can be mixed with any kind of polymer like alginate, carrageenan, etc. and then small beads are made by dripping into the liquid medium and then cultured into liquid medium with slow shaking condition.
7. **Feeder layer:** In many cases it is desirable to reduce the plating density, then a feeder layer consisting of X-irradiated non-dividing but living protoplasts are plated in agar medium and on this layer the isolated protoplasts are plated in a thin layer of liquid medium. Here the living but non-dividing protoplasts provide necessary growth requirement for the isolated less number of protoplasts.

20.3 PROTOPLAST FUSION AND SOMATIC HYBRIDIZATION

Isolated protoplasts are devoid of cell walls which make them easy tools for undergoing fusions *in vitro*. An important factor is that generally there is incompatibility

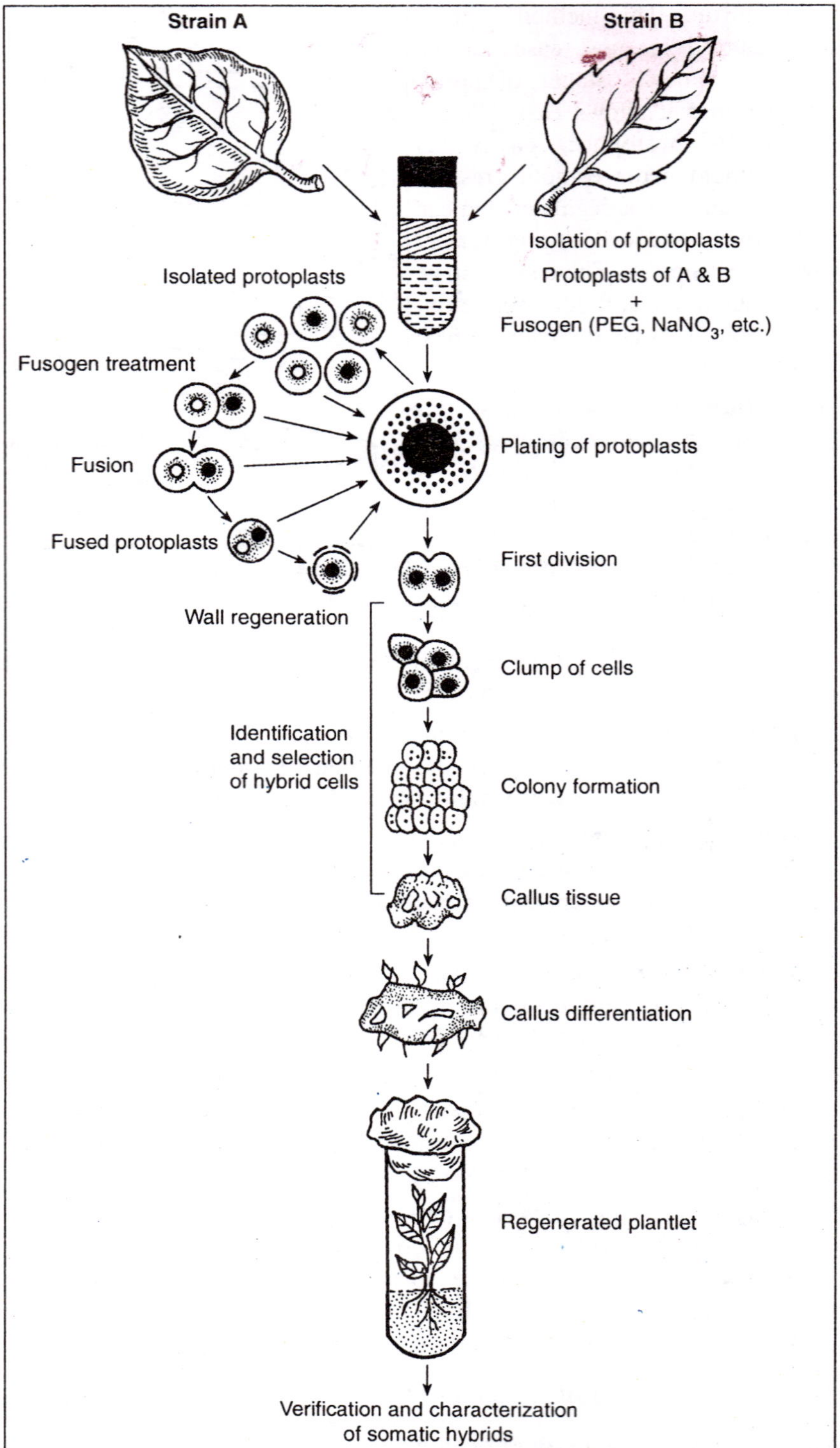

Fig. 20.5: Schematic representation of protoplast fusion and regeneration of plants

barrier between two protoplasts of different species or genera. The process of fusion may be spontaneous or it may be induced (Fig. 20.5).

A. Method of Protoplast Fusion

(a) Spontaneous Fusion Method

During enzymatic degradation of cell walls some of the adjacent protoplasts may fuse together to form homokaryons, sometimes more than two protoplasts fuse together and form multinucleate cells by a phenomenon where expansion and subsequent coalescence of the plasmodesmata led to such cases. However the spontaneous fusion products do not regenerate into whole plants except for undergoing a few divisions.

(b) Induced Fusion Method

Freshly isolated protoplasts can be induced to undergo fusion, irrespective of their origin with the help of different kinds of **fusogen** (fusion inducing agents) e.g., $NaNO_3$, lysozyme, High pH/Ca^{++}, polyethylene glycol (PEG), concavalin, polyvinyl alcohol, electrofusion, dextran sulphate, etc. Some of these treatments are discussed here.

(i) **$NaNO_3$ treatment:** Isolated protoplasts are suspended in mixture of 5.5% $NaNO_3$ and 10% sucrose soln. and incubated in a water bath at 35°C for 5 min. and centrifuged for 5 min. at 200 × g, following centrifugation the supernatant is decanted and the pellet is incubated at 30°C for 30 min. During this period most of the protoplasts undergo cell fusion. After sometimes these are washed with osmoticum and then plated properly in culture medium.

(ii) **Calcium ions at high pH:** Here the isolated protoplasts are incubated in a solution of osmoticum containing 0.05 M $CaCl_2$ at pH 10.5 (using 0.05 μ glycine—NaOH buffer) and at temperature 37°C. Aggregation of protoplasts generally takes place at once and fusion occurs within 10 mins. After this treatment, 20–50% of the protoplasts have been found to be involved in fusion.

(iii) **PEG treatment:** PEG has been found to be used as a fusogen in most of the successful cases of protoplast fusion. 1 ml of protoplast suspension is equally mixed with 1 ml of 28–56% PEG (1500–6000 MW) solution. The tube is then allowed to settle for 10 min to 40 min at room temperature. Both the mol. wt and the conc. of PEG used is critical in inducing successful fusions. After PEG treatment the elution of fusogen is done by using high pH/Ca^{++} containing solution which is most effective in enhancing the fusion frequency and survivability of protoplasts (Fig. 20.6A).

(iv) **Electrofusion:** Protoplasts are placed into a small cell culture containing electrodes and a potential difference is applied due to which protoplasts line up between the electrodes. In this fusion method, two step procedure is followed: a low voltage and rapidly oscillating AC field is applied, which causes the protoplasts to become aligned into chains by cell to cell contact. Second step is brief application of a high voltage DC pulse which induces reversible breakdown of the plasma membrane at the site of cell contact, leading to fusion and consequent membrane reorganisation. Heterokaryons produced by this electrofusion divide normally in culture medium and has the capability of regenerating the plantlet (Fig. 20.6B).

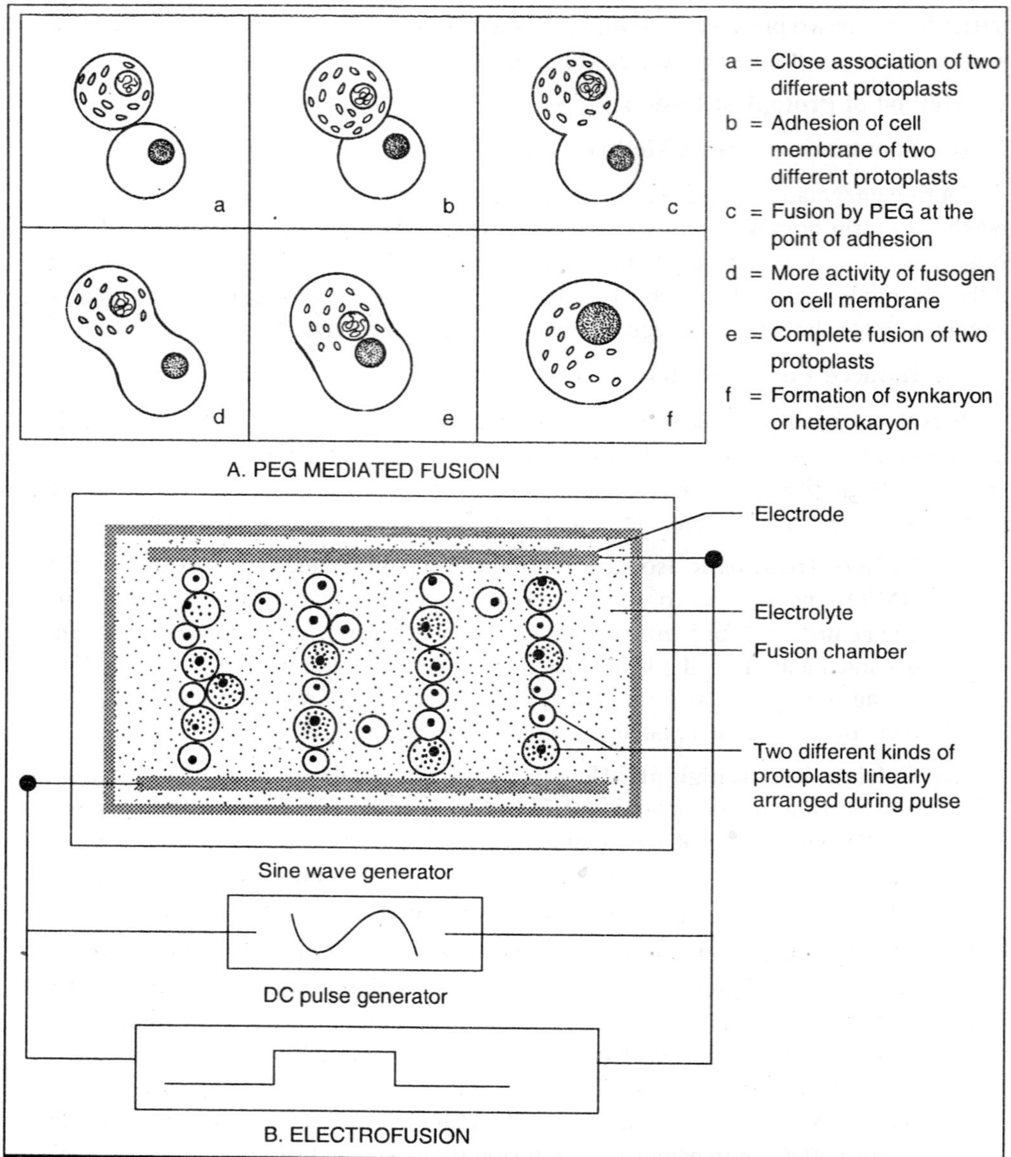

Fig. 20.6: Induced protoplast fusion: A. PEG mediated fusion; B. Electrofusion

B. Mechanism of Protoplast Fusion

Protoplast fusion consists of three main phases:

(i) **Agglutination or Adhesion:** Two or more protoplasts are brought into close proximity by using a variety of treatment like PEG, high pH, high conc. of Ca^{++} ions, etc.

(ii) **Fusion of plasma membrane at localised sites:** Membranes of protoplasts agglutinated by fusogen get fused at the point of adhesion, which results in the formation of cytoplasmic bridges between the protoplasts and fusion requires less than 10Å distance between two membranes. The high pH and high Ca^{++} ions have shown to neutralise the normal surface charge so that the agglutinated protoplasts can fuse due to intermingling of lipid molecules in membranes.

(iii) **Formation of heterokaryon:** Rounding off of the fused protoplasts occur due to the expansion of cytoplasmic bridges forming spherical heterokaryon or homokaryon.

C. Identification and Selection of Hybrid Cells

Following fusion treatment the protoplast population consists of parental type protoplasts, homokaryotic fused products and also heterokaryotic fusion products. The proportion of viable heterokaryotic fusion generally is lower, identification and recovery of protoplast fusion products have been based on observation of visual characters or hybrid cells may show genetic complementation for some growth requirements, etc.

(a) **Visual selection:** In most of the fusion programme generally the selection procedure involves the fusion between a non-green protoplasts of one parent and the green protoplast of another parent, as this facilitate the visual identification of heterokaryons under light microscope. The non-green protoplasts may be available from callus tissue and the green protoplasts from leaf tissue (Fig. 20.7A-B).

(b) **Selection by complementation:** If the parental protoplasts from the two parents can be identified by biochemical marker then the heterokaryons can be selected easily by using the proper growth requirement in media.

(i) **Complementation of resistance markers**—Dominant characters such as traits conferring resistance to antibiotics, amino acid analogues or toxic com-

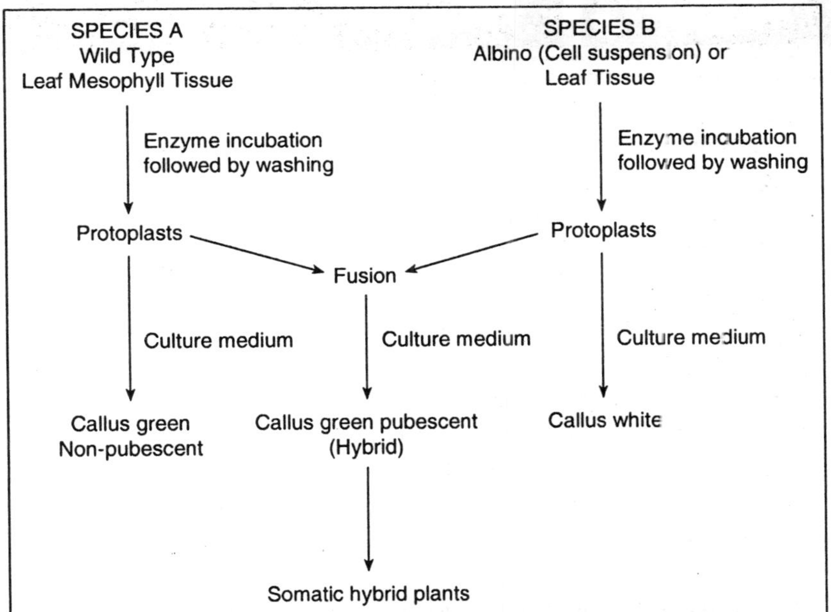

Fig. 20.7A: Selection scheme used in the intergeneric somatic hybridization of *Atropa belladonna* (wild type) with *Datura innoxia* (albino)

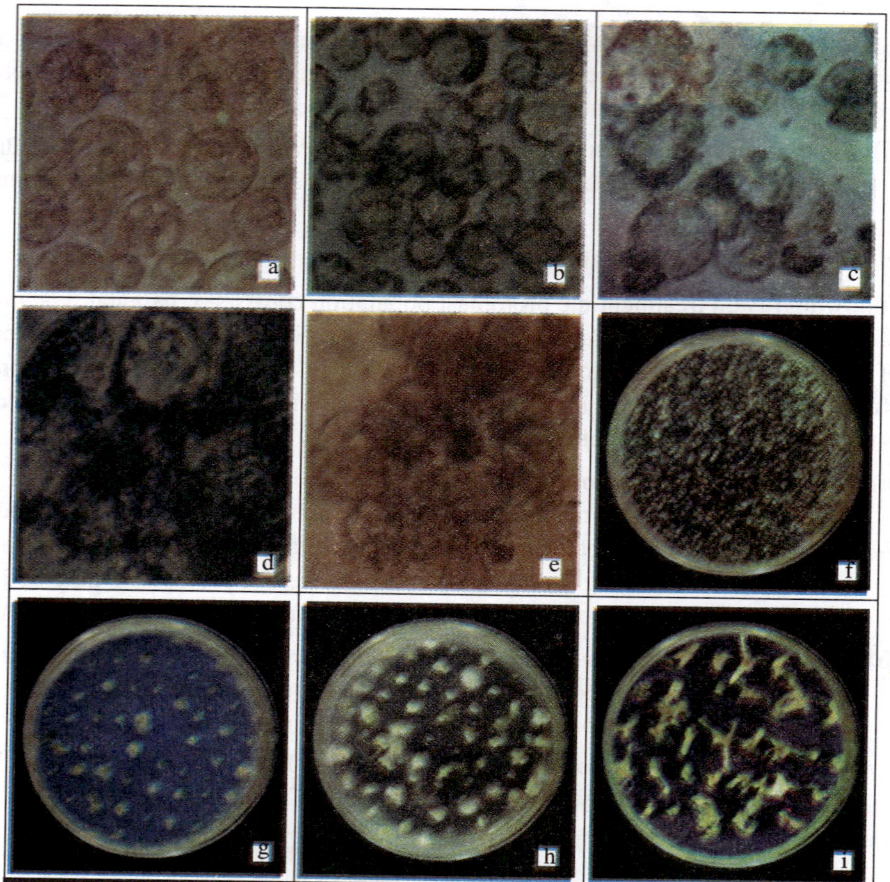

Fig. 20.7B: Plant regeneration from fused protoplast: (a) Callus protoplast, (b) Mesophyll protoplast, (c) Fused protoplast, (d–g) Division to calli, (h–i) Shoot bud regeneration

pounds have been selected as potent markers. When the protoplasts from two lines are being fused together then the fused product can be selected in presence of both the metabolites because of double resistance as compared to single parent.

(ii) **Use of metabolic inhibitors**—The parental protoplasts are treated with irreversible biochemical inhibitor such as iodoacetate or diethylpyrocarbamate and following treatment only hybrid cells will be capable of division.

(iii) **Auxotroph complementation**—Where two parental protoplasts are both mutated for the same enzyme but both of them are of two different mutational types. So the fused and hybrid protoplasts could be grown easily by complementing each other capable of producing active enzyme.

(iv) **Chlorophyll deficiency complementation**—This is the frequently used method for selecting the somatic hybrid where the normal plant protoplasts are allowed to fuse with an albino or mutated or chlorophyll deficient type of protoplast. The fused products or somatic hybrids must be able to produce the green colonies by complementation.

D. Verification and Characterisation of Somatic Hybrid

Somatic hybrids after regeneration should be verified through clear demonstration of genetic contributin of both the parents.

(i) **Morphological characters:** In most of the cases the somatic hybrids bear the characters from the parents or sometimes they have the somatic or sexual characters intermediate of both the parents. Such traits may be leaf size, leaf surface, pigment, flower shape, pollen character, etc.

(ii) **Isoenzyme analysis:** Electrophoretic banding patterns of isoenzymes have been extensively used to verify hybridity. Somatic hybrids may display isozyme banding of certain enzymes specific to both the parents or sometimes newer type of banding. The enzymes like esterase, isoperoxidase, phosphatase, alcohol dehydrogenase, etc. are studied.

(iii) **Chromosomal constitution:** Counting chromsomes in presumed somatic hybrid is a reliable and easy method of detection. Cytologically the chromosome number should be the sum of chromosome number of two parents. Besides the number, the size and structure of the chromosomes of both the parents are accounted for verification of somatic hybrids.

(iv) **Molecular technique:** With the availability of numerous molecular markers such as RFLP, AFLP, RAPD, microsattelites, etc. the somatic hybrid identification has become more easier. PCR technology are being utilised for hybrid identification. Specific restriction patterns of chloroplast and mitochondrial DNA have been used with great advantage to characterise the somatic hybrids.

20.4 CYBRID PRODUCTION

In sexual hybridization process the nuclear fusion takes place between two parents, but cytoplasm is contributed by the maternal parent only. Whereas in case of somatic hybrids due to cell fusion the nucleus as well as cytoplasm derived from both the parents.

But in the fusion programme it is possible to get the somatic hybrids where without nuclear fusion we can get the cytoplasmic hybrid plant only. Here for cytoplasmic hybrid or cybrid production the fusion is done between two protoplasts where one contributes only cytoplasm, the nucleus is inactivated and as a result the cytoplasmic material or genome get mixed or hybridized but the nuclear material is contributed by one parent only (Fig. 20.8).

There are **various approches** to achieve this type of fusion product or cybrids:

(i) The fusion of normal protoplast from one parent with the enucleated protoplasts from the other parent obtained by high speed centrifugation of protoplasts or by irradiation treatment.

(ii) Fusion between a normal protoplast and another protoplast which contains a non-viable nucleus (naturally produced).

(iii) Selective elimination of one of the nuclei from the fusion product just before heterokaryon formation or by selective elimination of chromosome of one parent just before first division.

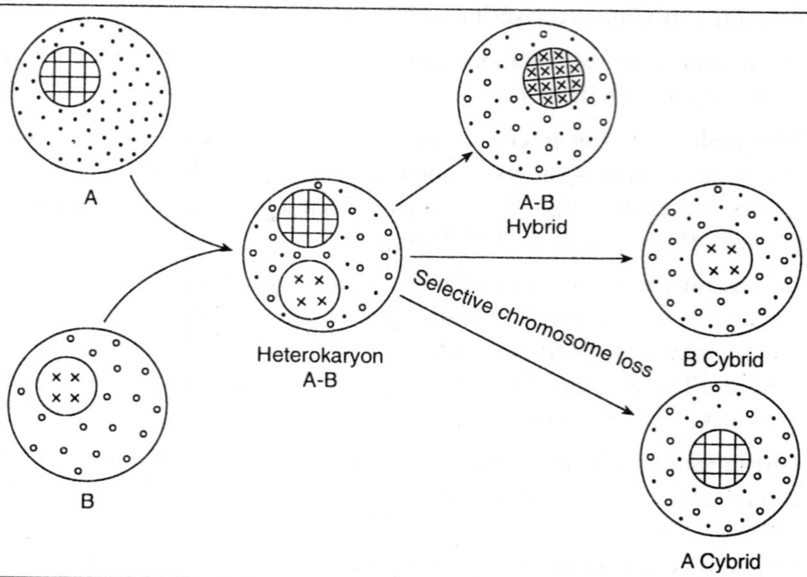

Fig. 20.8: Formation of a cytoplasmic hybrid "cybrid"

Cybridisation helps to achieve the alloplasmic constitution in a single step without following the method of back-crossing for 8–12 generations. Here the alloplasmic situations are created under the control of nucleus of one parent only. Application of cybrids would be directed to transfer the cytoplasmic male sterility character or herbicide resistance property to the recipient parent which are mainly governed by cytoplasmic genome.

Transfer of cytoplasmic male sterility character from 'OGURA' cytoplasm of *Brassica oleracea* to other species of *Brassica* sp. have been done successfully. Resistance to herbicide like atrazine has been transferred from *Brassica campestris* to *B. napus* by cybrid production.

20.5 PRACTICAL APPLICATIONS OF PROTOPLAST CULTURE

1. Production of Somatic Hybrids

To overcome the sexual incompatibility barriers the protoplast fusion programme may be taken to obtain the interspecific and intergeneric hybrids. For example, the fusion between *Lycopersicon esculentum* (tomato) and *Solanum tuberosum* (potato) have been possible to obtain the somatic hybrid '**Pomato**', at intergeneric level. Similarly attempts to overcome the conventional breeding barrier between many species of *Oryza* have been possible through this process. The interspecific somatic hybrids have been obtained through fusion with many wild species of *Oryza* like *O. brachyantha, O. eichingeri, O. officinales,* etc.

Many asymmetric hybrid have also been developed when there is partial fusion, i.e., these somatic hybrids do not contain the total complete set of chromosomes of both the parents, rather there are variation in chromosome number of different somatic hybrids obtained.

2. Somatic Hybrids with Specific Alien Gene

(a) **Gene for disease resistance:** There are many disease resistance genes available among the wild varieties of Solanaceae and Brassicaceae. So many of the protoplast fusion programmes have been done among the wild and cultivated varieties of Brassicaceae and Solanaceae successfully. For example, the disease resistance gene has been successfully transferred to *Lycopersicon esculantum* from a wild relative *Solanum orchranthum*, a woody vine yielding tomato like fruit. Resistance to blackleg disease (*Phoma lingam*) has been found in *Brassica nigra, B. juncea and B. carinata*, the symmetric and asymmetric hybrids between these gene donors and *B. napus*, the resistant somatic hybrids have been developed.

(b) **Abiotic stress resistance:** The fusion programme for transfer of abiotic stress resistance property has been mainly done on families like Fabaceae, Brassicaceae, Poaceae and Solanaceae. The somatic hybrids between *Solanum lycopersicoides* and *Lycopersicon esculentum* showed cold tolerance, the somatic hybrid between *Solanum tuberosum* and *S. acaule* developed disease resistant characters.

(c) **Quality characters:** Somatic hybrids between *Brassica* sp. and *Eruca sativa* i.e. *Erucobrassica* showed low erucic acid content which has been used further to transfer the low erucic acid gene into the cultivar variety by conventional breeding programme. The high nicotine content has been obtained in *Nicotiana tabacum* from *N. rustica* by this fusion programme.

3. Production of CMS Line

Cybrid production programme has become very much useful to transfer the agronomic traits which are encoded by cytoplasmic genome such as cytoplasmic male sterility character or the herbicide resistance property, etc. Cybridisation has been successfully used to transfer the CMS in rice, *Brassica* and *Chicory.* Resistance to herbicides (atrazine) and antibiotics as well as CMS has been successfully introduced in many cultivated species of *Brassica.*

4. Production of Unique Nuclear-Cytoplasmic Combination

Though the protoplast fusion programme results into the mixing of both nuclear and cytoplasm of both the parents initially but during developmental process sometimes some nuclear material may get eliminated as also the cytoplasm of one parent may predominate. In such cases the newer combination of nucleus and cytoplasm may develop some novel and unique germplasm which is not possible through conventional breeding programme.

5. Production of Autotetraploid or Allotetraploid

Somatic hybridisation can be used as an alternative way to obtain the tetraploid without the use of colchicine, i.e., fusion within the same species yield autotetraploid and between different species easily result into allotetraploid.

6. Production of Heterozygous Line

In case of vegetatively reproducing crop the protoplast fusion programme may be an useful way to produce heterozygous line, as there the general sexual breeding can not yield potential result.

7. Production of Fertile Diploid and Polyploid from Sterile Line

When sterile line is obtained through some conventional breeding programme then either by partial fusion, or by cybrid production the fertile diploid or by total fusion the polyploid line can be raised.

8. Production of Hybrid at the Juvenile Phase

Even before reaching the reproductive stage the protoplasts from the parental plants can be used to obtain the somatic hybrid or cybrid.

QUESTIONS

1. What is protoplast culture? Discuss protoplast culture methods.
2. Describe the methods of protoplast isolation and purification.
3. What are somatic hybrids? How is somatic hybridization performed? Mention its significance.
4. State the different methods of protoplast fusion. How do hybrid cells are identified and selected?
5. What are cybrids? How are they produced?
6. Discuss the practical applications of protoplast culture.

21 ANTHER-POLLEN CULTURE AND HAPLOID PLANTS

The significance of haploids in plant breeding and genetic study has been realised for a long time. Artificial production of haploids was attempted through distant hybridisation, delayed pollination, application of irradiated pollen, hormone treatment and temperature shock. In 1964, Guha and Maheswari first reported the haploid embryo and plantlet production from anthers of *Datura*.

Haploid may be grouped under **monoploids** (monohaploids), which possess half the number of chromosomes from a diploid species and **polyhaploids**, which possess half the chromosomes (gametophytic set) from a polyploid species.

One of the methods of haploid production is through androgenesis, where the haploids could be obtained by the culture of excised anther lobes (**anther culture**) from which the dividing microspore may give rise to embryo or by the culture of isolated pollen (**pollen culture**) (Fig. 21.1).

21.1 ANTHER CULTURE

Explants used for anther culture is very critical as the anther lobes bearing the **PMC** (pollen mother cell) of correct divisional stage signify for being divided to form the callus mass or the direct haploid embryo. Young flower buds with immature anther lobes are surface sterilised and the stamens are taken out with the fine forcep, the filaments are removed and then one of them is crushed in acetocarmine stain and checked for proper stage i.e., just released from the tetrad condition, released microspores are uninucleate and densely cytoplasmic preparing for the male gametophytic development. If it is found in the proper stage then the anthers are inoculated on proper media.

Within few days or weeks the anther walls may turn brown, the microspore within it divides to form either **callus** or the **embryos.** Due to the inside pressure of developing microspores, the anther-wall burst open and following the same tissue culture technique

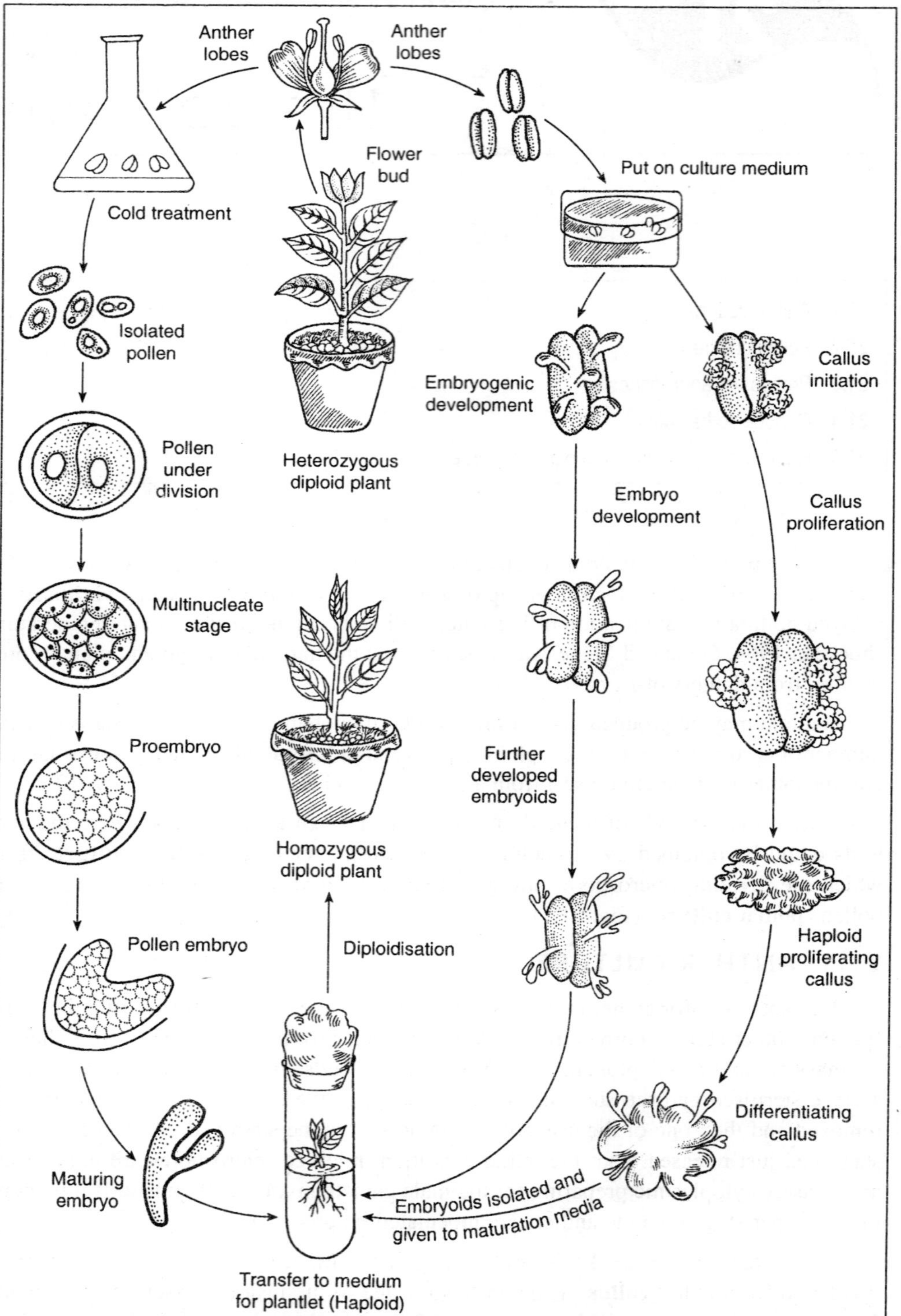

Fig. 21.1: Schematic representation of Anther and Pollen culture

the embryos or the plantlets are subcultured in proper media to get the whole plant in rooted condition to be transferred into soil (Fig. 21.2A–B).

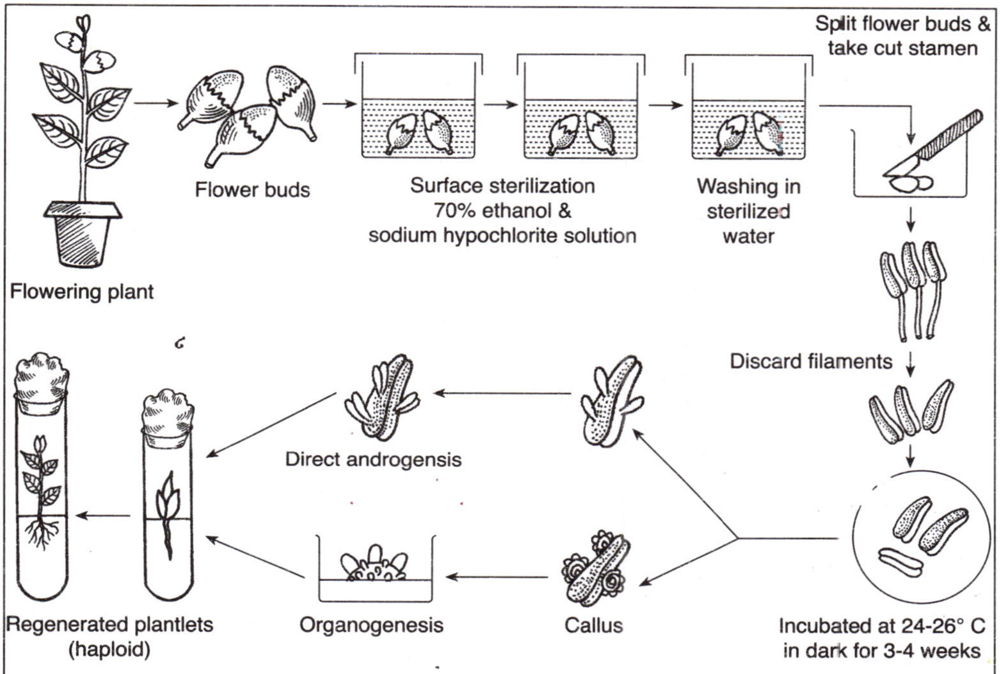

Fig. 21.2A: Anther culture

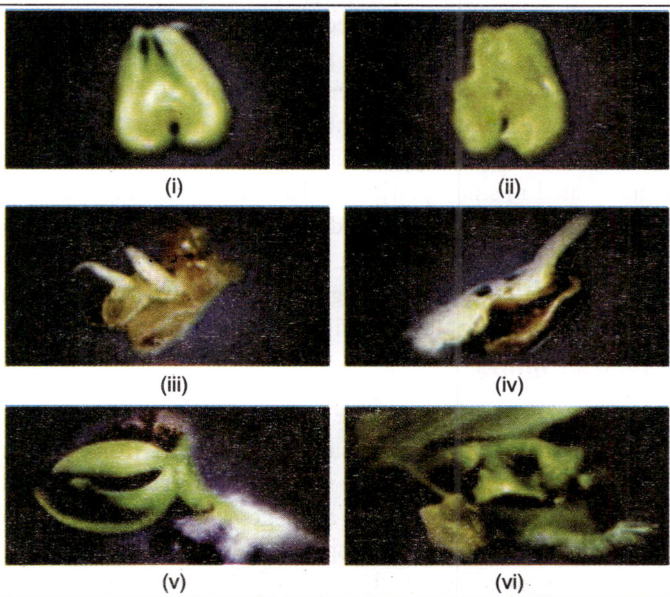

Fig. 21.2B: Haploid plant production through anther culture: (i) Anther put on the medium, (ii) Bursting of anther, (iii) Shoot primordia from callus, (iv) Root formation, (v–vi) Further growth of haploid plantlet

Flowchart for Anther Culture

(i) The young flower buds are collected and washed under running tap water to remove the dirts.

(ii) These are then surface sterilized by immersing in 70% ethanol or sodium hypochlorite soln.

(iii) Then washed in sterile water and transferred into a sterile petridish.

(iv) With the help of sharp scalpel and using forcep the buds are split open and anther lobes are taken out.

(v) One of the anther lobes of each bud is checked by crushing into acetocarmine stain under microscope for the proper stage of microspore development.

(vi) The filament portions are removed from the selected anther lobes.

(vii) The damaged anther lobes should be discarded and intact anther lobes are placed into proper media.

(viii) Incubated at 24°–28°C in dark for 3–8 weeks.

(ix) The haploid embryos or plantlets develop, come out by bursting the anther lobes.

(x) Individually these embryos or plantlets are removed and subcultured on suitable media to develop further and root development.

21.2 POLLEN CULTURE

Microspore or the immature pollen can be used as the explant to get the haploid plants directly. For pollen or microspore culture the flower buds are collected, surface sterilized and the anther lobes are dissected out from the flower buds as before. Then the anther lobes are squeezed with the help of a scalpel within a tube or small beaker to collect the microspore or pollen in nutrient media. Then the anther tissue debris is removed by filtering the suspension through a nylon sieve with a diameter slightly larger than the pollen size (40μ–100μ) allowing the microspore only to pass through it. Then the microspore-suspension washed and concentrated to a plating density. The microspores obtained are then mixed with an appropriate culture medium at a density of 10^3–10^4 microspore/ml, and plated in small petriplate. To ensure good aeration, the layer of liquid in the dish should be as thin as possible, and sealed with 'parafilm' to avoid dehydration. The responsive pollen will divide and form embryos or calli which directly or indirectly will form the haploid plantlet. By following the method of subculturing the whole plant suitable for soil transfer can be obtained (Fig. 21.3A–B).

Advantages of Pollen Culture

The main disadvantage of anther culture is that the plants or embryos or calli not only originate from microspore or pollen but may also originate from various other parts of anther lobes (anther wall, nucellus, tapetum) which are diploid tissues. Furthermore, anther wall or other tissue may affect the way of androgenesis by inhibiting the penetrance of media or stimulating effect of hormones supplied. This difficulty can be avoided by culturing the isolated microspores or pollen. This method has the following advantages:

(i) The explants i.e., microspores or pollens are all haploid cells.

(ii) The sequence of androgenesis can be observed starting from a single cell.

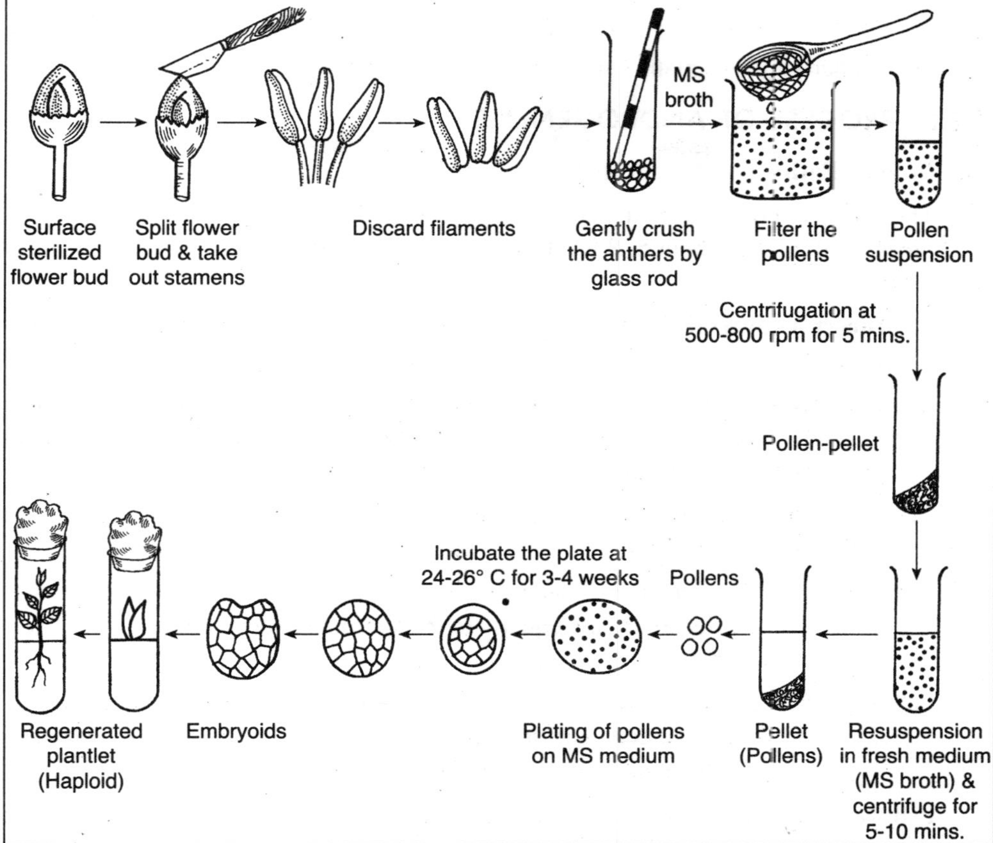

Fig. 21.3A: Pollen culture

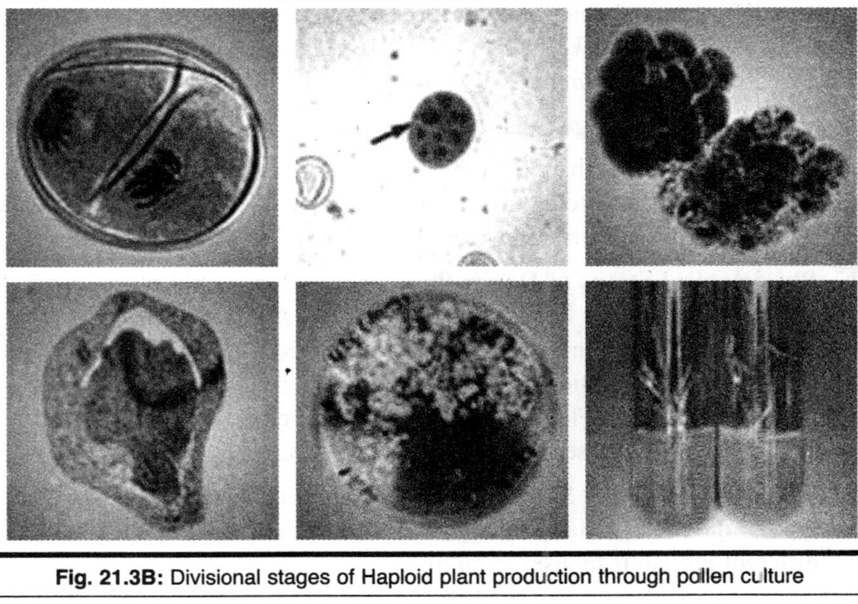

Fig. 21.3B: Divisional stages of Haploid plant production through pollen culture

(iii) The microspores are ideal for uptake, transformation and mutagenic studies, and the microspores are evenly exposed to chemicals and physical mutagens.

(iv) Higher yields of plants/anther could be obtained.

21.3 PROCESS OF ANDROGENESIS

A. *In Vitro* Pathways of Androgenesis

The anther or pollen normally starts undergoing in the pathway of androgenesis within 2 weeks and it takes about 5–8 weeks to obtain complete plantlets. Haploid plants are formed in two ways:

(i) **Direct Androgenesis:** Embryos are directly formed from the pollen/microspore without callus.

(ii) **Indirect Androgenesis:** Microspores undergoes division in unorganised fashion to give rise to callus and by embryogenic or organogenic induction the haploid plantlets may be obtained.

There are **four different pathways** to form the multicellular condition of pollen from the unicellular pollen (Fig. 21.4).

Pathway I: The microspore divides by an equal division and two identical daughter cells contribute equally to the sporophyte development e.g., *Datura innoxia*.

Pathway II: The uninucleate microspores divide unequally forming vegetative and generative cell. The sporophyte arises through divisions of vegetative cell, the generative cell gets degenerated e.g., *Nicotiana tabacum*.

Pathway III: The uninucleate microspore undergoes a normal unequal division but the pollen embryos are formed from the generative cell alone. The vegetative cell does not divide, e.g., *Hyocyamus niger*.

Pathway IV: The division of microspore is assymmetrical as in pathway II, but both the cells take part in embryo formation and sporophyte development e.g., *Datura metel*.

B. Factor Affecting Androgenesis

(a) **Stage of Pollen:** Anthers with microspores ranging from tetrad to the binucleate stage are responsive to culture. As soon as the starch deposition starts within the microspore there is no further development towards sporophyte.

(b) **Physiological Status of Donor Plant:** The anther should be collected from the flower buds of adult healthy plant and it is very important to use the healthy explant. The variation in response of anthers from plants grown under different environmental conditions may be due to the differences in endogenous level of growth regulators. Flowers from relatively younger plants, flowering at the beginning of flower-season are more responsive.

(c) **Genotype of Plants:** Success of anther culture is highly dependent on the genotype of the plant. It has been observed that various species and cultivars exhibit different growth responses in culture.

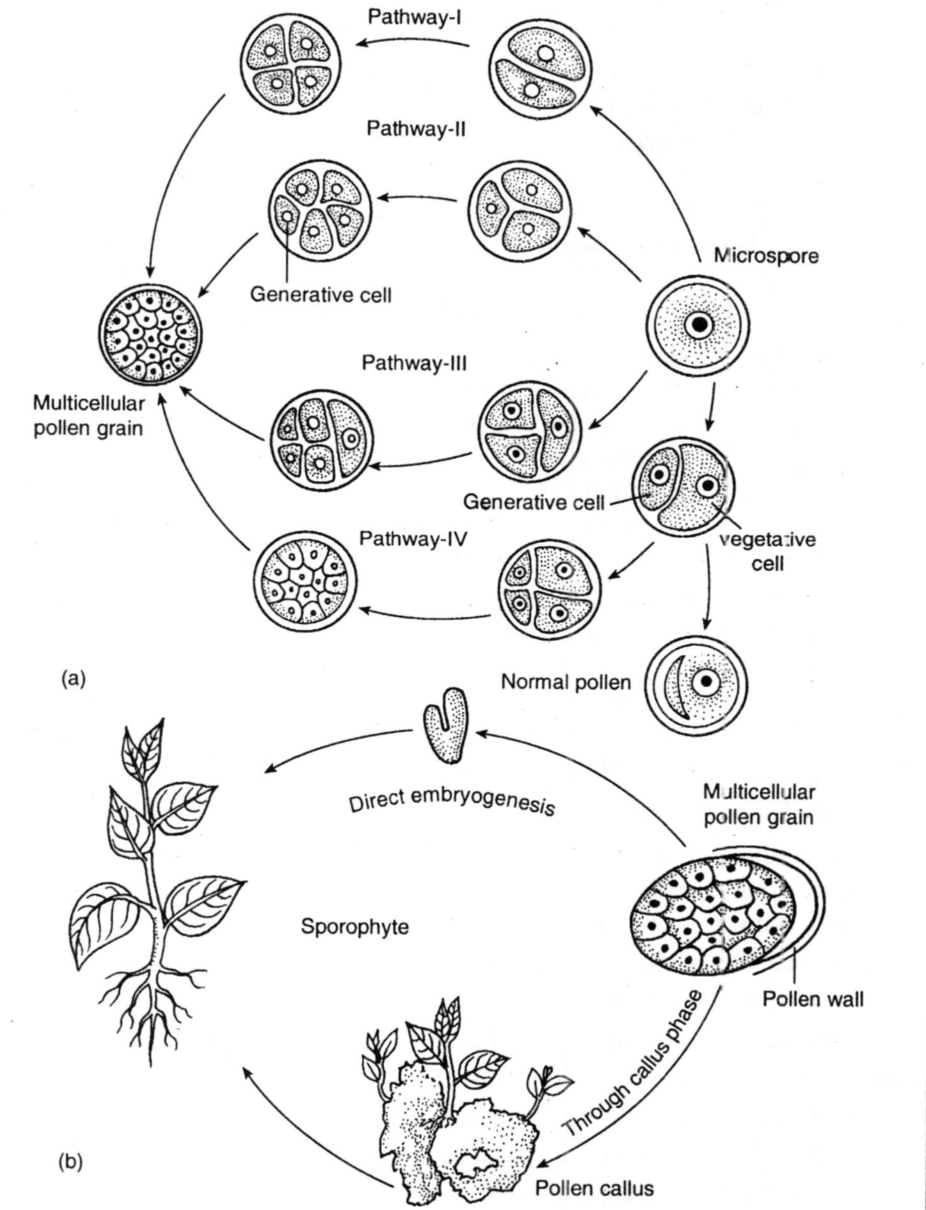

Fig. 21.4: (a) Pathways of microspore divisions leading to the formation of a multicellular pollen grain; (b) Multicellular pollen may directly form an embryo or produce sporophytes through a callus phase (after Bhojwani and Bhatnagar, 1990)

(d) **Pretreatment of Anthers:** The pathways towards androgenesis requires to stop the development of pollen cell towards gamete formation and to force to develop the multicellular condition. This induction can be given by different types of pretreatments.

(i) **Cold shock:** In general cold shock or cold treatment between 3–6°C for 3–15 days may be applied. Cold treatment helps the weak and non-viable anthers/ microspore to be killed, as a result the good material get more enrichment.

(ii) **Hot treatment:** In some species the hot water treatment (30–40°C for 24 hr–1 hr) helps more embryo development from the pollen. The temperature shock helps in dissolution of microtubules and causes abnormal division of microspores.

(iii) **Chemical treatment:** Chemicals like 2-chloroethyl phosphonic acid has a pronounced effect in increasing the haploid production in various species. It helps to develop the multinucleate condition with fewer starch grains.

(e) **Culture Media:** Composition of medium is one of the most important factors determining the success of anther culture. But it is difficult to draw a conclusion about the suitable media composition for different genes/species or even genotypes. Higher concentration of sucrose is essential for osmoticum and also androgenesis, chelated iron has been shown to induce embryogenesis. Minerals and growth regulators play important roles on embryogenesis but it totally depends on the endogenous level of hormones.

(f) **Culture Conditions:** The physical environmental conditions in which the cultures to be placed can enhance the differentiation. The cultures are incubated generally at 24–28°C. In the initial stages of induction of morphogenesis, darkness is normally more effective or cultures should be kept at low light intensity (500 lux). After formation of macroscopic structures, these can be transferred to a regeneration medium and kept at 14 hr. day light condition at 2000–4000 lux.

C. Diploidisation of Haploids

The ploidy level of plants derived from anther or microspore culture is highly variable. This variation may be due to endomitosis or fusion of various nuclei during the developmental stages of anthers at the time of excision and culture.

For obtaining homozygous diploid lines, the plants derived through anther culture must be analysed for their ploidy level and then the following methods can be applied:

Colchicine Treatment: As colchicine is a spindle inhibitor it is used to induce chromosome duplication in various ways:

(a) The regenerated plantlets can be treated with 0.5% colchicine solution.

(b) Repeated colchicine treatment to apical or axillary bud by cotton soaked in colchicine or by applying colchicine-lanolin paste.

(c) In case of cereals, the tillers can be put into colchicine solution cutting the crown and then again put back to soil.

(d) Anthers can be plated directly in colchicine supplemented media for a period and just after first division again the explants should be placed on colchicine-free media.

Endomitosis: Haploids are in general unstable in culture and have a tendency to undergo endomitosis to form diploid cells. This property can be exploited in some cases to obtain the homozygous plant.

21.4 GYNOGENIC HAPLOIDS BY OVULE CULTURE

Recent advances in plant tissue culture have developed the successful plant production from **ovary** or **ovule culture.** Here the megaspore or the female gametophyte of angiosperm is triggered to develop the *in vitro* haploid sporophyte. *In vitro* culture of unpollinated ovaries and ovules represents an alternative for the production of haploid and are called as **gynogenic haploids.** The first successful report of gynogenic haploid was made by San Noeum (1976) in Barley. Subsequently the gynogenic haploids have been obtained in various other crop plants like rice, wheat, maize, sunflower, sugar-beat, tobacco, etc.

21.5 HAPLOID PLANTS AND THEIR USE IN PLANT BREEDING

A. **Development of Pure Homozygous Line:** In plant breeding, it is very much essential to get the pure homozygous line which is generally obtained through selfing for 6–7 generations. But by the use of anther/pollen culture it can be reduced to few months or a year. These genetically pure homozygous lines are used for breeding as well as genetic research purpose. This technique is also helpful for breeding of these plants which have more elongated juvenile phase.

B. **Use in Hybrid Development:** As a result of complete homozygosity obtained from diploidisation, the traits are easily understood and can be fixed for further breeding programme for hybrid production.

C. **Significance in Early Release of Varieties:** As time taken for getting the homozygous line is less, so ultimately many new hybrid varieties can be obtained in less time applying this technique. Also the anther culture can be applied to the F_1 hyrbid plants which may not be stable in many cases but anther/pollen culture can help to get dihaploid plants with less effort which may be of great use.

D. **Hybrid Sorting in Haploid Breeding:** Haploid breeding involves the hybrid sorting and it is considered superior to pedigree and bulk method as the superior haploid plants are selected and hybridized. Haploids derived from anther culture in less time and with less effort give the advantage of selection of different superior gametes to be used for haploid breeding (Fig. 21.5).

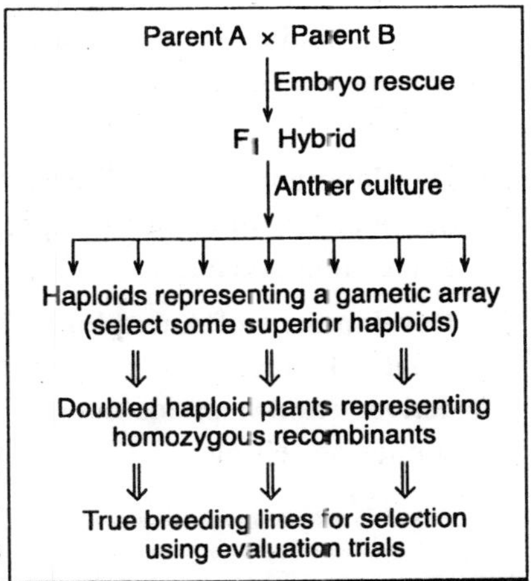

Fig. 21.5: Different steps in hybrid sorting method using anther culture technique

E. **Induction of Mutagenesis:** Haploid cell cultures are useful material for induction of mutations and to study the effect of mutation. This method can overcome the masking effect of presence of dominant gene. The screening method for detection of mutational effect is also easier in this technique.

F. **Induction of Genetic Variability:** The pollen/microspore are easy explant for production of genetically variable types by introducing the different foreign genes through different transformation procedure. These transformed or transgenic haploids can be used further in breeding programme.

Table 21.1: Some achievements of crop plants through hybrid sorting using anther culture

Crop	Varieties developed	Attributes
Rice	Tanfong-1, Huayn-2, Hua-03, Hua Han Zao, Xin Xion, Nanhua-II	Good quality, disease resistant, high yield
Wheat	Huapei-1, Jinghua-1, Lunghua-1, Yunhua-1	Rust resistant, cold resistant, high yield, large spikes, more tillers, resistant to rust, powdery mildew resistant
Tobacco	Tanyn 1, 2, 3	Disease resistant and mild smoke quality

G. **Generation of Exclusively Male Plants:** By haploid production of some dioecious male plant, the dihaploid plant containing both Y chromosomes can produce supermale. In case of *Asparagus*, male plants give more yield than female plants, so if haploids are produced from anthers of male (XY) plants and by chromosome doubling the Y chromosome containing plant will produce super male plant (YY) which can be subsequently vegetatively propagated.

H. **Development of Disease Resistant/Insect Resistant Varieties:** Haploid production can be used for the introduction of disease resistance genes into cultivars. An established cultivar is crossed with a donor disease resistant species and either F_1 or F_2 anthers are cultured. Haploids are developed and screened for resistance and then diploidized.

I. **Development of Salt Tolerant Varieties:** Salt tolerant breeding lines have been developed in different crop species which have been further integrated in conventional breeding procedure.

J. **Development of Aneuploids:** Haploids have been used in the production of aneuploids like monosomics in wheat, trisomics in potato. In tobacco nullisomics were derived from haploids obtained from monosomics which could not produce nullisomics on selfing.

QUESTIONS

1. Define haploid culture. State different types of haploid culture methods.
2. Describe the technique of anther and pollen culture to raise the haploid plants.
3. Distinguish between anther culture and pollen culture. Mention the advantages of pollen culture over anther culture.
4. What is androgenesis? Mention the *in vitro* pathways of androgenesis. Write the factors affecting androgenesis.
5. What are haploid plants? Discuss the use of haploid plants in plant breeding.
6. Write short notes on: (a) Diploidisation of haploids, (b) Gynogenic haploids (Ovule culture), (c) Hybrid sorting in haploid breeding.

22 RECOMBINANT DNA TECHNOLOGY AND GENETIC ENGINEERING

Biotechnology has come to the forefront of scientific disciplines simply because of tremendous development of techniques for genetic engineering. The genes which are responsible for all characters can now be chemically resolved, isolated, synthesized and subjected to manipulation forming a major aspect of biotechnology otherwise termed *Recombinant DNA Technology*.

Genetic engineering involves manipulation of the genetic material towards a desired and in a directed and predetermined way. This method aims at isolating DNA fragments and recombining them; i.e., two DNA molecules are isolated and cut into fragments by one or more specialized enzymes and then the fragments are joined together in a desired combination and restored to a cell for replication and reproduction. When the recipient organism is a microbe, then as the microbe multiplies, it is possible to obtain millions of copies of a specific region of DNA by allowing the cell to multiply.

22.1 PURPOSE AND BASIC STEPS

Purpose: The interest in genetic engineering principally is due to its varied applications (Fig. 22.1):

1. Production of varieties of plants having particular desirable characteristics (e.g., resistance or tolerance to disease, drought, development of CMS line, etc.).

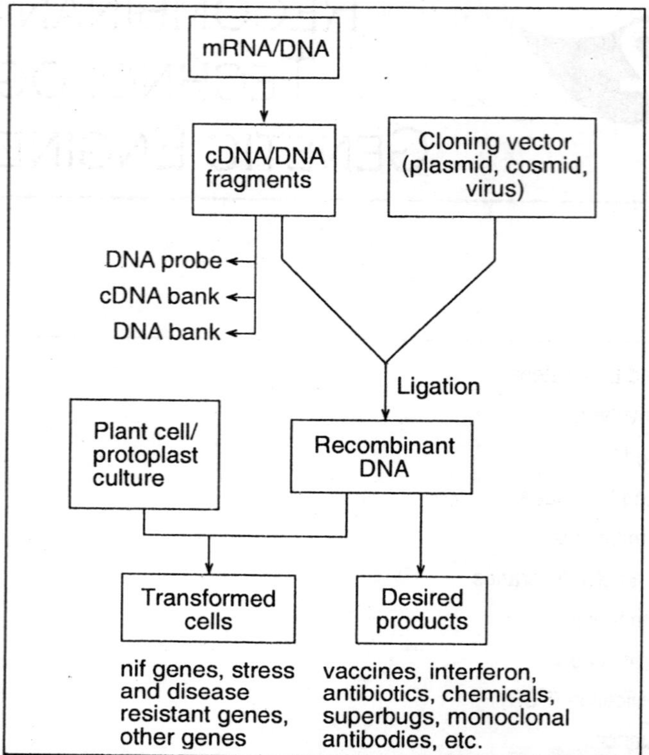

Fig. 22.1: Area of application of genetic engineering techniques

2. Improvement in the production of biochemicals and commercially important organic chemicals.
3. Correction of genetic disorder in higher organisms.
4. The sequencing of gene, the prerequisite for mapping the genome as well as utilizing the gene for horizontal transfer involved in raising transgenic organism.

Basic Steps: In principle the recombinant DNA technology involves certain basic steps, such as (Fig. 22.2):

1. Isolation of the gene (*target DNA*) to be cloned.
2. Insertion of the gene into another piece of DNA called a *vector* which allows it to be taken up by the recipient cell and replicated.
3. Transfer of the *recombinant vectors* into cells of recipient organisms, either by transformation or by infection using viruses.
4. Selection of those cells which contain the desired recombinant vectors.
5. Growth of the transformed organism.
6. Expression of the gene to obtain the desired product.

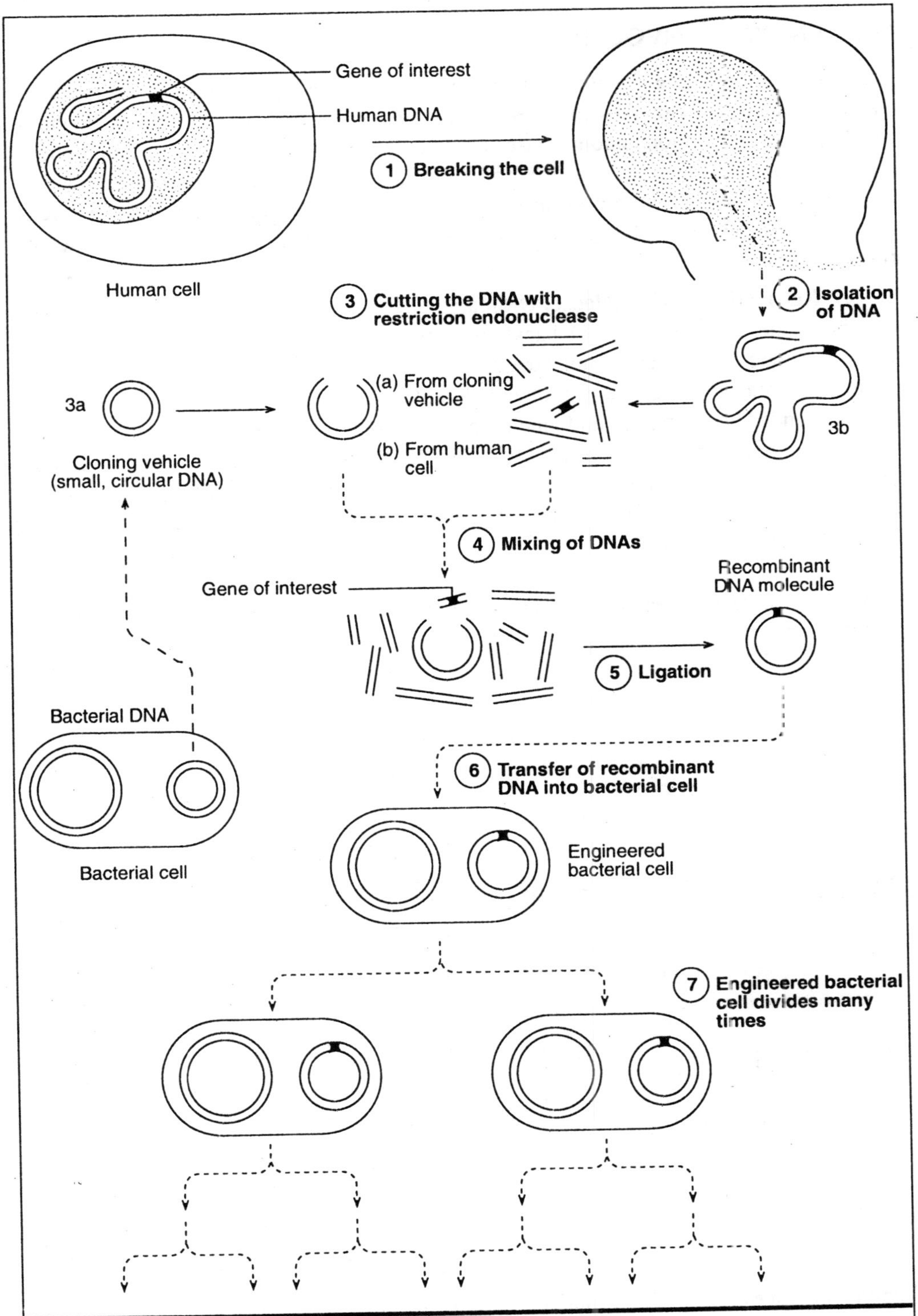

Fig. 22.2: Basic methods of genetic engineering and gene cloning

22.2 ENZYMES INVOLVED

Genetic engineering became possible with the discovery of mainly two types of enzymes: the cutting enzymes called *restriction endonucleases* and the joining enzymes called *ligases*. Restriction endonucleases or restriction enzymes, as they are called popularly, recognize unique base sequence motifs in a DNA strand and cleave the backbone of the molecule at a place within or, at some distance from the recognition site. Whereas ligase is the enzyme that joins a 5′ end of a DNA with a 3′ end of the same or of another strand.

(a) **Restriction Endonuclease:** Ordinary nucleases are endonucleases or exonucleases. The former cleaves the DNA backbone between two nucleotides, i.e., it cleaves the double stranded DNA at any point except the ends, but it involves only one strand of the duplex. The latter remove or digest one nucleotide at a time starting from 5′ or 3′ end of a DNA strand. The restriction endonucleases cleave only at specific regions in a particular DNA, so that discrete and defined fragments are obtained at the end of total digestion. The name 'restriction' endonuclease originated from an observation of a system of restriction of the growth of the phage lambda in particular strains of the *E. coli* host cell (Meselson and Yuan, 1968).

Table 22.1: Source of restriction enzymes, cleavage sites and productions of cleavage

Microorganisms	*Restriction enzymes*	*Cleavage sites*	*Cleavage products*	
Bacillus amyloliquefaciens H	Bam HI	5-GGATCC-3 3-CCTAGG-5	5-G 3-CCTAG	GATCC-3 G-5
B. globigii	Bgl II	5-AGATCT-3 3-TCTAGA-5	5-A 3-TCTAG	GATCT-3 A-5
Escherichia coli RY13	Eco RI	5-GAATTC-3 3-CTTAAG-5	5-G 3-CTTAA	AATTC-3 G-5
Haemophilus influenzae Rd	Hin dIII	5-AAGCTT-3 3-TTCGAA-5	5-A 3-TTCGA	AGCTT-3 A-5
H. parainfluenzae	Hpa I	5-GTTAAC-3 3-CAATTG-5	5-GTT 3-CAA	AAC-3 TTG-5
Klebsiella pneumoniae OK 8	Kpn I	5-GGTACC-3 3-CCATGG-5	5-GGTAC 3-C	C-3 CATGG-5
Streptomyces albus G	Sal I	5-GTCGAC-3 3-CAGCTG-5	5-G 3-CAGCT	TCGAC-3 G-5
Serratia marcescens	Sma I	5-CCCGGG-3 3-GGGCCC-5	5-CCC 3-GGG	GGG-3 CCC-5

Most restriction enzymes recognize only one short base sequence in a DNA molecule and make two single strand breaks, one in each strand, generating 3′OH and 5′P groups at each position. The sequences recognized by restriction enzymes are often *palindromes*, i.e., inverted repetition sequences which are symmetrical. Restriction enzymes can cut DNA in two ways to generate *blunt ends* (cut precisely at opposite sites, e.g., HpaI) and *staggard ends* (cut at asymmetrical position, e.g., Eco RI) with short single stranded over-hangs at each end. A large number of restriction enzymes have been identified and classified into three categories (type I, II, III) on the basis of their site of cleavage.

Restriction enzymes have three important *features:*

1. Restriction enzymes make breaks in palindromic sequences.
2. The breaks are usually not directly opposite to one another.
3. The enzymes generate DNA fragments with complementary ends.

The commonly employed restriction enzymes are listed in Table 22.1.

(b) **DNA Ligase:** Ends of DNA strands may be joined by the enzyme polynucleotide ligase, called '*glue*' of the recombinant DNA molecule. The enzyme catalyses the formation of a phosphodiester bond between the 3′OH and 5′P terminals of two nucleotides. The enzyme is thus able to join unrelated DNA, repair nicks in single strand of DNA and join the sugar phosphate backbones of the newly repaired and resident region of a DNA strand (Mertz and Davis, 1972).

The enzyme which is extensively used for covalently joining restriction fragments is the ligase from *E. coli* and that encoded by *T4 phage*. As the main source of DNA ligase is T4 phage, hence, the enzyme is known as T4 DNA ligase.

The ligation reaction is controlled by several factors, such as pH, temperature, concentration and kinds of sticky ends, etc. As ligase uses the ends of DNA molecules as substrates rather than the entire DNA, the kinetics of joining depend on the number of ends (concentration) available for joining.

(c) **Alkaline Phosphatase:** The broken fragments of plasmids, instead of joining with foreign DNA, join the cohesive end of the same DNA molecules. The treatment with alkaline phosphatase prevents recircularisation of plasmid vector and increases the frequency of production of recombinant DNA molecule.

(d) **DNA Polymerase and the Klenow Fragment:** The DNA polymerase that is generally utilized is either the DNA Pol I from *E. coli* or the T4 DNA polymerase encoded by the phage gene. The *E. coli* enzyme is basically a proof-reading and repairing enzyme. It is composed of 3 subunits each with a specific activity. They are: 5′–3′ polymerase, 3′–5′ exonuclease and 5′–3′ exonuclease. The enzyme is useful for synthesizing short length of a DNA strand, especially by the nick translation method. The 5′–3′ exonuclease activity may be deleted, this edited enzyme is referred to as the klenow fragment. The T4 DNA Pol possesses, like the klenow fragment, only the polymerase and proofreading (3′–5′ exonuclease) functions.

(e) **Reverse Transcriptase:** Retroviruses (possessing RNA) contain RNA dependent DNA polymerase which is called *reverse transcriptase* (Mizutani, Temin, Baltimore, 1970). This produces single stranded DNA, which in turn functions as template for complementary long chain of DNA. This enzyme is used to synthesize the copy DNA or complementary DNA (cDNA) by using mRNA as a template. The enzyme is very useful for the synthesis of cDNA and construction of cDNA clone bank and to make short labelled probes.

22.3 VECTORS USED

Cloning Vector: By cloning, one can produce unlimited amounts of any particular fragment of DNA. In principle, the DNA isolated and cut pieces are introduced into a suitable host cell, usually a bacterium such as *Escherichia coli*, where it is replicated, as the cell grows and divides. However, replication will only occur if the DNA contains a sequence which is recognized by the cell as an *origin of replication*. Since such sequences are infrequent, this will rarely be so, and therefore, the DNA to be cloned, has to be attached to a carrier, or *vector DNA* which does contain an origin of replication.

Criteria of an Ideal Vector

Vectors are those DNA molecules that can carry a foreign DNA fragment when inserted into it. A vector must possess certain minimum qualifications to be an efficient agent for the transfer, maintenance and amplification of the passenger DNA.

1. The vector should be small and easy to isolate.
2. They must have one or more origins of replication so that they will stably maintain themselves within host cell.
3. Vector should have one or more unique restriction sites into which the recombinant DNA can be inserted.
4. They should have a selectable marker (antibiotic resistance gene) which allows recognition of transformants.
5. Vector DNA can be introduced into a cell.
6. The vector should not be toxic to host cell.

Types of Vector

Based on the nature and sources, the vectors are grouped into bacterial plasmids, bacteriophages, cosmids and phagemids (Fig. 22.3).

(a) **Plasmid:** Plasmids are the extrachromosomal, self replicating, and double stranded closed and circular DNA molecules present in the bacterial cell. A number of properties are specified by plasmids such as antibiotic and heavy metal resistance, nitrogen fixation, pollutant degradation, bacteriocin and toxin production, colicin factors, etc.

Plasmids have following *advantages* as cloning vehicle (Cohen et al., 1973):

1. It can be readily isolated from the cells.
2. It possesses a single restriction site for one or more restriction enzymes.
3. Insertion of foreign DNA does not alter the replication properties.
4. It can be reintroduced into cell.
5. Selective marker is present.
6. Transformants can be selected easily by using selective medium.
7. Multiple copy numbers are present in a cell.

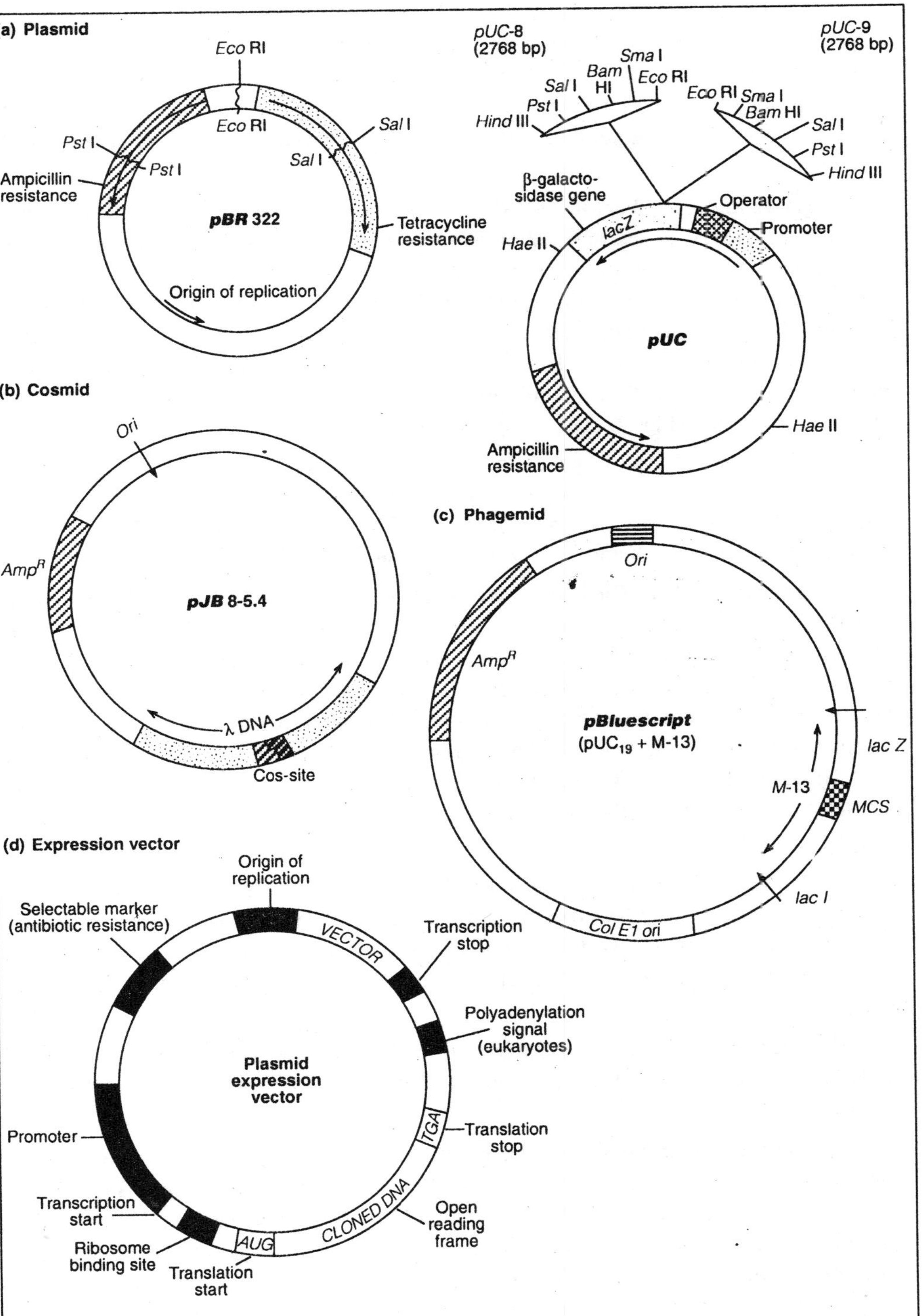

Fig. 22.3: Vectors (a–plasmid, b–cosmid, c–phagemid, d–expression vector)

Some plasmid vectors are pBR 322, pBR 327, pUC vectors, yeast plasmid vector and Ti, Ri plasmids. Ti and Ri Plasmids are widely used in plant system for genetic transformation. Among higher plants, Ti plasmid of *Agrobacterium tumefaciens* or Ri plasmid of *A. rhizogenes* are the best known vectors. T-DNA, from Ti or Ri plasmid of *Agrobacterium*, is considered to be very potential for foreign gene transfer in cloning experiments with higher plants.

pBR 322 and pUC Vectors: pBR322 is a derived plasmid from a naturally occurring plasmid col E1, composed of 4362 bp DNA and its replication may be more faster. The plasmid has a point of origin of replication (*ori*), two selectable marker genes conferring resistance to antibiotics, e.g., ampicillin (amp^r), tetracycline (tet^r) and unique recognition sites for 20 restriction endonucleases. Tetracycline resistance gene has a cloning site and insertion of foreign segment of DNA will inactivate the tet^r gene. The recombinant plasmid will allow the cells to grow only in presence of ampicillin but will not protect them against tetracycline (Fig. 22.3a).

Another plasmid used in gene cloning is pUC vector available in pairs with reverse orders of restriction sites relative to lac^Z promoter. This is a synthesized plasmid possessing ampicillin resistance gene (amp^r), origin of replication from pBR322(*ori*) and lac^Z gene from *E. coli.* pUC 8 and pUC 9 make one such pair (Fig. 22.3a).

(b) **Bacteriophage:** The bacteriophage has linear DNA molecule, a single break will generate two fragments, foreign DNA can be inserted to generate chimeric phage particle. But as the capacity of phage head is limited, some segments of phage DNA, not having essential genes, may be removed. This technique has been followed in λ (Lambda) phage vectors to clone large foreign particle. Plasmid can clone up to 20 to 25 kb long fragments of eukaryotic genome. The examples of different Lambda phage vectors are λ gt 10, λ gt 11, EMBL 3, etc. M-13 is a filamentous bacteriophage of *E. coli* whose single stranded circular DNA has been modified variously to give rise M-13 series of cloning vectors.

(c) **Cosmid:** Cosmids are plasmid particles, into which certain specific DNA sequences, namely those for *cos* sites are inserted which enable the DNA to get packed in λ particle. Like plasmids, the cosmids perpetuate in bacteria without lytic development. The cosmids have high efficiency to produce a complete genomic library (Fig. 22.3b).

(d) **Phagemid:** These are prepared artificially by combining features of phages with plasmids. One commonly used phagemid is pBluescript IIKs derived from pUC-19 (Fig. 22.3c).

(e) **Plant and Animal Viruses:** A number of plant and animal viruses have also been used as vectors both for introducing foreign genes into cells and for gene amplification. Cauliflower Mosaic Viruses (CaMV), Tobacco Mosaic Viruses (TMV) and Gemini Virus are three groups of viruses that have been used as vectors for cloning of DNA segments in plant system. SV 40 (Simian Virus 40), human adenoviruses and retroviruses are potential as vectors for gene transfer into animal cells.

(f) **Artificial Chromosomes:** Yeast Artificial Chromosome (YAC) or Bacterial Artificial Chromosome (BAC) vectors allow cloning of several hundreds kb pairs which may represent the whole chromosome. It can be cloned in yeast or bacteria by ligating them to vector sequences that allow their propagation as linear artificial chromosome.

(g) **Transposons:** Transposable elements like Ac-Ds or Mu-1 of Maize, P-element of *Drosophila* may also be used for cloning vector and transfer of gene among eukaryotes.

Expression Vector: A vector that has been constructed in such a way that inserted DNA molecule is put under appropriate promoter and terminator sequences for high level expression through efficient transcription and translation. Example: Use of promoters ('nos' from T-DNA) or expression cassettes (pRT plasmids) (Fig. 22.3d).

Shuttle Vector: There are plasmids capable of propagating and transferring genes between two organisms (e.g., *E. coli* and *A. tumefaciens*). It has unique origins of replication for each cell type as well as different selectable markers. It can, therefore, be used to shuttle gene from prokaryotes to eukaryotes. Example: pBin19.

22.4 GENE CLONING TECHNIQUE

Gene cloning is the technique which involves the insertion of a specific fragment of foreign DNA into a cell through suitable vector, in such a way that inserted DNA replicates independently and transferred to progenies as a result of cell division.

The technique involves a collection of fragments obtained by digestion with a restriction enzyme and can be made to anneal with a cleaved vector molecule, yielding a number of hybrid vectors containing different fragments of foreign DNA. A particular DNA segment of interest must be isolated from the whole set of all vectors possessing

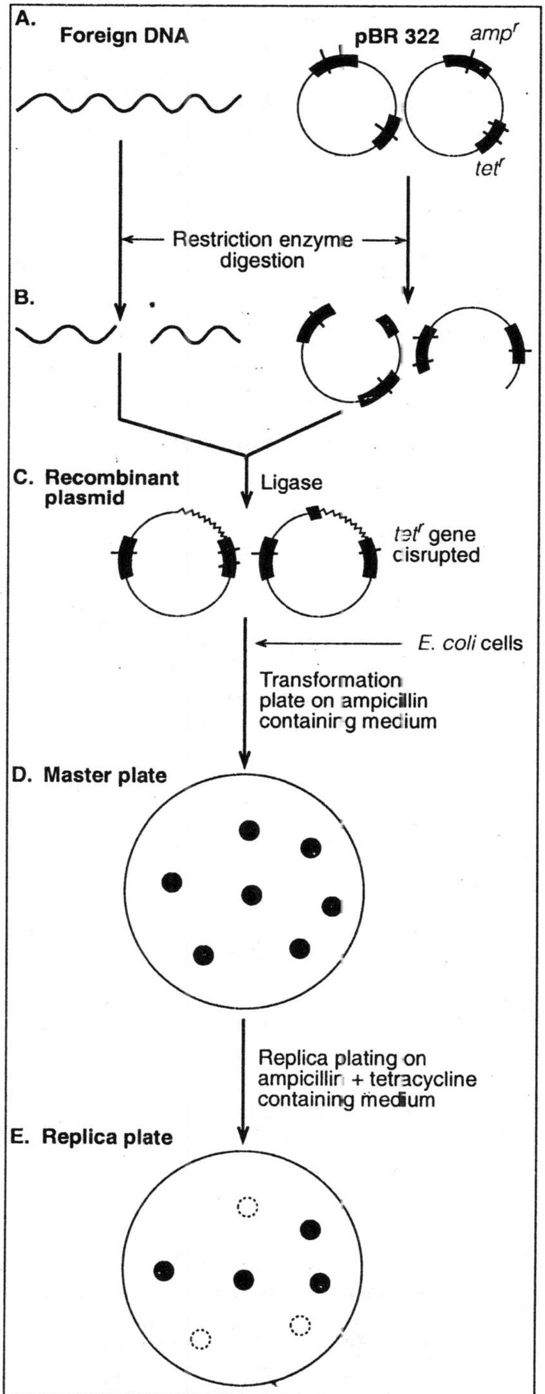

Fig. 22.4: Diagrammatic presentation of gene cloning

foreign DNA through gel electrophoresis. Now the recombinant DNA molecule with vector is transferred to a bacterium through transformation technique and the transformed cell line of bacteria is then perpetuated and selected for the particular clone (Fig. 22.4).

I. **Isolation of DNA to be Cloned:** The DNA of interest, i.e., target DNA may be genomic DNA or complementary DNA or synthetic DNA. The genomic DNA of interest if contained in a particular restriction fragment, that can be isolated from gel after electrophoresis. Otherwise, a complementary DNA (cDNA) fragment is prepared directly by using mRNA as template. The polyadenylated mRNAs are separated from other types of RNAs through affinity column chromatography. These mRNAs are then copied to cDNAs with the help of reverse transcriptase. In these cases as the cDNA is obtained from mRNA, so it must contain the uninterrupted coding sequence of gene and the recombinant DNA molecule will synthesize the eukaryotic gene product in prokaryotic cell. One can also synthesize the desired DNA fragment by machine.

II. **Insertion of Foreign DNA Fragment into a Vector:** The cDNA thus isolated above or obtained from gene bank is fragmented by using the specific restriction enzyme to develop specific cohesive ends. The cloning vector is also treated with the same restriction enzyme, so that the cohesive ends are generated (Fig. 22.5). For insertion of double stranded cDNA into a cloning vector, it is necessary to add to both termini single stranded DNA sequence which should be complementary to a tract of DNA at the termini of linearized vector. In order to get efficient formation of recombinant DNA molecules, addition of sticky ends on both termini is necessary.

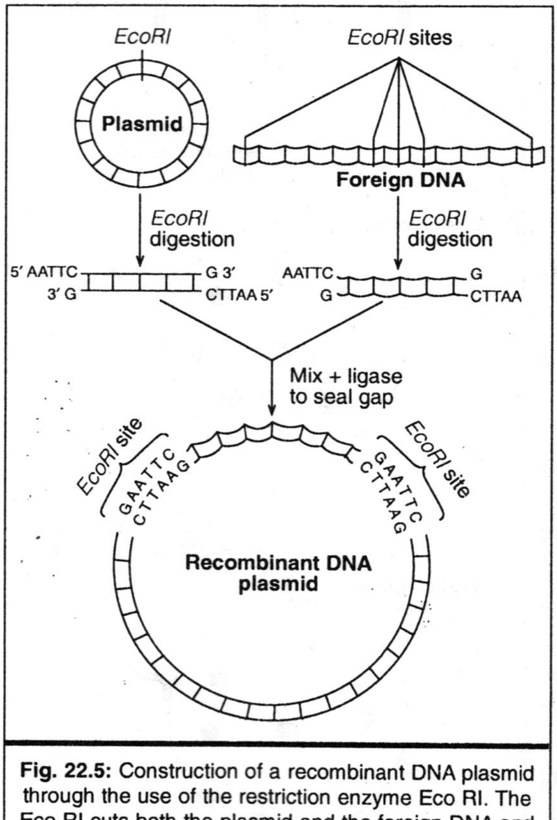

Fig. 22.5: Construction of a recombinant DNA plasmid through the use of the restriction enzyme Eco RI. The Eco RI cuts both the plasmid and the foreign DNA and by mixing, we get recombinant DNA plasmid (after Russel, 1987)

There are two methods for generation of cohesive ends on the double stranded cDNA: (i) use of linkers (ii) homopolymer tails. Linkers are the chemically synthesized double stranded DNA oligonucleotides containing on it one or more restriction sites for cleavage by restriction enzymes (Fig. 22.6). Linkers are ligated to blunt end by T4-DNA ligase. Using terminal transferase the synthesis of homopolymer tails of the defined length at both 3′ termini of double stranded DNA and vector is possible. If the polyT tail is added at the termini of foreign DNA, then polyA tail is added at the restriction site of the vector, so that the complementary sticky ends are formed and they get annealed by T4-DNA ligase (Fig. 22.7).

IIIA. **Transfer of Recombinant DNA into Bacterial Cell:** Before the recombinant DNA can be bulked up by cloning, it must be taken up by a suitable bacterial host cell, which is then said to be transformed, i.e., a host bacterial cell must accept the plasmid with the foreign gene, get it incorporated into its genome and start transcribing that gene. The event of entering the plasmid with foreign DNA into the cell is known as transformation. A mild heat shock is given to the mixture which results in the uptake at higher frequency of the DNA. The selection of transformed cells is done by allowing the bacteria to grow in antibiotic selection medium.

IIIB. **Cloning in Eukaryotes:** In eukaryotes the nucleus is separated from the rest of cell through nuclear membrane, many of the genes are split genes with exons and introns. As such genetic engineering with eukaryotes needs special methods. When eukaryotic genes are cloned in prokaryotes, the split genes cannot be correctly expressed, because prokaryotes do not have the machinery for splicing out the RNA transcribed from the introns of a gene. So the eukaryotic cells are needed for cloning and expression of cloned eukaryotic genes.

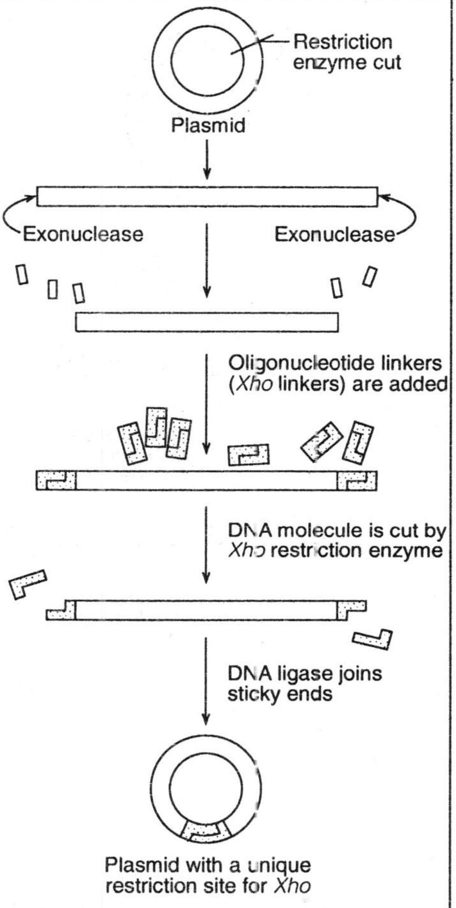

Fig. 22.6: Addition of a "linker" (carrying a restriction site) to a vector molecule

Among eukaryotes, DNA cloning has been done in yeast, mouse and in higher plant species. In yeast, a 2μ plasmid DNA is an appropriate cloning vehicle, which can be transferred through efficient transformation method. This involves protoplast production followed by PEG directed introduction of DNA into protoplasts.

IV. **Detection of Recombinant Clone:** From the large number of colonies produced by transformation to select or screen out the few colonies which contain the recombinant plasmid—the use of antibiotics is one of the most easy and useful methods for this purpose. The transformed cells can be plated on selection medium containing different antibiotics. The colonies which grow, can be said to have a plasmid, as the antibiotic resistance gene of plasmid enables the bacteria to grow. For example, the plasmid pBR322 contains genes for ampicillin resistance (*amp*^{r}) and tetracycline resistance (*tet*^{r}). Thus the transformants can be detected by their plating potential on medium containing either (or both) of these antibiotics.

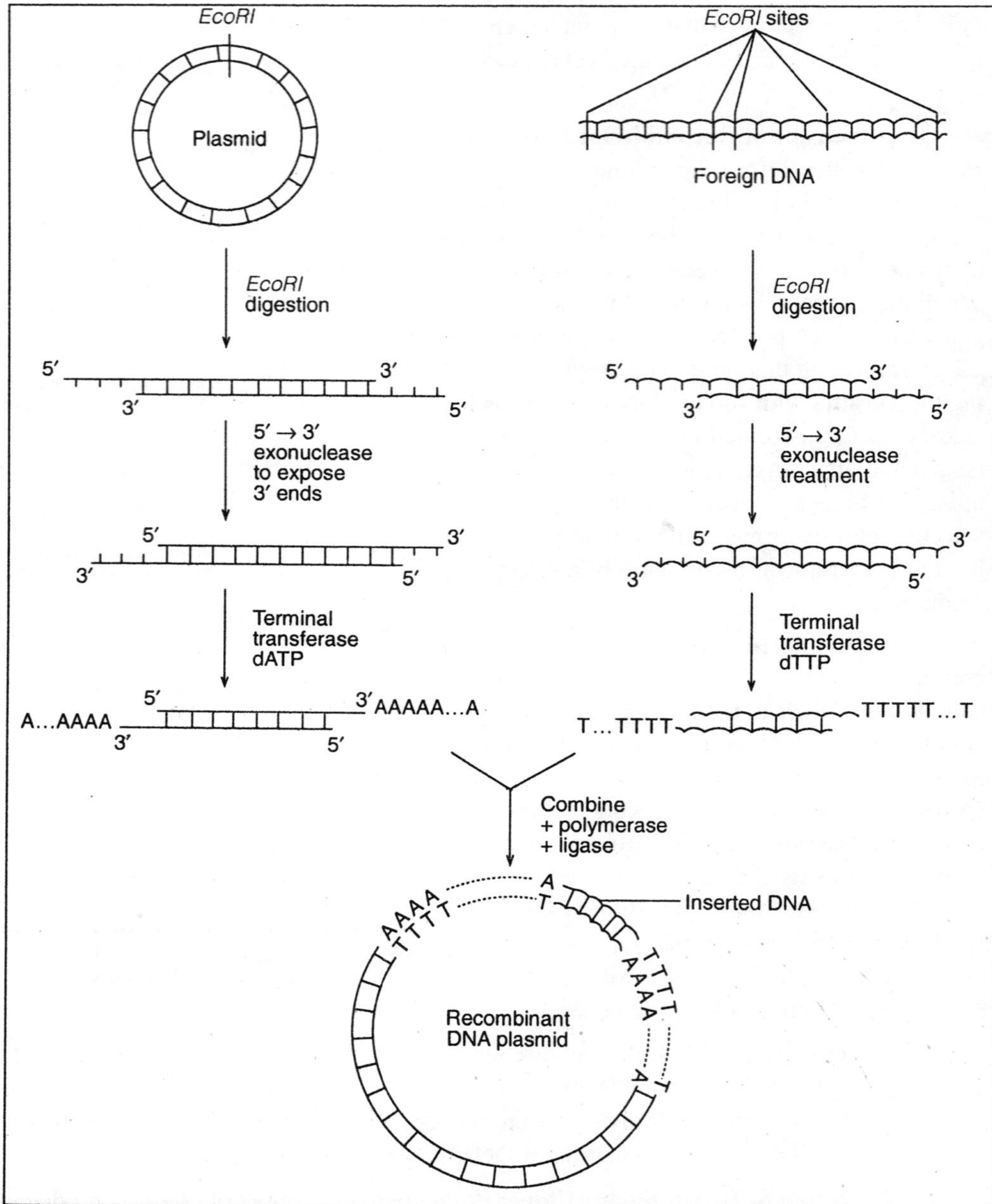

Fig. 22.7: Construction of a recombinant DNA plasmid using the enzyme terminal transferase to synthesize complementary ends on the linearized plasmid and a restriction enzyme generated fragment of foreign DNA (after Russel, 1987)

The presence of cloned DNA fragments can be detected by insertional inactivation of suitable genetic system. For example, DNA fragment of interest can be inserted into one of the antibiotic resistance genes (*tet^r^*) of pBR322, inactivating that gene (*tet^s^*) and other remains active (*amp^r^*). To selectively kill cells with antibiotics, the original master plate with ampicillin in medium is subjected to replica plating method with both ampicillin and tetracycline. Bacteria with recombinant plasmid do not grow on replica plate, only with non-recombinant plasmid will grow. Recombinant colonies are thus identified and selected from master plate (Fig. 22.8).

The detection of recombinant clones can also be done by using chromogenic substrates. The most popular system uses X-gal, a colourless substrate on cleavage by β–galactosidase, a blue coloured product is formed, then the expression of *lac Z* gene can be detected easily. Host cells that are Lac^- are used, so that the Lac^+ phenotype will only arise when the vector is present. Furthermore, if a DNA fragment is cloned into *lac Z* gene (Eco RI site of Charon 16A), any recombinants will be *lac* Z^- and therefore will not produce β-galactosidase and plaques will remain colourless in presence of X-gal.

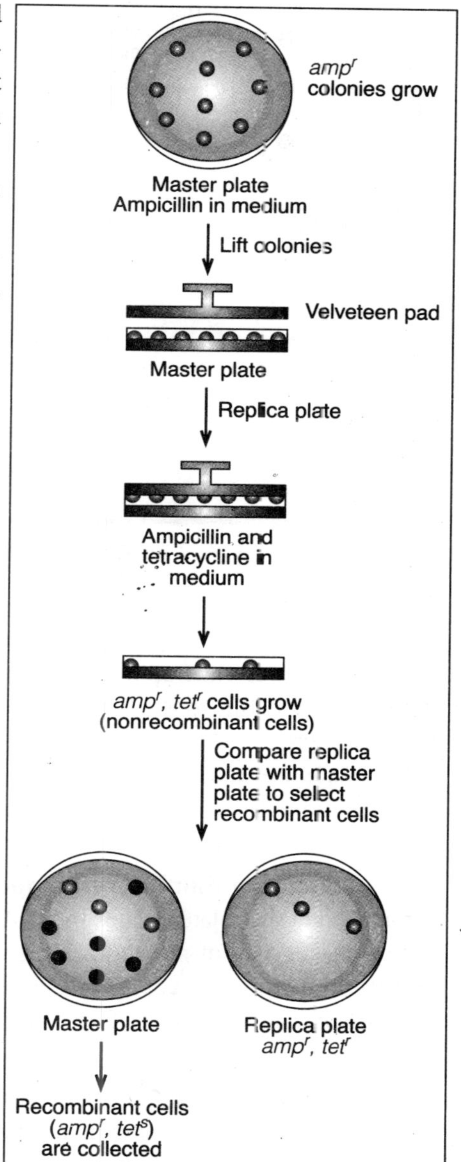

Fig. 22.8: Selecting for recombinant cells after transformation (after Barnum)

22.5 SCREENING TECHNIQUE

Screening is a procedure by which a population of viable cells is subjected to some sort of analysis that enables the desired sequences to be identified.

Colony Hybridisation Technique

This technique has been developed by Gaustein and Hogness (1975). It requires the radioactively labelled DNA probe with a sequence complementary to atleast one part of the desired DNA. The probe may be partially pure mRNA, a chemically synthesized oligonucleotide or a related gene, which identifies the corresponding recombinant DNA.

Colonies to be tested are replicated from a solid medium on to a nitrocellulose filter paper. A portion of each colony remains on the medium, which is called master plate. The paper is treated with NaOH, which simultaneously breaks open the cells and denatures the DNA. The paper is then saturated with ^{32}P labelled mRNA, complementary to the gene being sought, and DNA-RNA renaturation occurs. After washing to remove unbound (^{32}P) mRNA, the position of the bound radioactive phosphorus usually detected by autoradiography, desired colonies are located (Fig. 22.9). If the protein product of a gene of interest is synthesized, immunological techniques allow the protein producing colonies to be identified using radioactive labelled antibodies.

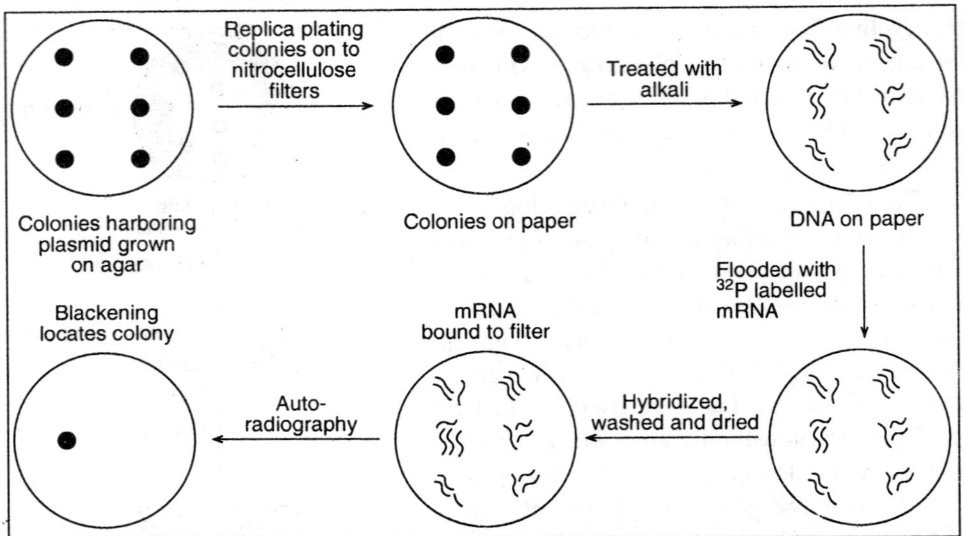

Fig. 22.9: Colony hybridization (after Hartl, Freifelder, Snyder, 1988)

Blotting Techniques

A mixture of DNA, RNA or protein fragments of different molecular weight can be separated by gel-electrophoresis due to their different mobility and can be visualized directly in gel. But to confirm the identity of these bands with a known and available molecular probe, it is possible to hybridize these bands with a labelled probe. When the DNA bands are thus identified by blotting, it is called Southern blotting; when RNA bands are identified, it is called Northern blotting; and when the protein bands are detected, the technique is called the Western blotting.

(a) **Southern Blotting Technique:** This method was developed by Southern (1975) for analysing the related genes in DNA restriction fragment. This technique helps to provide physical map of restriction sites within a gene located normally on a chromosome, reveals the number of copies of the gene in the genome, the degree of similarity of the gene when compared with other complementary genes.

In the method, the genomic DNA is digested with one or many restriction enzymes and the digested sample is gel-electrophoresed, consequently the DNA fragments are then denatured by alkali treatment; the gel is then laid on top of a buffer saturated filter paper, placed on a solid support; upper surface of the gel is covered with nitrocellulose filter paper overlaid with dry filter paper and then stack of paper-towel. The dry filter paper when draws the buffer through the gel, the single stranded DNA molecules get attached to the nitrocellulose membrane; after baking at 80°C, DNA fragments are permanently fixed to the membrane; the filter is then allowed to hybridize with radio-labelled probe (DNA/RNA) of known sequence which is complementary to blot-transferred DNA. After hybridization the filter is washed thoroughly in buffer; the hybridized band regions are detected autoradiographically by placing the nitrocellulose filter in contact with photographic film. The images showed the hybridized DNA molecules (Fig. 22.10).

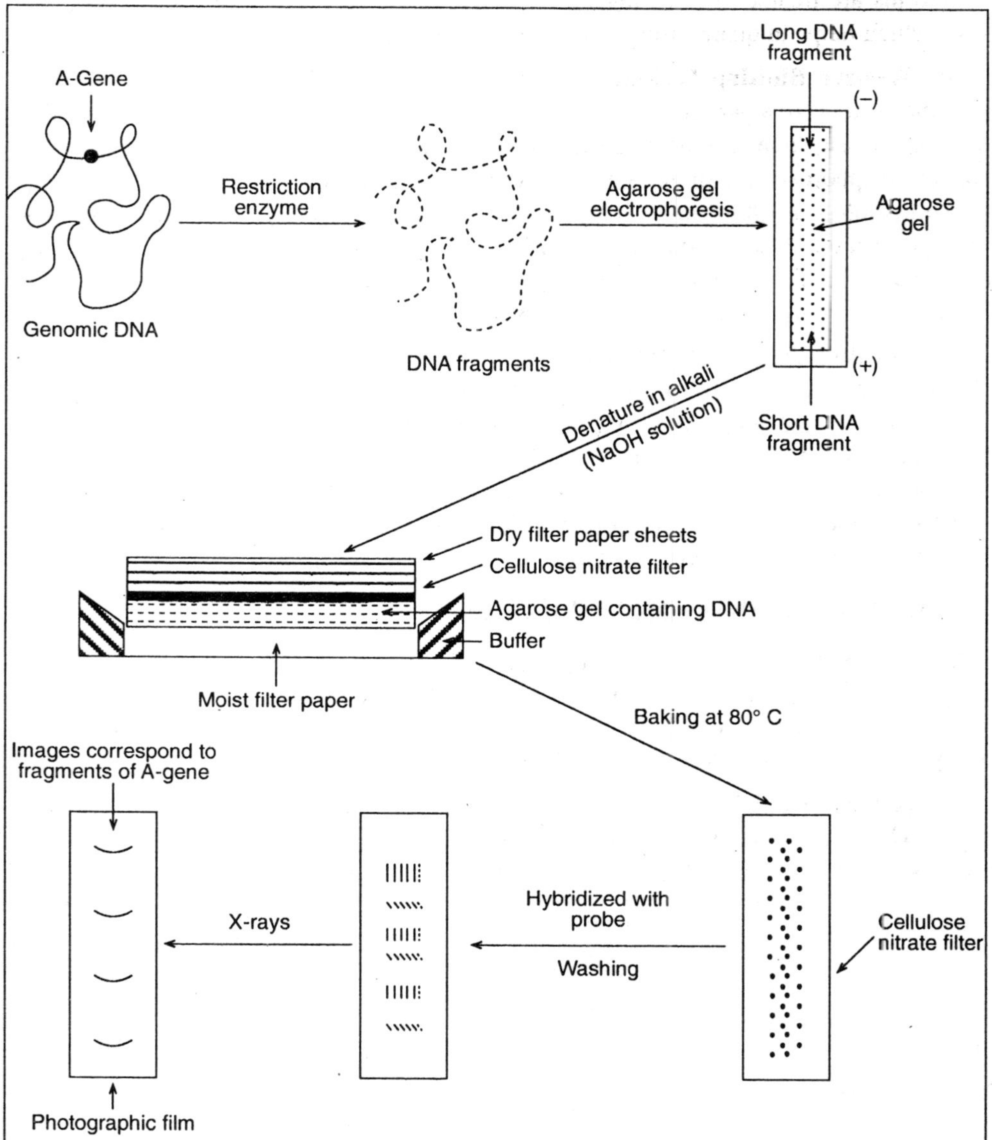

Fig. 22.10: Procedure of Southern blotting technique

(b) **Northern Blotting Technique:** Alwine et al. (1979) devised the technique in which RNA bands are blot-transferred from the gel onto chemically reactive paper (aminobenzyl oxymethyl cellulose). The hybridized bands are found out by autoradiography following the same method like Southern. But the name Northern blotting came as it is the extended method of Southern. In this technique the blot-transfers are reusable because of the firm covalent bonding of RNA to the reactive paper. Thomas invented the technique that under appropriate condition the mRNA can also be blot-transferred to nitrocellulose paper and are then hybridized by single stranded probe (RNA or DNA).

The hybrids are treated with SI nuclease to remove the single stranded non-hybridized probe, which helps in quantitative estimation of mRNA.

(c) **Western Blotting Technique:** Towbin et al. (1979) developed the technique to detect the proteins of a particular specificity. When the transferred gene expresses the protein in alien cell, then it is detected through this technique. The technique follows the steps: electrophoresis of cell extract (protein) in polyacrylamide gel; blotting of proteins on to nitrocellulose filter paper; hybridization of proteins by using radiolabelled antibodies (I^{125}antibodies) of known structure; and detection of hybridized sequences by autoradiography. If the radioactive label is not used, the bound antibody may be detected by a second antibody tagged with an enzyme.

(d) **Dot Blot or Slot Blot Technique:** In this technique pure extracted genomic DNA to be tested are spotted on a nitrocellulose membrane, where an apparatus is used for placing spots on the membrane through slots, thus the spots are made in the form of oblong slots rather than round blots. There is no step involving digestion with enzyme; gel electrophoresis or transfer from gel to membrane is needed. This technique is considered much more convenient, when no restriction sites are needed to be studied, only to detect the presence of foreign DNA clone in the cell/tissue.

22.6 GENOMIC AND cDNA LIBRARIES

Genomic DNA Library: A genomic library is a collection of independently isolated vector linked DNA fragments derived from a single organism. It contains at least one copy of every DNA sequence in the genome. An ideal library is one that represents all of the sequences with smallest possible number of clones.

The genomic DNA libraries can be prepared by the complete digestion of the total genomic DNA with a restriction enzyme and the fragments are inserted into a suitable vector like λ phages (Fig. 22.11). The drawback of this method is that sometimes the

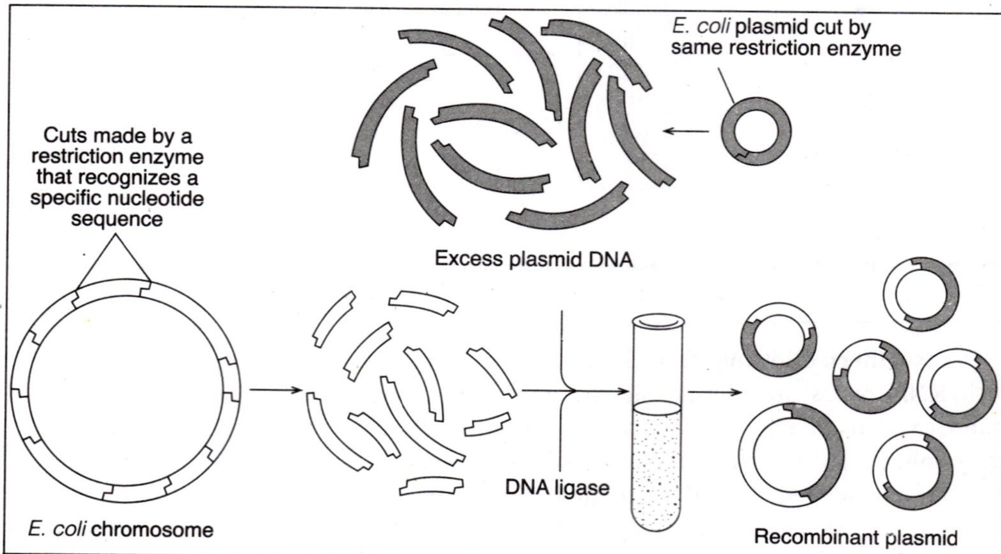

Fig. 22.11: Construction of a genomic library using recombinant DNA techninque

sequence of interest may contain multiple restriction sites, so digestion with RE results into two or more pieces. In this method the eukaryotic DNA is broken up into smaller fragments, thus an entire library would necessarily contain a large number of phages, and screening of which is very laborious. The problems of this method can be avoided by random shearing of total DNA and cloning of large fragments. This method ensures that sequences are not excluded from the cloned library simply because of the distribution of restriction sites. In this procedure the randomly fragmented DNA is partially digested with RE which has short recognition sites. The fragments of desired size are collected through agarose gel electrophoresis, so the population of overlapping fragments that are close to random can be cloned directly.

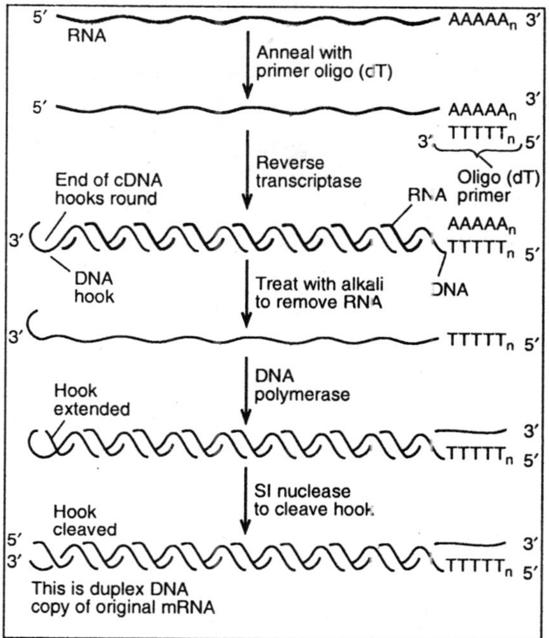

Fig. 22.12: Synthesis of cDNA from mRNA using reverse transcriptase enzyme

cDNA Library: Complementary DNA (cDNA) libraries can also be prepared by isolating mRNAs from tissues which are actively synthesizing proteins, like roots and leaves in plants, ovaries or reticulocytes in mammals, etc. The mRNAs are used for copying them into cDNAs through the use of reverse transcriptase. Then the cDNA molecule can be made double stranded and cloned (Fig. 22.12).

However, cDNA clones will differ from genomic clones in lacking the introns present in split genes, and have the advantage of being capable to be expressed in bacteria, which do not have the machinery to process the eukaryotic mRNA. There are far less number of cDNA clones in a bank than in a genomic library, which makes easier to look for a desired gene. Screening of cDNA bank also provides fairly unambiguous results.

22.7 MOLECULAR MAPPING

Preparation of linkage map based on recombination data is always handicapped due to nonavailability of mutants for many genes. This limitation has largely been overcome in recent years by molecular mapping through ISH (*in situ* hybridization), Restriction mapping, RFLP (Restriction Fragment Length Polymorphism) analysis, RAPD (Random Amplified Polymorphic DNA) technique, Minisatellites and microsatellites (VNTR) study. Molecular maps are both physical and genetic.

Molecular Physical Map

In molecular physical map, actual distances are measured in terms of nucleotide sequences.

(a) ***In situ* hybridization (ISH: FISH & GISH):** For localization and mapping of different segments of chromosomes and gene loci at the microscopic level, application of molecular hybridization *in situ* is now widely adopted (Fig. 22.13A). *In situ hybridization (ISH)* principally uses probe sequences, tagged with radioisotopes or fluorescent compounds (or a chemical reporter). The initial step is denaturation of the target which is followed by hybridization with probe of the complementary sequences to undergo reannealing or pairing. The complementary sequences of the probe bind selectively at the target site. The hybridized sites are localized either through autoradiography or immunofluorescence as well as counter staining with specific stains detected cytologically. The *in situ* hybridization technique was developed initially by Pardue and Gall and later modified by different authors.

The **fluorescent in situ hybridization (FISH)** is the most powerful technique at present, through which the target loci at the chromosome is hybridized with complementary probe sequence, tagged with fluorescent compound. There are two approaches, namely direct or indirect, of FISH technique in plant chromosome. In the indirect method, the probes are tagged with reporter molecule, such as biotin, digoxigenin and finally they are located by fluorochrome conjugated antibodies such as avidin. The main principle of this method is to make the probe-target sequence as antigenic so that, it can be detected through antibody. The common fluorochromes are FITC (fluorescein-isothiocyanate) as well as rhodamine. In the direct method, the probe is directly labelled by fluorochrome-labelled antibodies. The direct labelling of fluorochrome with the probe is the rapid method of ensuring good resolution.

Due to lack of karyomorphological markers, metaphase chromosome analysis cannot distinguish parental genomes in hybrids. If ISH technique is applied with total genomic probes where the plant has multigenomic constitution, the parental chromosome can be directly identified in the hybrids, such as *Triticum* and *Secale* in *Triticale* or in the hybrid of *Gasteria* and *Aloe* (Fig. 22.13B). Thus method of total genome *in situ* hybridization is otherwise termed as **GISH (genomic *in situ* hybridization)** technique. Since its first demonstration in identification of parental genomes in hybrid between *Hordeum chilense* and *Secale africanum* by Schwarzacher et al., the technique has been extensively applied to elucidate ancestry of hybrids and polyploids. GISH remains a very effective tool in genome identification, their orientation and in establishing genomic relationships between species. The use of GISH in meiosis helps in understanding intergenomic homologies as well as in elucidating the possible transfer of chromosome segments through intergenomic recombination.

The FISH technique, using different colour combinations by different probes, is now being applied to detect simultaneously different genomes, or chromosome segments by extension of the technique—otherwise termed as **multicolour FISH**. This method has been used to distinguish three genomes in hexaploid wheat and to detect several translocation sites and insertions in polyploid species of *Triticum* and *Aegilops*. The multicolour FISH technique is now regarded as a powerful tool for gene mapping as well as detection of abnormalities, including insertion and breakage points with chromosome specific or genome specific dispersed probes. The term **chromosome painting** is used in FISH technique where chromosome specific dispersed probes are used to detect the location of

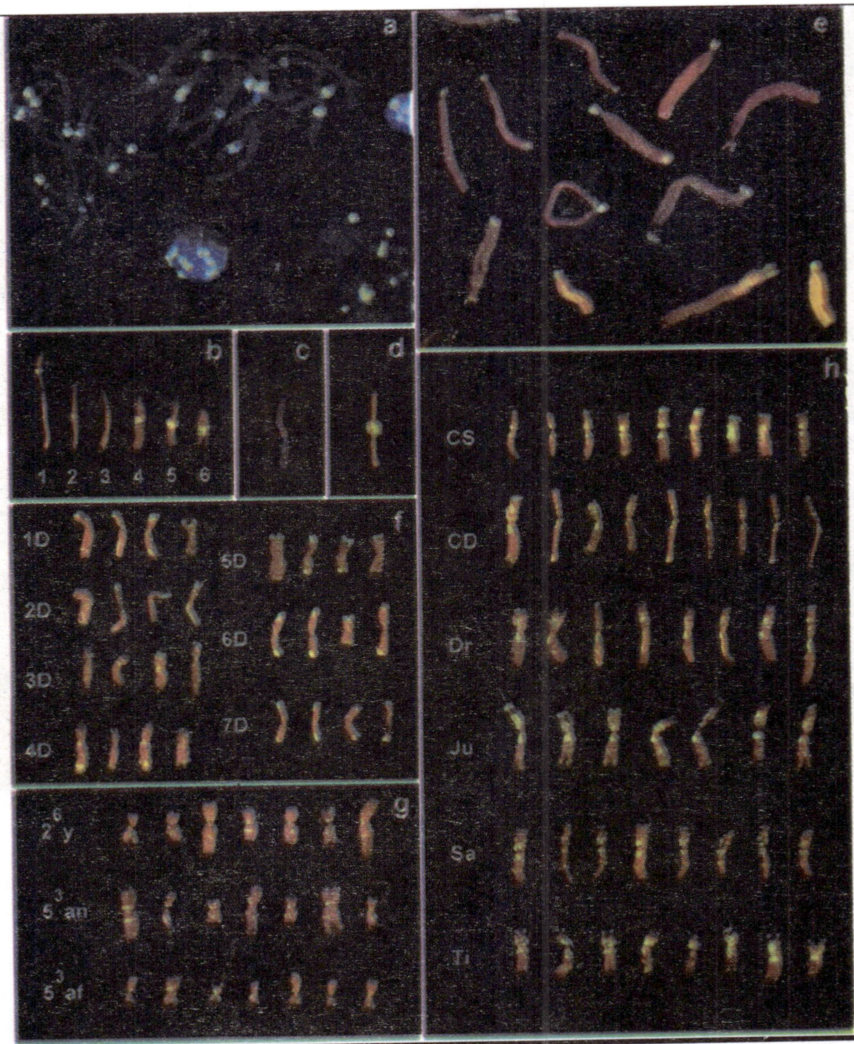

Fig. 22.13A: Localization of repetitive DNA sequences following ISH (*in situ* molecular hybridization) (from Sharma and Sen)

(a) Metaphase plate and interphase nuclei of field bean
(b) Localization of repeats on field bean chromosomes
(c+d) Visualization of repeats on field bean-acrocentric chromosomes
(e) Localization of telomeric repeats on field bean chromosomes
(f) Images of group D wheat chromosomes after localization of repeats
(g) Localization of GAA repeats in bareley chromosome 5 isolated from two translocation lines and barley chromosme 5^3 isolated from translocation line T3-5af
(h) Polymorphism in banding in wheat chromosme 3B

complementary target sequences in the complement. The FISH technique in recent years has further been modified to locate single copy or tandem sequences, utilizing primer mediated extension and amplification *in situ* by PCR method. The application of this method lies also in confirmation of location of foreign gene at the chromosome level in transgenic individuals.

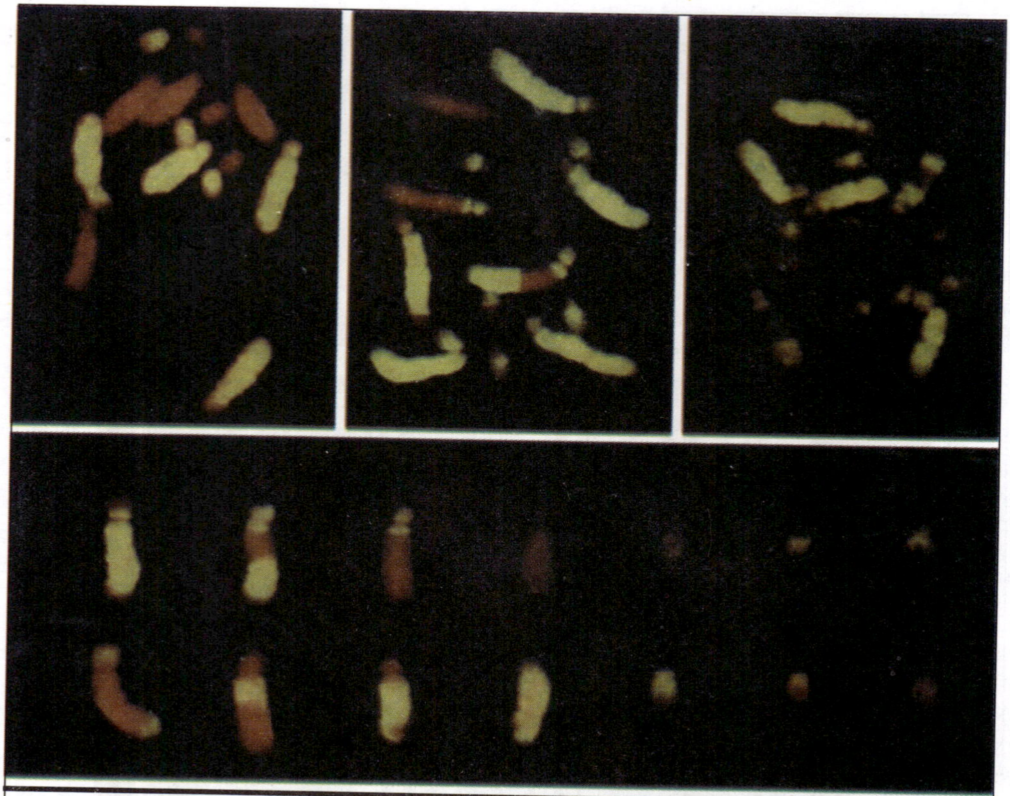

Fig. 22.13B: Mitotic chromosomes of *Gasteria lutzil* X *Aloe aristata* F_1 hybrid and back-crosses with *G. lutzii* following GISH (genomic *in situ* hybridization). The larger *Gasteria* genome (four long and three short chromosomes) fluoresces yellow; the smaller *Aloe* genome fluoresces orange. In each, the complete *Gasteria* genome fluoresces uniformly yellow while the recombinant chromosomes from the F_1 show yellow and orange segments, indicating cross-over points (from Sharma and Sen)

(b) **Restriction Mapping:** Determination of nucleotide sequence of a gene is done by cleavage of corresponding DNA at specific sites with the help of restriction endonucleases. These sites of cleavage can be identified and mapped to give to a restriction map. On a restriction map, is found a linear sequence of sites, each for a specific enzyme and the distances between them are measured as number of base pairs of DNA (Fig. 22.14). The data of digestion of a DNA molecule by more than one endonucleases can be utilized to arrange the sites of breakage in a definite order. This process involves double digests (enzyme A and B) to determine the cleavage positions of DNA due to one enzyme with respect to another enzyme. The results of reciprocal digests (A followed by B and B followed by A) show the same fragments. The original DNA sample is also digested by a mixture of both the enzymes to confirm the results of individual successive digests. Overlapping region of different digests (A and B) can be detected which allow to prepare the restriction map.

(c) **Other Methods:** Following methods are also used for this purpose.

(i) **Chromosome walking:** It is a technique where different sequential clones from a genomic library are identified by successive hybridization and restriction maps are pre-

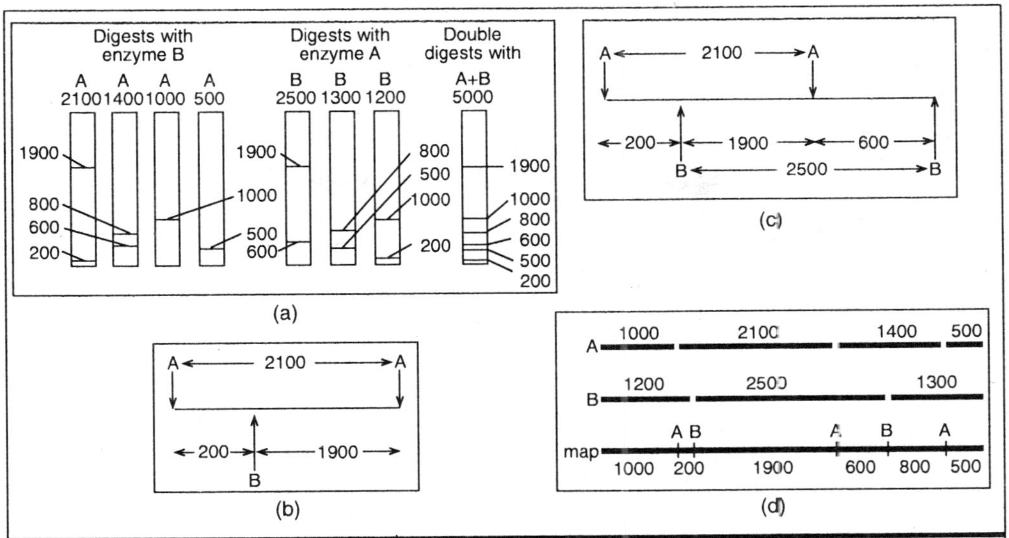

Fig. 22.14: (a) Technique of double digests to determine cleavage positions of DNA due to one enzyme with respect to those due to another enzyme (the two enzymes are A and B); the four gels on the extreme left show the result of electrophoresis after digesting with enzyme B each of the four fragments obtained after digestion with A; three central gels represent results of digesting with enzyme A each of the three fragments obtained after digestion with enzyme B and the solitary gel on the right shows the results of digesting the intact DNA with both the enzymes simultaneously. (b) The sites of cleavage by enzymes A and B in one of the four fragments, 2100 bp long obtained due to cleavage by enzyme A, the fragment has a site for enzyme B at 200 bp length from one end. (c) Reconstruction of restriction map showing restriction sites of enzymes A and B in two overlapping fragments (2100 bp fragment obtained due to enzyme A and 2500 bp fragment obtained due to B). (d) A restriction map, in the form of linear arrangement of cleavage sites, the upper two figures show sites due to individual enzymes and the bottom figure shows arrangement of cleavage sites of both enzymes, sites of more enzymes can be added to this map by same technique

pared separately. As a result the whole chromosomal regions are characterized and restriction map of entire chromosome can be prepared following the technique of chromosome walking.

(ii) **Chromosome jumping:** Chromosome jumping approach to construct the physical map is utilized in order to bring the molecular marker close to the gene of interest. This technique will minimize the size of the DNA segments carrying the gene to be cloned in comparison to the used linked molecular marker.

(iii) **Radiation hybrid mapping:** The radiation hybrids are somatic cell hybrids in which an irradiated fragments of human DNA are randomly integrated into the rodent chromosomes. After screening a panel of hybrid clones with a human marker allows the marker to be mapped.

(iv) **QTL mapping:** A QTL (Quantitative trait loci) may be defined as a region of the genome that is associated with an important quantitative agronomic trait. The QTL mapping is an advanced method using RFLP maps. It exploits the complete linkage maps by interval mapping of 'QTL' and identifies the crosses for QTL mapping. The effect of each genome segment located between a pair of marker loci (but not QTL associated with a single RFLP) is assessed by interval mapping.

Molecular Genetic Map

These maps are prepared utilizing the following molecular markers.

I. Non-PCR based

(a) **Restriction Fragment Length Polymorphism (RFLP):** When the genomic DNA from several individuals (genetically related) are digested with same restriction enzymes and separated on gel through electrophoresis, blotted and hybridized with a labelled DNA clone of known sequence, polymorphism in hybridization pattern reveals the relatedness among different individuals, and such variation is termed as 'Restriction Fragment Length Polymorphism' (RFLP) (Fig. 22.15).

RFLP is used to infer phylogenetic relationship in both plants and animals. This technique is also used to prepare chromosome maps in human, mice, fruitfly, or in plants like maize, tomato, lettuce and rice. The inheritance or linkage relationship also can be studied through this technique analysing the parent, F_1 and F_2 population.

II. PCR based

(b) **Random Amplified Polymorphic DNA (RAPD):** If artificially synthesized primers with arbitrary sequences are used for PCR (Polymerase chain reaction) amplification, DNA segments to be amplified will be selected at random which will thus provide a truly random sample of DNA markers, which are described as 'Random Amplified Polymorphic DNA' (RAPD). The steps involved are:

(i) Amplification of extracted DNA by PCR using random primers.

(ii) Separation of amplified DNA on agarose gel.

(iii) Visualization of markers on the gel and its photography.

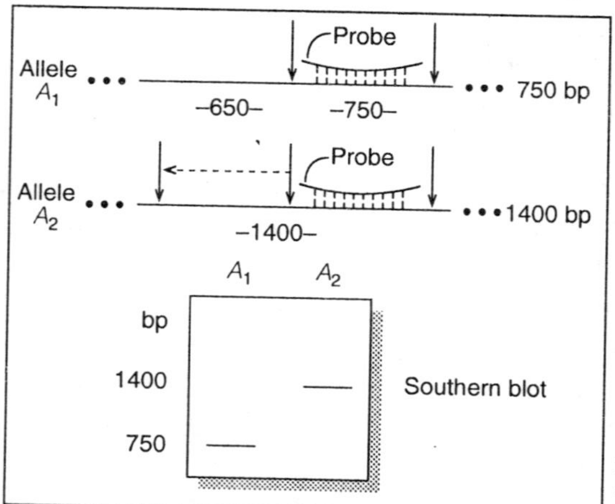

Fig. 22.15: Restriction fragment length polymorphism (RFLP) analysis. Allele A1 is a gene segment, 750 bp long, identified by probe binding. In allele A2, the restriction site to the left of allele A1 has been changed. The probe thus recognizes a 1400 bp fragment instead of the 750 bp fragment. During Southern blotting, two different bands show up from the two alleles. Arrows indicate restriction sites

(c) **Amplified Fragment Length Polymorphism (AFLP):** This technique involves the ligation of adaptors to DNA fragments obtained after RE digestion. Then these fragments are allowed for PCR. The primers used here are generally sequence of adaptors which help to amplify the whole length between the two restriction sites. Then the different fragment lengths are separated on gel.

(d) **SCAR (Sequence Characterised Amplified Region:** Amplified RAPD products are cloned, sequenced and primer sequences are ascertained from the end of band, the primers are generally longer which gives greater degree of specificity.

(e) **Variable Number Tandem Repeats (VNTR):** The DNA differences between two individuals can be detected by variable number of tandem repeats which are short repeat sequences arranged in tandem order. VNTRs are divided into two groups—*Minisatellites* (10–25 bps) and *Microsatellites* (1–5 bps). A DNA polycore probe of such micro- or minisatellites can detect simultaneously a large number of variable loci containing tandem DNA repeat sequences.

(f) **Sequence-Tagged Sites (STS):** Sequence tagged sites are DNA length of 100–500 bp that are unique in genome. They are created by polymerase chain reaction amplification of primers obtained by sequencing segments of the genome.

(g) **Single Nucleotide Polymorphism (SNP):** The map can also be marked off in differences among individuals that amount to changes in single base pairs. These differences are called SNPs.

22.8 DNA FINGERPRINTING

Each individual has an unique DNA pattern. No two individuals are absolutely alike, excepting in monozygotic twins. This uniqueness of difference in DNA sequence is the basis of the technique of *DNA fingerprinting* (Jeffreys, 1986). Digestion of DNA with specific enzymes, such as restriction enzyme as mentioned earlier, electrophoresis, Southern blotting and detection through hybridization with specific nucleotide probes, have made DNA fingerprinting a very powerful tool in the study of biodiversity and revolutionized forensic science.

The DNA differences between two distinct individuals or polymorphism of the genome, can be detected by *microsatellites* or *minisatellites*. There are variable numbers of short repeat sequences (SRSs) arranged in tandem order. Minisatellites are repeats of 10–25 bp while microsatellites are repeats of 1–5 bp of DNA. A DNA polycore probe of such mini- or microsatellites can detect simultaneously a large number of variable loci containing tandem DNA repeat sequences. Such a complex fingerprint pattern is very specific to each individual. The analysis of this short tandem repeats (STR) have made the fingerprint test very sensitive, permitting DNA typing of degraded plant organs as well.

At present short single stranded *oligonucleotide probes* can be synthesized. As such their use has facilitated DNA fingerprinting research, because they can yield information of high degree of polymorphism. In general in plants, short tandem repeats are very common. The number of short tandem repeats (STRs) are species specific. In *Petunia* it is one in 11 kb, in *Brassica* it is one in every 25.4 kb, whereas in *Arabidopsis*, maize and barley it is one in every 42 kb, 58 kb and 156 kb respectively. Rice has variety of STRs, but only AT type of STR is found in barley.

22.9 GENE AMPLIFICATION: POLYMERASE CHAIN REACTION (PCR)

PCR provides a simple and ingeneous method for exponential amplification of specific DNA sequences by *in vitro* DNA synthesis, i.e., this technique has made it possible to synthesize large quantities of DNA fragments without cloning it. Kary Mulis in 1985 developed the technique based on the use of an enzyme which is named as Taq DNA polymerase. The PCR technique has now been automated and is carried out by a specially designed machine.

Technique: The technique involves the following three steps (Fig. 22.16):

I. **Denaturation of DNA Fragment:** The target DNA containing sequence to be amplified is heat denatured (around 94°C for 15 sec) to separate its complementary strands, this process is called melting of target DNA.

II. **Annealing of Primers:** Primers are added in excess and the temperature is lowered to about 68°C for 60 sec., as a result the primers form the hydrogen bonds and anneal to the DNA on both sides of the DNA sequence.

III. **Primer Extension:** Finally different nucleoside triphosphate (dATP, dGTP, dCTP, dTTP) and a thermostable DNA polymerase (Taq polymerase from *Thermus aquaticus* and Vent polymerase from *Thermococcus litoralis*) are added to the reaction mixture, it helps in polymerization process of primers and, therefore, extends the primers (at 68°C) resulting in synthesis of multiple copies of target DNA sequence.

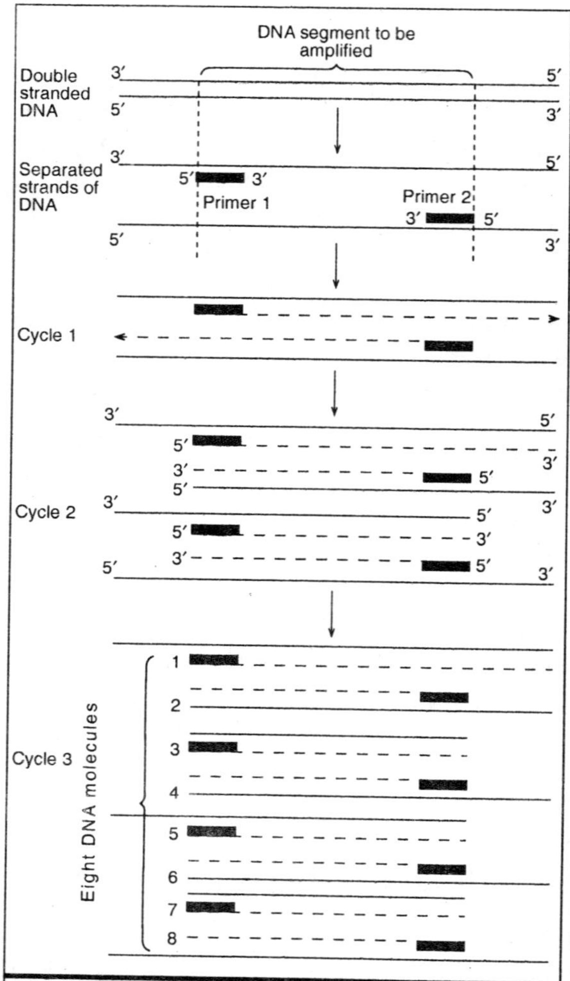

Fig. 22.16: Basic reaction of the PCR (only three cycles of PCR are shown; in each cycle primers are shown by solid boxes, template strands are shown by continuous lines and newly synthesized strands are shown by broken lines) (after Dharmalingam, 1990)

After completion of all these steps in one cycle, again the second cycle is repeated following the same process. If 20 such cycles occur, then about one million copies of target DNA sequence are produced. Recently this technology has been improved much more, where instead of Taq polymerase the rTth polymerase is used which transcribe RNA to DNA, and thereafter amplify the DNA.

Modified Forms of PCR: The conventional PCR is the symmetrical PCR technique. There are some other modified forms of PCR which are used for various purposes:

AP-PCR (Arbitrarily Primed Polymerase Chain Reaction): It requires only a single primer of relatively much smaller length compared to the primers used in PCR. This technique is used for DNA profiling, in animal and plant biotechnology as well as in forensic medicine.

Asymmetrical PCR: Target sequences of one strand may be amplified in several order of magnitude more as compared to its complementary strand. This approach is particularly useful for generating single stranded DNA fragment to be used for sequencing of DNA.

IPCR (Inverted Polymerase Chain Reaction): In this method it allows the amplification of DNA flanking a known DNA sequence, the primers are facing outwards. Using the inverse PCR, the unknown sequences flanking known sequences can be readily amplified.

RT-PCR (Reverse Transcriptase Polymerase Chain Reaction): Although the PCR amplification is generally performed on the DNA template but using this technique the RNA also can be used for amplification. This technique is particularly useful for studying the expression of genes and for monitoring the obscure species of mRNA.

Nested PCR: Nested PCR primers are ones that are internal to the first primer pair. The larger fragments produced by the first round of PCR is used as the template for the second PCR. This technique eliminates any spurious non-specific amplification products.

Application of PCR in Biotechnology: PCR has many fold applications.

1. *The amplification of gene fragments as fast alternative of cloning:*
   (a) Inserts of bacterial plasmids can be amplified with primers.
   (b) DNA from known sequence can be obtained by designing primers.
   (c) PCR helps in identification of homologous sequences from related organisms.
   (d) Using RT-PCR the 3′ end of cDNA can be amplified (RACE: Rapid Amplification of cDNA Ends).
   (e) Reverse PCR helps to know the flanking sequences of a known DNA clone.
2. *Modification of DNA fragments:* Site directed mutagenesis using oligonucleotides as PCR primers provides a powerful approach to study structure-function relation.
3. *Diagnosis of pathogenic microorganism:* DNA from the infected parts of a person or animal may be subjected to PCR with primer specific gene of the pathogen and diagnosis can be done on amplification of DNA.
4. *DNA analysis of archaeological specimens:* As DNA is relatively stable and remain intact for a long period of time, PCR can help in analysis of DNA from those embedded materials.
5. *Detection of mutation relevant for inherited diseases:* Any point mutation, a deletion or an insertion and expanded tandem trinucleotide repeat can be detected by PCR. Somatic mutations in oncogenes or tumour repressor genes can also be detected by PCR with primers flanking the insertions or deletions.
6. *Analysis of genetic markers for forensic applications,* for paternity testing and for the mapping of hereditary traits.
   (a) Amplification of SSR.
   (b) RAPD (Random Amplified Polymorphic DNA) with arbitrary, often short (10 bp) primers.
7. *Species-specific amplification of DNA segments* between interspersed repeat elements (IRS) using the primer based on the SINE sequence (Short Interspersed Nuclear Elements).

8. *Genetic Engineering using PCR:* Using PCR we can incorporate alteration or mutation in the ultimate product by choice altering, removing or adding sequences to the primer at the 5′ end.

   By recombinant PCR technique, it is possible to join two DNA fragments at a specific site through complementary overlaps (This technique is termed as splicing). By synthesizing two mutagenic primers, spanning the internal site to be changed, it is possible to introduce mutations within a fragment.

22.10 GENE SEQUENCING

When a particular gene of interest or a DNA fragment is isolated, the sequencing of that piece of DNA becomes essential. The sequence of DNA refers to the order of nucleotide bases along its sugar phosphate backbone. No technique can determine the sequence of bases in an entire gene in a single experiment, so it is necessary to cut the whole gene into fragments of manageable size (few hundred base pair long) and purify each fragment. There are two different methods which are now routinely used for determination of DNA sequences.

(a) **Maxam & Gilbert's Chemical Degradation Method:** In this method the following steps are involved (Fig. 22.17)

1. The 3′ ends of DNA fragments are labelled.
2. The labelled strands are then separated, of which both strands are labelled at 3′ ends.
3. The mixture is divided into four samples, each treated with a different reagent having the property of destroying either only G or only C or A and G, T and C; the concentration is adjusted in such a way that 50% of target base is destroyed.
4. Fragments of different sizes having ^{32}P are produced.
5. Electrophoresis is done using each of the four samples in four different lanes of the gel.
6. Autoradiograph of the gel helps to determine the sequence from the position of bands in four lanes.

(b) **Sanger and Coulson's Dideoxynucleotide Synthetic Method:** This method is based on utilization of DNA polymerase, 2′–3′ dideoxynucleotides, and different dNTPs. Here the dideoxynucleotide terminates the chain reaction of DNA synthesis which produces the different length of chains. The method involves the following steps:

1. Four reaction tubes are set up each containing single stranded DNA sample to be sequenced, all four dNTPs (radioactively labelled) and an enzyme for DNA synthesis (DNA polymerase I).
2. Each tube also contains a small amount of (much smaller amount relative to four dNTPs) one of the ddNTP, so that four tubes have each different ddNTP, bringing about termination at a specific base – Adenine (A), Cytosine (C), Thymine (T), and Guanine (G).
3. The fragments generated by random incorporation of ddNTP leads to termination of reaction and so the different fragments are produced which can be

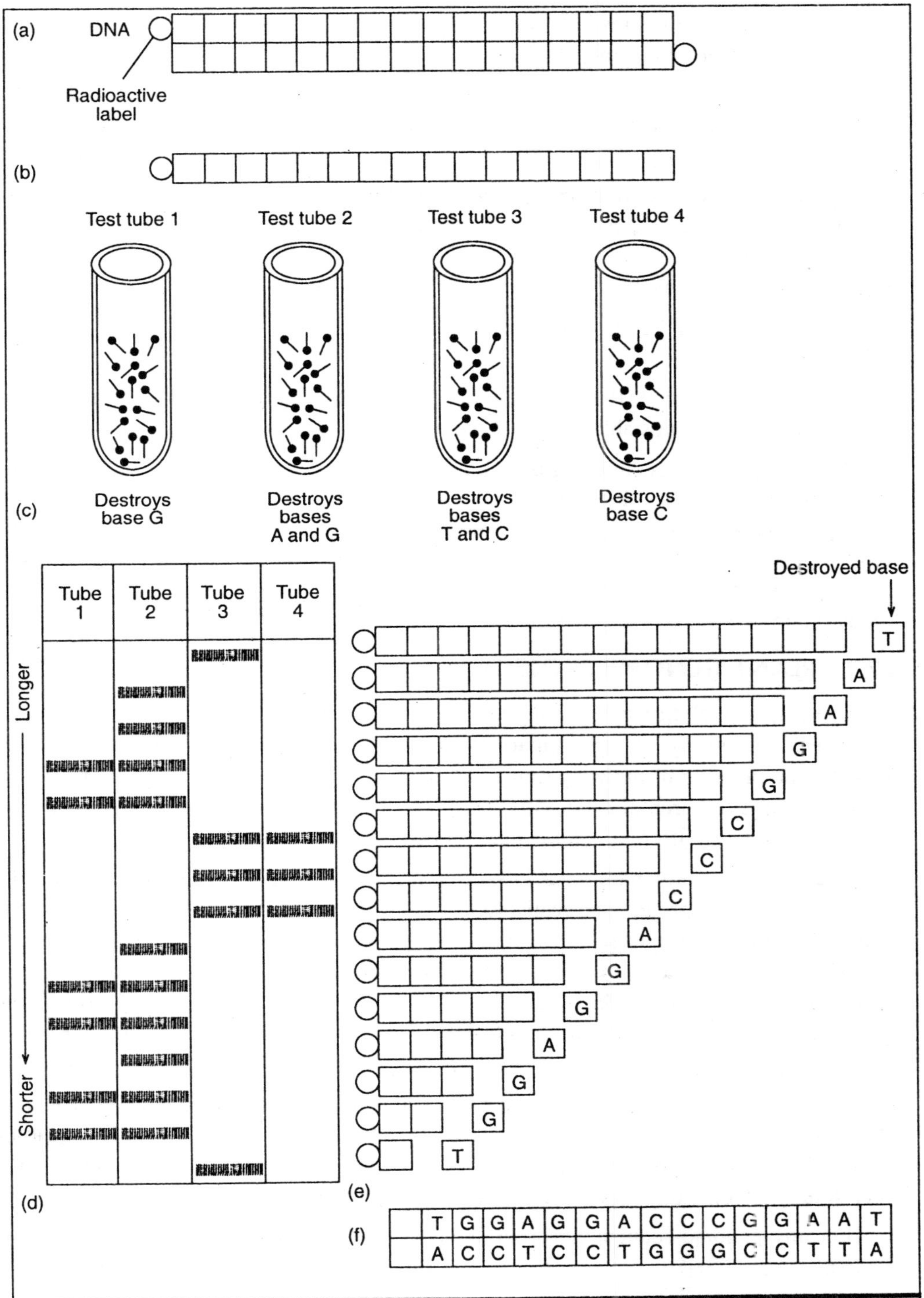

Fig. 22.17: Different steps involved in Maxam and Gilbert's method of DNA sequencing

separated by high resolution polyacrylamide gel, four adjoining lanes are loaded by four different samples.

4. The gel is then autoradiographed, the position of different bands in each lane can be visualized, and based on the position of the bands, the DNA sequence can be read out very easily.

(c) **Direct DNA Sequencing using PCR:** PCR is also used for sequencing the amplified DNA product. This method is more reliable, less time consuming and can utilize either the whole genomic DNA or cloned fragments for sequencing a particular DNA segment. The sequencing in this method involves only two steps:

1. Generation of sequencing templates (double stranded or single stranded using PCR).
2. Sequencing of PCR products either with the thermolabile DNA polymerase or with the thermostable Taq DNA polymerase.

So, here the DNA sequencing does not require any cloning vector. In both the cases of enzyme, Sanger's dideoxy method or Maxam-Gilbert's chemical method can be used. In Sanger's method, as usual, the sequencing primer is labelled with ^{32}P and the mixtures with amplified DNA, Taq polymerase and appropriate buffer are incubated at 70°C for 5 min. The reaction is stopped by addition of formamide and mixtures are allowed to run polyacrylamide sequencing gel, which can be read by computer or manually.

22.11 PLANT GENOME PROJECT

The major revolution in the study of genomes of different species was brought about due to the availability of recombinant DNA and PCR technologies. These techniques helped in preparation of molecular maps of many plant and animal genomes. The objective of genomic research in any species is to sequence the whole genome and to decipher functions of all the different coding and non-coding sequences.

The technology for largescale DNA sequencing has enable scientists to undertake genome sequencing project in a realistic time scale. Since the time of first 'large' genome sequencing in bacteriophage λ in 1983, the projects on different groups have been completed. Some notable examples include the bacterium *Escherichia coli*, the yeast *Saccharomyces cerevisae,* the weed *Arabidopsis thaliana,* the rice *Oryza sativa,* the nematode *Caenorhabditis elegans,* the fruit fly *Drosophila melanogaster*, the mouse *Mus musculus,* the primate chimpanzee *Pan troglodytes,* the human *Homo sapiens sapiens.*

In crop science genomics has been promoted using *Arabidopsis*, a weed plant found all over the world, easy to grow, short span of life cycle, one of the smallest genome among dicot plants and *Oryza sativa* (rice), one of the members from monocot having very simple genome organisation and similarity with other major cereal plants.

For any genome sequencing programme, the **following steps** are undertaken (Fig. 22.18):

1. **Construction of Linkage Maps with Molecular Markers:** Different kinds of molecular markers like RFLP, RAPD, microsatellites, AFLP, etc. are used to construct the actual physical linkage map. The RFLP markers are then hybridised with vectors

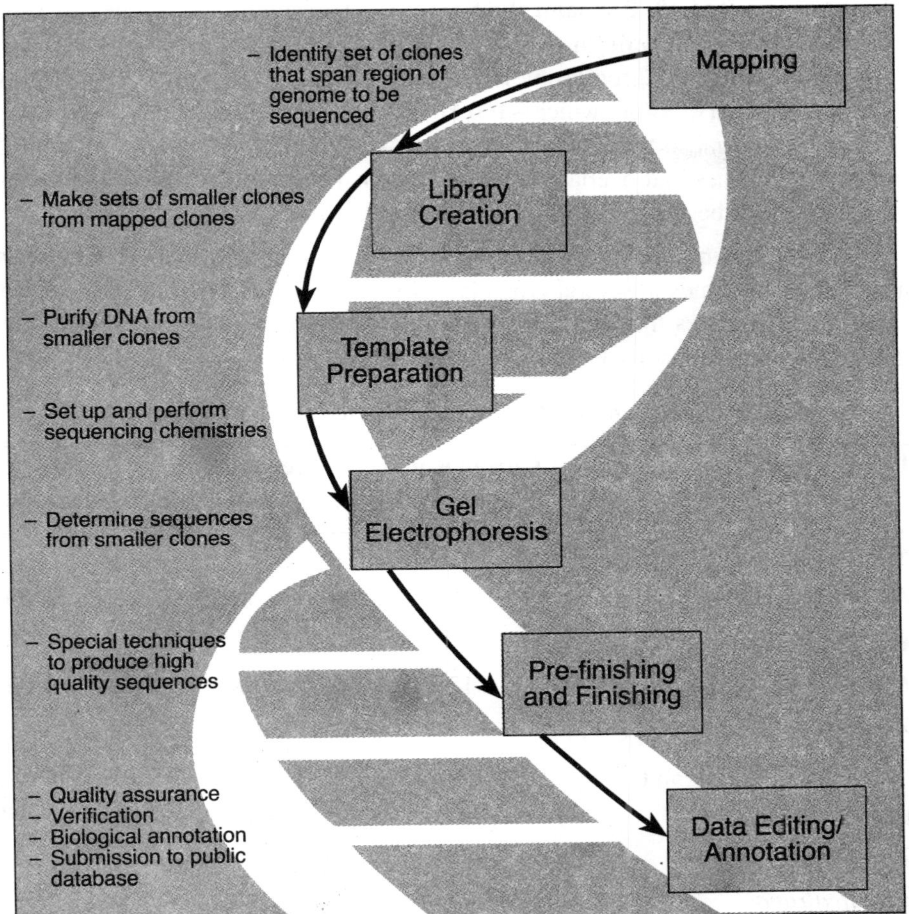

Fig. 22.18: DNA sequencing process

containing the DNA fragments that are to be sequenced, enabling the positioning of various cloned DNA fragments along the chromosome.

2. **Gene Libraries:** Construction of gene libraries using restriction endonuclease and then by cloning the sheared DNA into vectors like cosmid or YAC is essential to identify the overlapping clones.

3. **Screening of Libraries and Constructing Contigs:** STS (sequence targetted sites) or the EST (expressed sequence tags) help a lot in sequencing process by facilitating the task of annotation of the final sequence. When the ordered overlapping clones are available then the contigs are constructed, which represent stretches of contiguous sequence ready clones.

4. **Sequencing:** After identification of YAC clones these are further fragmented by restriction endonuclease and again subcloned into cosmid. After identification of overlapping 'cosmid' clones these are then sheared and cloned in M 13 vectors or pUC plasmid and used for sequencing using the Sanger method.

5. **New Vectors BAC & PAC and the Shot Gun Approach:** YAC clones are found to be unstable and chimeric in nature, so the use of alternative vectors such as BAC (Bacterial Artificial Chromosome) or the PAC (Pl derived Artificial Chromosome) became necessary, within which ~100 kb can be cloned easily. With the aid of computers and special software ends the every BAC are sequenced and then matched with other BAC clones and overlapping are detected. This process reduces the number of BAC clones to be sequenced fully and it is more powerful technique.

The genome sequencing in *Arabidopsis thaliana* was followed by the efforts of sequencing the genomes in several crop plants like cereals, oilseeds, legumes, vegetables, etc. The plant genomes that are sequenced or targeted to be sequenced during the first decade of the present century have been listed in Table 22.2.

Table 22.2: Plant genomes sequenced/targeted to be sequenced by 2010

Plant	Chromosome number (n)	Genome Size (Mb)	Year of completion
Weeds			
Arabidopsis thaliana	5	125	2000
Cereals/Millets			
Oryza sativa	12	389	2005
Zea mays	10	2865	—
Sorghum bicolor	10	760	—
Oilseeds			
Brassica oleracea	9	600	—
Brassica rapa	10	500	—
Legumes			
Lotus japonicus	6	472	2001
Medicago truncatula	8	500	—
Vegetables			
Lycopersicon esculentum	12	950	—
Trees			
Populus trichocarpa	19	550	2004

Arabidopsis Genome Project

Three American groups namely Meyerowitz, Somerville and Goodman—at three different universities took the first step leading to the project on sequencing of the *Arabidopsis* genome later known as **AGI**. The first RFLP map was created by Meyerowitz, second by the Goodman group and later integrated. Initially the conventional '**clone-by-clone**' was dominant approach but later **shot gun approach** was followed using cDNA libraries and ESTs.

By 1999 the first report came on chromosome 2 and 4, but now the whole genome has been sequenced with all the informations. Few of them are as follows:

- The genome has approx. length of 145 Mbp, of which genes are present (coding region is 2–2.5 kb) at each 4–5 kb interval.
- Telomere and centromere regions are full of repeated DNA.
- Centromeric region has transposons and pseudogenes.
- Sometimes entire stretches of DNA and genes are duplicated between chromosomes (Chr. 4 has 37 genes duplicated on Chr. 5).
- Approx. 20% of the genes have signal sequences and target products into organelles such as chloroplasts or mitochondria. Entire mitochondrial genome of *Arabidopsis* is represented on chromosome no. 4.
- It has about 20,000–50,000 genes of various functional groups.

Rice Genome Project

Scientists from Japan started the Rice Genome Programme (**RGP**) in 1991. In 1998 the second phase of RGP was launched. Now 10 countries are participating in the **international Rice Genome Sequencing Programme** (IRGSP). *Monsanto* (2000) produced the draft of rice genome of Japonica variety which have explored the following:

- The rice genome is estimated to comprise of 420–466 Mbp of DNA.
- Monsanto produced 399 Mb of sequences from 3391 BAC clones in 2000 (not available to public).
- In 2002 the draft sequence of Indica rice variety was published in Beijing Genomic Institute and Syngenta published a draft sequence of Japonica variety.

  Similar to *Arabidopsis*, the early work began with random genomic clones, but in recent years cDNAs and ESTs have supplied a large number of markers. IRGSP has made the progress in preparing contig maps employing YAC and BAC.
- From the sequence data available, it has been reported that the total number of genes in rice is not much larger than in *Arabidopsis* and most of the genes of two plants show homology.
- According to available data on rice from Gen Bank, 28,282,731 bases of sequences has been submitted from the rice EST sequencing projects.
- Average rice gene is 2.2 kbp containing 3.9 exons and 2.9 introns. The density of rice gene is one gene per 5.7 kbp.

QUESTIONS

1. What is genetic engineering? Give a brief account of method of genetic engineering with reference to gene cloning.
2. Define recombinant DNA technique. Mention the enzymes involved and vectors used in this technique.

3. What are the criteria of an ideal vector? Write about pBR 322 and pUC vectors. Distinguish between expression and shuttle vectors.
4. Write a brief note on the applications of genetic engineering.
5. Write short notes on following techniques:
   (i) Colony hybridisation technique
   (ii) Blotting techniques
   (iii) Polymerase Chain Reaction (PCR)
   (iv) DNA sequencing
   (v) DNA fingerprinting
   (vi) Molecular mapping
   (vii) Restriction mapping
   (viii) Restriction Fragment Length Polymorphism (RFLP).
6. Write a short note on Plant Genome Project with special reference to *Arabidopsis* and rice.

23 BIOINFORMATICS

Bioinformatics is currently defined as the study of information content and information flow in biological systems and processes. It serves as the bridge between observations (data) in diverse biologically-related disciplines and the derivations of understanding (information) about how the systems or processes funcion and subsequently the application (knowledge). Though Hwa Lim, Father of Bioinformatics, coined the word 'bio-informatique' in 1987, but Temple Smith used the term '**Bioinformatics**' in 1991.

***In Silico* Biology**, a new area of Biology, has been developed in recent years because of generation of data in the field of genetics at an unprecedented exponential rate; the management and use of which requires the increasing use of computers and the relevant software.

Computational Biology, another term often used interchangeably with bioinformatics, although the former typically focuses on algorithm development and specific computational methods, while the latter focuses more on hypothesis testing and discovery in the biological domain.

Systems Biology, another area of research, emerged due to availability of enormous amount of molecular data and bioinformatics tools creating unprecedented opportunities to assemble and integrate this data into networks of genes, proteins and biochemical pathways.

Bioinformatics involves collection, storage, retrieval and analysis of biological data that has a lot of applications in pharmaceutical, agricultural and food industries, and in molecular genetics research. Biological data are generated from various genome sequencing projects, obtained by different techniques like DNA sequencing (genome and EST), 2D gel electrophoresis, mass spectroscopy (MS, MALDI, LC-MS), protein crystallization, microarrays (e.g., cDNA, oligos, peptide), molecular markers (e.g., RFLP, RAPD, AFLP, SNP). Thus bioinformatics is an interface of biological sciences, mathematics, physical sciences and computer sciences, i.e., the integrated field of biology and information technology.

23.1 SCOPE AND AREAS OF BIOINFORMATICS

The *scope of bioinformatics* has been depicted in Fig. 23.1A.

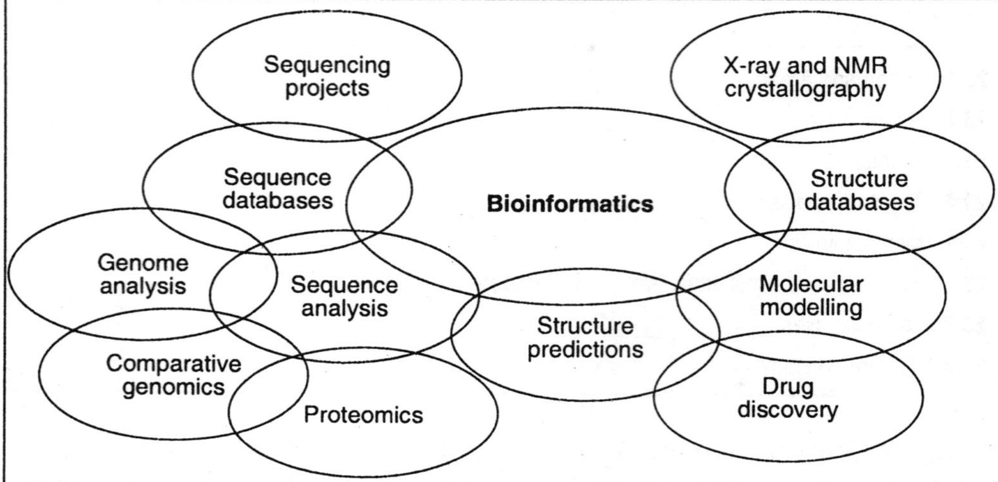

Fig. 23.1A: Scope of Bioinformatics

The subject bioinformatics is based on the following *areas* (Fig. 23.1B):

Genomics: Whole genome analysis and sequences; experimental analysis involving thousands of genes simultaneously.

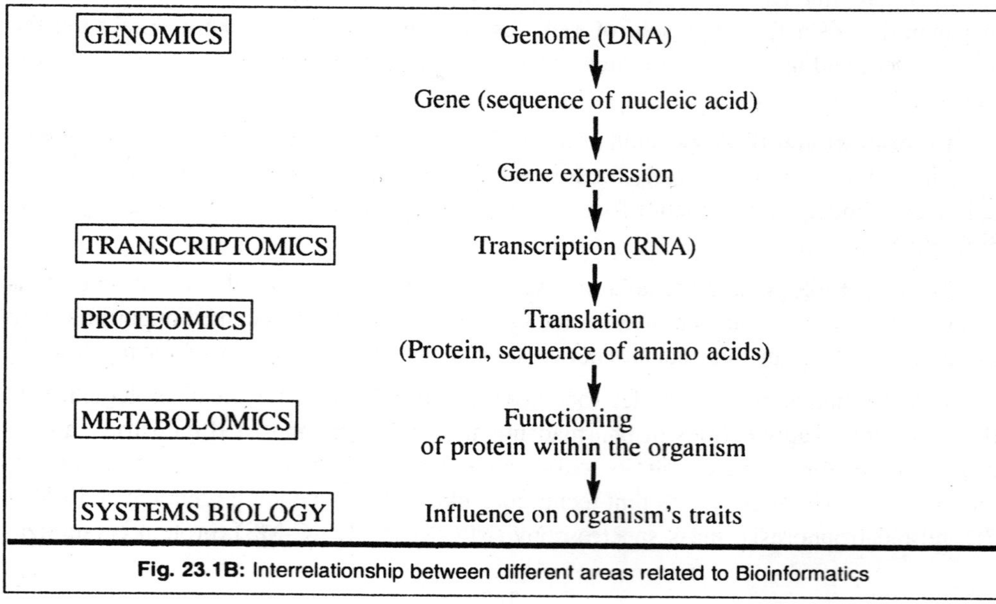

Fig. 23.1B: Interrelationship between different areas related to Bioinformatics

Transcriptomics: DNA chips and array analysis; expression arrays, comparative analysis between species and strains.

Proteomics: Proteome analysis of an organism.

Metabolomics: Analysis of different types of metabolite molecules.

Pharmacoinformatics: Medical application to genetic diseases.

Chemoinformatics: Other application as in forensic.

23.2 GENOMICS

The complete genetic content of an organism is **genome**, and the DNA obtained is called genomic DNA. This genomic DNA of prokaryote contains all the coding region and can be sequenced, whereas the DNA of eukaryotes includes both intron and exon sequences (coding sequence) as well as noncoding regulatory sequences such as promoter, and enhancer sequences. The subject genomics is the complete analysis of the entire genome of a chosen organism which involves the study of physical structure of the organism's genome or the genetic make up of an organism to know the number of genes present and the type of genes, i.e., to study the function of different genes.

Genome Sequencing Method: The method for sequencing gene or DNA segments through Maxam and Gilbert's chemical degradation method or Sanger's dideoxynucleotide synthetic method have been discussed in previous chapter. The method for *whole genome sequencing* through *shotgun assembly* has also been discussed in earlier chapter.

Whole Genome Sequence Data: Complete nucleotide sequences of nuclear, mitochondrial and chloroplast genomes have already been worked out in large number of prokaryotes and several eukaryotes. By the year 2005, among prokaryotes, approx. 1400 viral genomes, 250 bacterial genomes (230 eubacteria and 20 archaea), 500 mitochondrial genomes, 35 chloroplast genomes have been fully sequenced.

Among the eukaryotes namely the whole genome of *Saccharomyces cerevisiae* (yeast), *Coenorhabditis elegans* (nematode), fruitfly (*Drosophila melanogaster*), Human (*Homo sapiens*), Crucifer weed (*Arabidopsis thaliana*) and rice (*Oryza sativa*) have been sequenced already and data available for annotation studies.

The sequence data of eukaryotic nuclear genome is an important source of identification, discovery and isolation of important genes. This data is very much helpful in variety of application relevant to animal, plant and microbial biotechnology.

Functional and Structural Genomics: Once the whole genome sequence becomes available, the next step is to assign the function to different regions of genome. *Functional genomics* is the subject which is based on the use of genetic information to delineate protein structure, function, pathways and networks. Function may be determined by 'knocking out' and 'knocking in' expressed genes in model organisms such as worm, fruitfly, yeast or mouse.

Structural genomics involves solving the experimental structures of all possible protein folds which is playing an important role in high throughput function assignment.

Significance of Genomics: All the informations require input in probability theory, database management and manipulation, and computer science. This will help in

(a) identification of open reading frame sequences, (b) gene splicing sites (introns), (c) gene annotation (intergenomic comparisons) and (d) determination of sequence patterns of regulatory sites and gene regulations.

23.3 PROTEOMICS

The entire protein component of a given organism is called '**proteome**', the term coined by Wasinger in 1995. A proteome is a quantitatively expressed protein of a genome that provides information on the gene products that are translated, amount of products and any post translational modifications. Proteomics is an emerging area of research in the post-genomic era, which involves identifying the structures and functions of all proteins of a proteome. It is sometimes also treated as structural based functional genomics.

Methods of Proteome Analysis: Resolution and identification of proteins are possible by 2D-PAGE (Polyacrylamide Gel Electrophoresis) and Mass Spectrometry; comparative 2-D gel approach or protein chip approach helps to identify the proteins in up or down regulated system. A variety of other techniques are used for protein identification, the most common being matrix assisted laser desorption ionization time-of-flight mass spectrometry (MALDI-TOF MS). The hybrid electrospray ionization (ESI) method of quadrupole TOF-MS with its increased mass accuracy, is becoming increasingly established.

Scope of Proteomics: Proteomics deals with significant problems like:

(a) Identification of functional domains in protein sequences.

(b) Single, multiple protein alignment (homology).

(c) Determining sequence-structure, sequence function relationships (structural bioinformatics).

(d) Discovery of protein pattern and providing the framework for the analysis of signalling networks.

Reverse Genetics: The research in proteomics has made it possible to get the knowledge of all the proteins produced in an organism which may or may not be directly responsible for any phenotypic trait, but this may be helpful to know the functions of all the genes in that organism. This has made the approach of reverse genetics feasible because from the study of proteins, one can deduce the function of gene and the trait.

Significance of Proteomics: The knowledge of proteomics is complementary to genomics and has become a major thrust area of genetics, molecular biology and biotechnology research. From the whole genome sequence, functional genes are identified as open reading frames (ORFs) having initiation and termination codon, but ORF always does not represent any functional gene. Verification of gene product by proteome analysis serves a very useful purpose for 'annotation of the genome'. Post translational modification and expression of proteins, functions—all are regulated by various activities of the cellular metabolism—but all these are also due to proteolysis or protein-protein interactions.

Yeast Proteome: The complete sequence of the whole genome of yeast has been worked out in 1996, nearly 6200 genes are present in this small organism. In 2001, the functions of 93% of the proteins (5800 proteins) encoded in the genes were also elucidated. Later (2002-2005) networks involving the interactions have also been studied. Study of yeast proteome in great details will be very useful for the study of functions of genes of higher organisms including human, since yeast is the simplest eukaryote.

23.4 TRANSCRIPTOMICS

After the genome sequences are being completed, the new questions arise about the functional roles of different genes; the cellular processes in which they participate; mechanism by which the genes regulate the interaction of genes and gene products; changes in level of gene expression in different cell types and states.

To answer all these questions, the new area of science has got emerged which is transcriptomics. The transcription of genes to produce RNA is the first stage of gene expression. Although mRNA is not the ultimate product of a gene, but it is the first step of gene regulation and information about the transcript levels which is needed for understanding gene regulatory networks. The **transcriptome** is the complete set of mRNA transcripts produced by the genome at any one time. Unlike the genome, the transcriptome is extremely dynamic, all the cells of an organism contain same genome but the transcriptome varies considerably in different cells at different circumstances due to different patterns of gene expression.

Techniques for Transcriptome Analysis: High throughput techniques based on DNA chip/ microarray technology (i.e., cDNA microarrays, oligo microarrays), cDNA-AFLP (cDNA-amplified fragment length polymorphism) analysis, SAGE (serial analysis of gene expression) and a new technique MPSS (massively parallel signature sequencing) are used for transcriptome analysis. The cDNA microarray technique is based on the ability of the mRNA molecule to bind specifically to or hybridize to, its original DNA coding sequence in the form of a cDNA template spotted on an array. *DNA chip* is prepared on a silicon or glass based surface with regions of known sequence of chosen target DNA, which can hybridize with an unknown labelled DNA sample. Besides using cDNA clones as probes on an array, oligonucleotides of around 20 nucleotides can also be used as probe. Microarray experiments allow for comparison of gene expression profiles between two mRNA samples (e.g., treatment vs. control, or treatment 1 vs. treatment 2). The most important advantage of microarray-based technology is that large data sets from different experiments can be combined together in a single database, which allows gene expression profiles from either different samples or samples from different treatments to be compared with each other and analysed together.

Significance of Transcriptomics: As the transcriptome includes all mRNA transcripts in the cell, it reflects the genes that are being actively expressed at any given time, with the exception of mRNA degradation phenomenon such as transcriptional attenuation. The study of transcriptomics examines the expression level of mRNA in a given cell population.

Many DNA sequences that have been isolated shown to have no known function. However, if they show similar expression patterns to a characterized gene, it is likely that their functions are similar. It is sometimes possible to identify conserved regulatory sequnces of such genes. Ultimately, these studies promise to expand the size of existing gene families, reveal new patterns of coordinated gene expression across gene families and uncover entirely new categories of genes.

Furthermore, the product of any one gene usually interacts with those of many others, therefore, transcriptomics will provide precise knowledge on coordination among genes and their inter-relationships. It will also help to understand the integration of gene expression and function at the cellular level, revealing how multiple gene products work together to produce physical and chemical responses to both static and changing cellular needs.

23.5 METABOLOMICS

Genomics is concerned with the total complement of genes and proteomics, the analysis of the entire set of proteins, metabolomics has been defined as the qualitative and quantitative measurement of all low molecular weight **metabolites** in a given sample, cell or tissue and the integration of the data in the context of gene function analysis. In the 'post-genomic era', the genome expression profiling methods have come up at the level of the transcriptome, proteome and the metabolome. The comprehensive measurements of the working parts of biological systems at these different levels of organisation will allow a full and global comparison of the differences between cell types, tissues, organs and whole organisms (plants, animals and microbes) to prove unknown aspects of gene function, physiology and metabolism.

Areas of Metabolomics: Metabolic analysis can be divided into four general areas:

(a) *Target compound analysis:* The quantification of specific metabolites.

(b) *Metabolic profiling:* Quantitative and qualitative determination of a group of related compounds or of members of specific metabolic pathways.

(c) *Metabolomics:* Quantitative and qualitative analysis of all metabolites.

(d) *Metabolic fingerprinting:* Sample classification by rapid global analysis, without extensive compound identification.

Techniques for Analysis of Metabolites: As there are different kinds of metabolites available in cellular system so there is no single analytical method to detect the metabolite present in the extract. The detection also depends on the solvent used to get the tissue extract. A mixture of techniques such as gas chromatography, high pressure liquid chromatography and capillary electrophoresis are used to separate different metabolites according to various chemical and physical properties. Proton (1H) NMR can detect any metabolite containing hydrogen, gas chromatography (GC) provides high resolution compound separations and can be used in conjunction with a flame ionization detector (GC/FID) or a mass spectrometer (GC/MS). HPLC, with UV detection is a common method used for targeted analysis of plant materials and for metabolic profiling of individual classes. LC/MS and LC/NMR are powerful instruments for the structure determination. Mass analyser like fourier transform ion cyclotron resonance instruments (FT-ICR-MS) can help to obtain the 'mass profiles' from the crude extract without any chromatographic separation.

Significance of Metabolomics: Metabolomics is a relatively new discipline, and techniques for high throughput metabolic profiling are still under development. The advantage of metabolomic analysis is that the biochemical consequences of mutations, changes in the environment and treatment with drugs can be observed directly.

23.6 BIOINFORMATICS—INSTITUTES, WEBSITES, DATABASES, TOOLS

Institutes: Major public domain bioinformatics facilities are (Fig. 23.2):

(a) NCBI – National Centre for Biotechnology Information, USA.

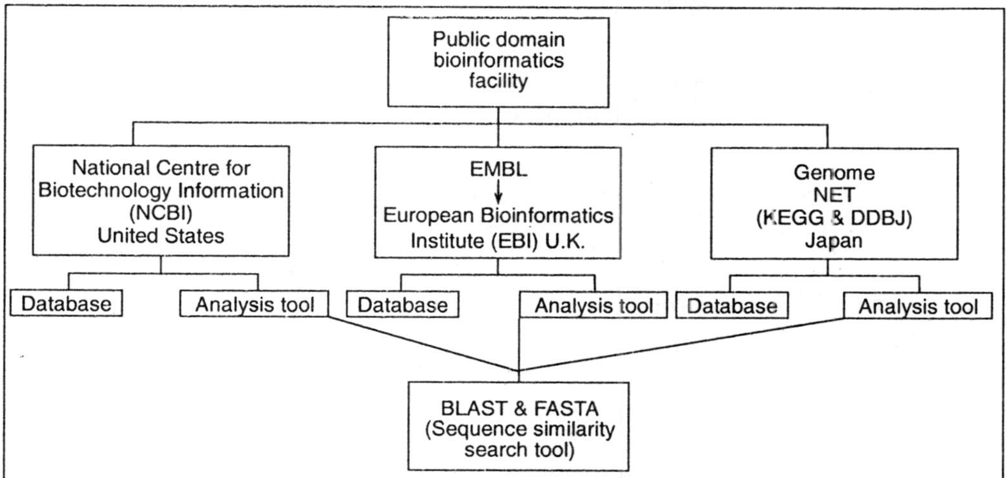

Fig. 23.2: Major publicly available databases and data mining tools

Table 23.1: Some important websites commonly used for bioinformatics (From P. K. Gupta)

Subject	Source	Link
Nucleic acid sequence	GenBank	http://www.ncbi.nih.gov:80enterz/query/fcgi? bd-Nucleotide
Genome sequence	SRS at EMBL/FBI	
	Entrez Genome	http://srs.cbiac.uk
	TIGR database	http://www.ncbi.nlm.nin.gov:80/entrez/query. bd=Genome
Protein sequence	GenBank	http//www./tigr.org/tbl/
	SWISS-PORT at ExPASY	http://www.ncbi.nlm.nin.gov:80/entrez/query.fcgi? bd=Protein
	PIR	http://www.expasy.ch/spro/
Protein structure	Protein Data Bank	http://www.ndrf.georgetown.edu
Post translational modifications	RESID	http://www.rcsb.org/pdb/
Biochemical and biophysical	ENZYME	http://www.ndrf.georgetwon.edu/pirwww/search/ textresid.html
	BIND	http://www.expasy.ch/enzyme
Biochemical pathways	Path DB	http://www.ncbi.nlm.nih.gov:80/entrez/query.fcgi? db=Structure
	KEGG	http://www//ncgr.org/software/pathdb
	WIT	http://www.genome.ad.anl.jp/eegg/
Microarray	Gene Expression Links	http://www.wit.mcs.anl.gov/WIT2/
Other interesting sites	European Bioinformatics	http//industry.ebi.ac.uk/valan/MicroArray
	Institute	http://www.ebi.ac.uk
	DNA Database of Japan	http://www.nig.oc.jp/home.html

(b) EBI – European Bioinformatics Institute, UK.

(c) SIB – Swiss Institute of Bioinformatics, Switzerland.

(d) Genome NET (KEGG & DDBJ), Japan.

Websites: Some important websites commonly used for bioinformatics are depicted in Table 23.1.

Databases: Bioinformatics is involved in storing the sequence information in different nucleic acid and protein databases which can be assessed by people all over the world through network technology.

Protein databases: The major protein databases are: PDB, SWISS-PROT, PROSITE, ExPASy, PIR, PRINTS, BLOCKS, PRODOM, Pfam, Inter Pro.

Nucleic acid databases: The major nucleic acid databases are: GenBank, DDBJ, Ref Seq, dbEST, NDB, CSD, EMBL.

Tools for Genetic Studies: In order to deal with molecular data, a wide range of softwares are now available which facilitated analysis of data in user-friendly manner. These tools are classified into four classes:

1. *Statistical analysis tools* (Table 23.2)
2. *Genome analysis tools* (Table 23.3)
3. *Sequence alignment tools* (Table 23.4)
4. *Genome annotation tools* (Fig. 23.3).

Table 23.2: Two major tools for statistical analysis of data for genetic studies

Name	Features
SPSS	Produces results in understandable/presentable formats (in tabular and graphical form), and also allows sharing of results with others using a variety of reporting methods.
SAS	Used for a variety of statistical analyses, information storage and retrieval, data modification and programming, report writing, simple and advanced statistics, and file handling. The SAS/STAT software used for all kinds of statistical analyses; SAS/ETS software is used for time series analysis and SAS/IML is used for interactive matrix manipulation.

Table 23.3: Tools for genome analysis (From P. K. Gupta)

Purpose	Tools and their features
(i) *Tools for molecular-marker development*	
SSR mining	TROLL (tandem repeat occurrence locator); TRF (tandem repeat finder); SSRFinder; MSL (microsatellite locator); SSRIT (SSR identification tool); SSRSEARCH; MISA (Microsatellite); Sputnik; SSR Primer, etc.
SNP mining	SNP pipeline (PHRED, PHRAP and DEMIGLACE); PolyPhred; PolyBayes; AutoSNP; SEAN; PARSESNP; SniPpER; SNPF (Single Nucleotide Polymorphism Finder); TRACE_DIFF, etc.

Contd.

Purpose	Tools and their features
Primer designing	(i) Normal PCR - Primer3, Prima; (ii) site-directed mutagenesis - Primer Generator, PrimerX; (iii) RNAi - RNAi Design.
(ii) *Tools for large-scale genome sequencing*	
Base calling/quality assignment	PHRED: (i) calls bases (base calling means assigning bases to peaks of different colours) and (ii) assigns quality values to the bases.
Fragment assembly	PHRAP: (i) assembles shotgun DNA sequences, constructs contigs, and provides extensive information about assembly including quality values for contig sequences.
Vector trimming	Cross match; compares a set of sequences to a set of vector sequences and trims vector sequences to produce vector-masked versions of these sequences.
Viewing and editing contigs	Consed/Autofinish: allows viewing, editing, and finishing sequence assemblies created with PHRAP.
Sequence assembly program	CAP3: clips 5′ and 3′ low-quality regions of sequences; uses base quality values (Q-values) in computation of overlaps between sequences; multiple sequence alignments, and generation of consensus sequences.
(iii) *Tools for map construction, map alignment and QTL/LD analysis*	
Map preparation	MAPMAKER, JoinMap, etc.
QTL analysis	(i) single locus analysis: QTL Cartographer, PLABQTL; (ii) two locus analysis: QTLMapper/QTLNetwork, PseudoMarker, etc.
LD analysis	LDA, TASSEL, STRUCTURE, GDA, (LD estimation); LDMAP, ALLASS (LD mapping); QTDT (association tests and TDT), etc.
MAP comparison	cMAP: for comparing the order of loci in different maps for the same species or in the maps of related genera (for comparative genomics).
Anchoring of BAC contigs to genetic maps	iMAP: allows unambiguous assignment of BAC contigs to the genetic map, i.e., assists in preparing genetically anchored BAC contig maps.

Table 23.4: List of sequence alignment tools and their specific features (From P. K. Gupta)

Name of tool	Features
(i) *Pair-wise sequence alignment tools*	
FASTA family	(i) [fastaa] and [ssearch], compare a protein sequence against a protein database or a DNA sequence against a DNA database. (ii) [fastx/fasty], translates a DNA sequence into amino acid sequence of a protein and compares such a protein against a protein database. (iii) [tfastx/tfasty], compares a protein sequence against a DNA sequence translated in three forward and three reverse frames.

Contd.

Name of tool	Features
PatMatch	Searches for short nucleotide or peptide sequences (3-30 nt or amino acid).
BLAST (earlier versions)	(i) BLASTp compares protein sequence against a protein sequence; (ii) BLASTn compares a nucleotide sequence against a nuclotide sequence; (iii) BLASTx compares a nucleotide sequence translated in all reading frames against a protein sequence; (iv) tBLASTn compares a protein sequence against a nucleotide sequence translated in all reading frames; (v) tBLASTx compares the six frame translations of a nucleotide sequence against the six frame translations of a nucleotide sequence; (vi) BLASTz compares long stretches of nucleotides (>2 kb).
BLAST (later versions)	(i) VecScreen (for match with sequence of plasmids, phage, cosmids, BACs, PACs and YACs). (ii) IgBLAST (for analysis of immunoglobulin sequences). (iii) MegaBLAST (for comparing a set of ESTs with a set of genes, and grouping them in clusters). (iv) SNP BLAST (for match query sequence with sequence carrying SNPs available in the SNP database, dbSNP). (v) PowerBlast (for alignment of query sequence with all sequences rather than one to one alignment).
ParAlign	Exploits parallelism involving division of a major task into small tasks that are performed in parallel to perform a very rapid computation.
Protein Engine and Transeq	Used first for translating DNA sequences into proteins, which are then used for further similarity searches and for the study of 3-D structures encoded.
(ii) *Multiple sequence alignment tools*	
ClustalW	Multiple sequence alignments of divergent sequences; evolutionary relationships can be seen via viewing Cladograms or Phylograms.
Clustal X	New windows interface for the ClustalW multiple sequence alignment program.
MSA	Multiple Sequence Alignment: for alignment of several protein or nucleic acid sequences.
SAGA	Sequence Alignment by Genetic Algorithm: for multiple sequence alignments.
PHYLIP	Following are some of the 30 programmes, which PHYLIP carries for phylogenetic analysis: (i) PROTPARS and PROTDIST use protein sequences; (iii) DNAPARS uses DNA sequence; (iv) NEIGHBOR uses distance matrix prepared using raw data. (v) DRAWGRAM draws a tree based on output from other phylogeny programme.

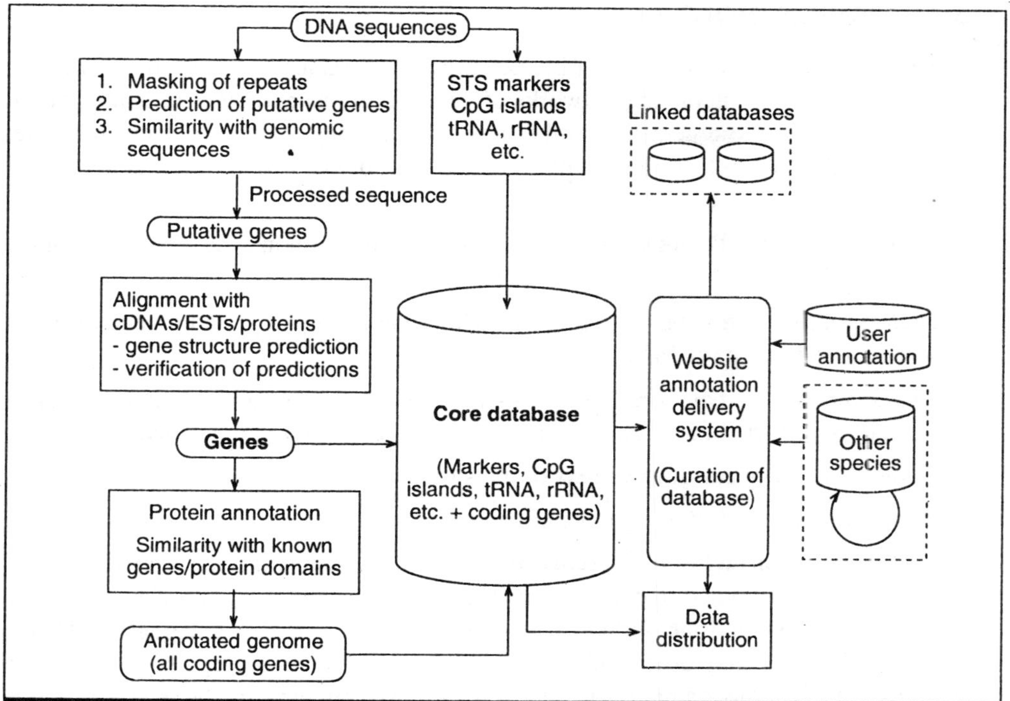

Fig. 23.3: Basic outline of an automatic genome annotation pipeline and delivery system. The diagram can be partitioned into (i) analysis pipeline stage (shown on the left), and (ii) delivery system (shown on the right); the data is distributed between the two stages as the central core database. The sequence data enters the pipeline and flows down through the gene prediction analysis before being stored in the core database. The website delivery system is a key feature as it serves to integrate the gene predictions and features from the core database alongside multiple external databases (From P. K. Gupta)

23.7 BIOINFORMATICS IN INDUSTRY

Bioinformatics has a great impact in agriculture, health-care and on environment which will bring bioindustrial revolution.

Food Industry: Functional genomics is playing a major role in food biotechnology industry. The complete genome sequence information available in different databases generates information that can be used for finding metabolic pathways, improving cell factories and development of novel preservation methods.

Agricultural Industry: Crops are improved by producing plants that have disease resistant genes to pathogens like fungi and bacteria. Homology searches, finding conserved motifs and molecular modelling is useful in identifying disease resistant genes. Fungicides that can efficiently kill the pathogens are designed by molecular modelling.

Pharmaceutical Industry: Chemoinformatics is playing a key role in pharmaceutical industry to design new drug targets from genomic data at a very faster rate. Disease causing genes are identified using the tools of genomics and proteomics. Drug lead identification and drug optimization became easy using the tools of genomics and proteomics. Pharmaceutical industry is also using the sequence information in the production of vaccines and therapeutic proteins.

23.8 SYSTEMS BIOLOGY

Systems biology is a new area of biology where an organism is viewed as an assembly of integrated and interacting networks of genes, proteins and biochemical reactions, which put life into the organism. Unlike molecular biology which focus on molecules, such as sequence of nucleotides and proteins, systems biology focus on systems that are composed of molecular components.

Scope of Systems Biology: To understand the whole system of an organism following are the keys:

1. *Understanding structure of the system,* such as gene regulatory and biochemical networks, as well as physical structures.
2. *Understanding of dynamics of the system,* both quantitative and qualitative analysis as well as construction of theory/model with powerful prediction capability.
3. *Understanding of control methods* of the system.
4. *Understanding of design methods* of the system.

Goals and Approaches of Systems Biology: The fundamental characteristics of systems biology are embodied in data and knowledge integration, the comprehensiveness of the data acquisition, and the ability to digitalize biological output. **Integrative Biology** and **Digital Biology** may be considered synonymous. Systems biology seeks to explain the biological phenomenon not on a gene-by-gene basis, but through the interaction of all the cellular and biochemical components in a cell or an organism. Systems biology is still

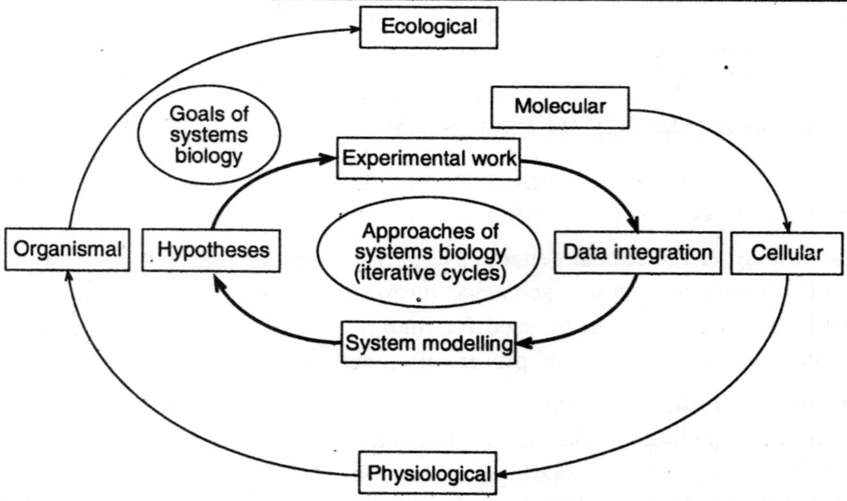

Fig. 23.4: A diagrammatic representation of goals and approaches of systems biology. (i) the outer thin spiral depicts the goals of systems biology, which involve describing processes across all layers of biological organization including molecular, cellular, physiological, orgnismal and ecological, and (ii) the inner thick circle shows iterative process of experimentation at a global level, which includes the following: data integration, system modelling, and generation of hypotheses, which lead to the design of new experiments that start a new round of the cycle. Each iteration refines the model and deepens our biological understanding of the system. Ultimately, such models may be used in a predictive mode and applied to human health-care and agriculture (From P. K. Gupta)

in its infancy but that has to be explored and the area that we believe to be the main stream in biological sciences in this century (Fig. 23.4).

The Institutes for Systems Biology, USA; Ottowa Institute of Systems Biology, Canada, have been established to cater the needs of this new emerging discipline.

Steps in Systems Biology: There are five simple steps to understand the systems biology –

1. Defining of input and output of a system.
2. Defining the relevant parts of the system under study.
3. Different states of inputs are used and the relationship between these states, and the output are quantified.
4. The changes in system states are related to output using one of a variety of mathematical tools, such as principal component analysis.
5. With the help of a computer, the original network map is prepared on the basis of the results of the previous experiments.

Systems Biology in Plant Science: The entire genome sequence of *Arabidopsis* and rice has been published in the year 2000 and 2005 respectively. The large scale c-DNA sequencing projects are rapidly progressing for plants like *Medicago*, corn, wheat, soyabean, sugarcane, poplar, etc. Functional genomics addresses the question—what is the function of all these genes? Computational biology integrates data and ultimately helps us to understand the functioning of whole biological systems. Plant systems biology deals with the aspects like cell cycle, root development, cell death, bud formation, flower formation, leaf development, plant-microbe interactions, etc. Each system is highly complex, and biologists now have the tools at hand to view the global behaviour of their preferred model systems and to better select the genes that are likely to play key roles in the regulation of entire processes. Furthermore, cloning of ORFs (Open Reading Frames),

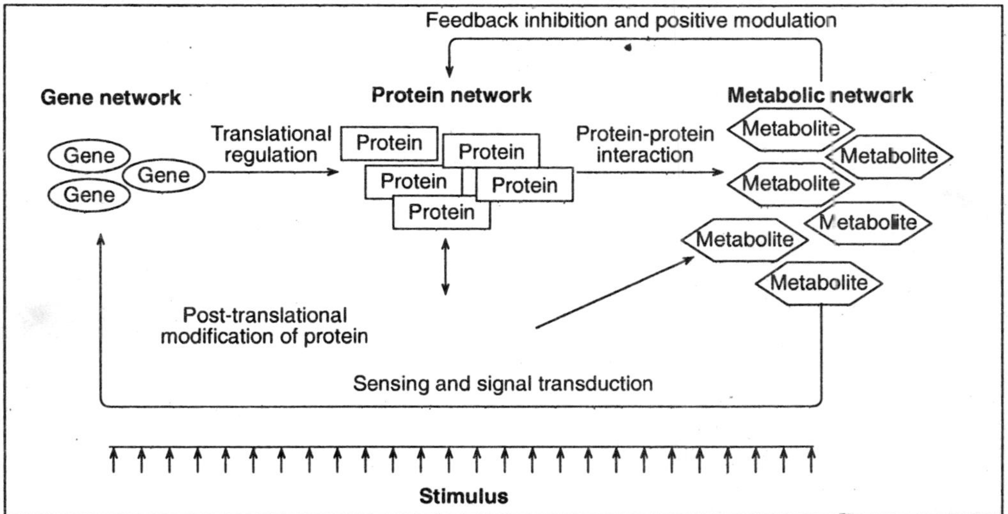

Fig. 23.5: An outline of a metabolic network system, where responses to a developmental and/or environmental stimulus involve (i) feedback regulation and (ii) regulation of gene expression due to signal transduction (From P. K. Gupta)

promoters and making of new constructs can change or improve the gene expression which may be helpful.

Significance of Systems Biology: With the development of new tools and techniques the complex system between genes, proteins and other metabolites can be understood; they do not work in isolation, they interact among themselves in highly organized and complex ways. Systems biology deals with these complex interactions, which may be the biosynthetic pathways or translational modifications of proteins or the signal transduction pathway leading to expression of a set of specific genes (Fig. 23.5). For example, the immune system is not the result of a single mechanism or a set of genes; instead, the interactions of numerous genes, proteins, metabolisms and the organism's external environment, produce immune responses to fight infections and diseases. Thus, a study of all aspects of any system in an organism makes the subject of systems biology.

QUESTIONS

1. Define Bioinformatics. Mention the areas on which it is based. State principal websites, databases, tools in bioinformatics.
2. Give an account of Genomics, Transcriptomics, Proteomics, Metabolomics.
3. What is systems biology? Discuss its scope and significance.

24 GENE TRANSFER AND TRANSGENIC PLANTS

The aim of plant genetic engineering is to transfer a foreign gene to a recipient species to alter the genome of the recipient and to give a new characteristic of the foreign gene. The plant with the introduced foreign gene is termed as '**Transgenic**' and the process is termed as '**Transgenesis**'. The crop thus genetically modified is called '**GM' (genetically modified) crop**.

In case of higher plants, cells or protoplasts can be cultured and used for regeneration of whole plants. The alien gene of interest can be delivered in different systems to obtain the transgenic plants. Besides cultured cells, protoplasts, meristem cells, pollen or zygotes can also be used. So for gene transfer in plants it should be decided first which will be target cell/tissue for gene transfer experiment. Several gene transfer methods are available which may be direct or indirect. The gene to be transferred has to be attached with a carrier i.e. vector.

24.1 VECTORS OF GENE TRANSFER

The common features of vectors used for transformation of eukaryotic cells include — (i) multiple cloning sites (unique restriction site), (ii) bacterial origin of replication, and (iii) selectable marker gene which will allow to select the proper transformed cell.

Ti and Ri Plasmid: The potential use of the *Agrobacterium* Ti plasmid (*A. tumefaciens*) and Ri plasmid (*A. rhizogenes*) as vectors arises from the ability of the bacterium to somehow transfer and stably integrate a piece of the plasmid DNA into the plant nuclear genome. The transferred DNA is known as T-DNA and carries several genes which are expressed within the plants and becomes integrated into the plant chromosomal DNA. This integration seems to depend on the presence of two repeated sequences located at either end of T-DNA. Other genes on the Ti plasmid include those for attachment of the bacterium to plant cell walls for transfer of T-DNA into the plant cell and for the uptake and catabolism of the appropriate opine. The non-integrating region of the Ti plasmid which is

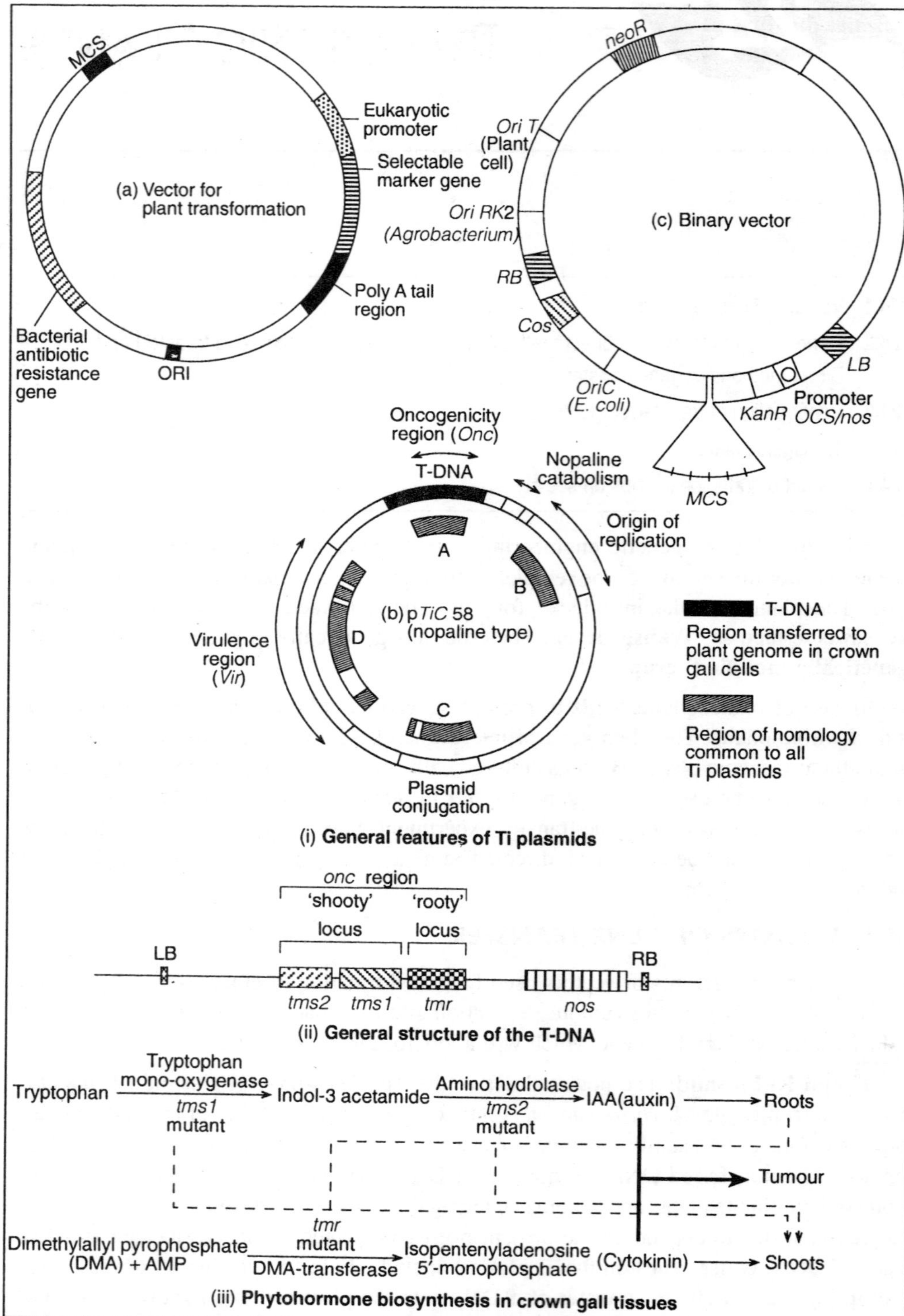

Fig. 24.1: Vectors for plant transformation

essential for transfer and integration of the T-DNA is the *vir* region, which is located near to the T-DNA.

Structure of Ti plasmid: Most Ti-plasmids have four regions in common (Fig. 24.1):

Region A (T-DNA): Responsible for tumour induction; comprising of T-DNA, flanked by 25 bp repeat sequence (left and right borders).

T-DNA consists of

(i) *Onc* region of three genes, representing 'shooty locus' and 'rooty locus', responsible for auxin and cytokinin synthesis.

(ii) *Os* region, responsible for synthesis of opines (octopine and nopaline).

Region B (ori): Responsible for origin of replication.

Region C (con): Responsible for conjugation.

Region D (vir): Responsible for virulence, also called *vir* region, comprising of six operons (A, B, C, D, E and G).

Plant Viruses as Vector: There are two groups of plant viruses which contain DNA—the cauliflower mosaic viruses which have double stranded DNA and the gemini viruses which have single stranded DNA. The Cauliflower Mosaic Virus (CaMV) is often cited as the most likely potential vector for introducing foreign genes into plants. CaMV has one useful feature is that the naked DNA being infective, is able to enter plant cells; inside the cells, the DNA is replicated and encapsidated within virus particles, which then invade the rest of the plant. As the CaMV DNA does not become integrated into the chromosomal DNA, so it is uncertain that it goes to all cells.

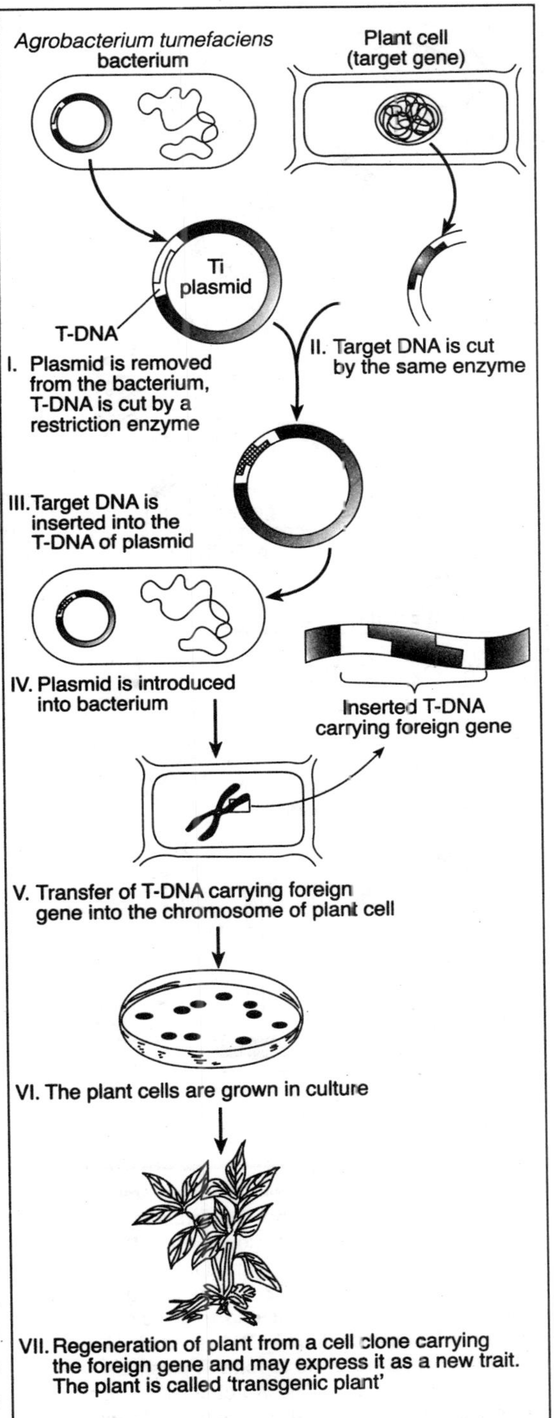

Fig. 24.2: Ti plasmid used to transfer foreign genes into plants

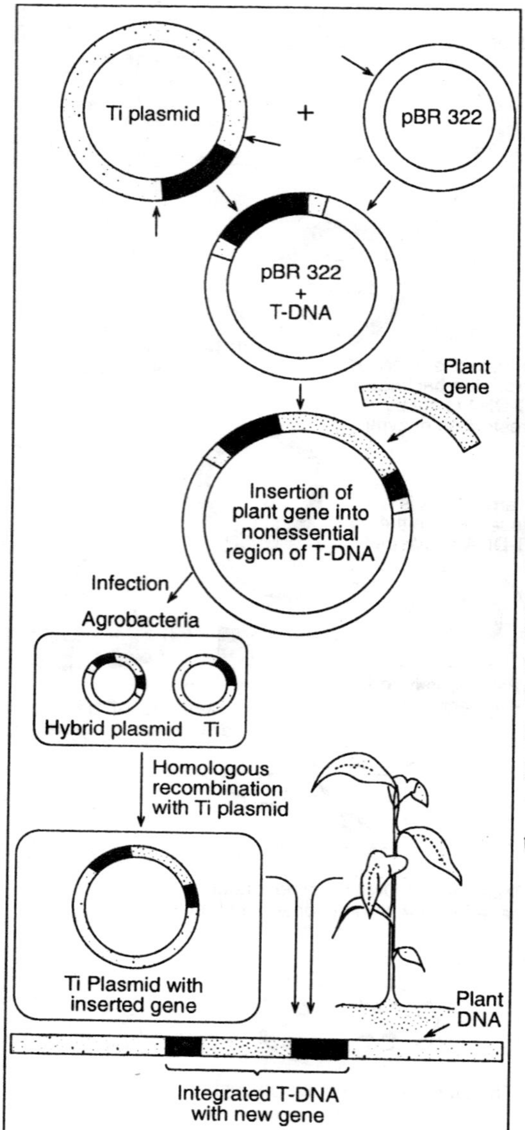

Fig. 24.3: A procedure for using the Ti plasmid as a vector. First the T-DNA of the Ti plasmid is cut out with restriction enzymes and cloned into pBR 322. Next a foreign plant gene is inserted into the cloned T-DNA region in the pBR 322. The resulting hybrid plasmid is mixed with Agrobacteria colonies containing normal Ti plasmids; their T-DNA recombines with that of the hybrid plasmids to form Ti plasmids carrying the foreign gene. The Agrobacteria carrying the foreign gene are used to infect plants, which incorporate the modified T-DNA into their chromosomes (after Watson, Tooze, Kurtz, 1983)

24.2 GENE TRANSFER METHODS

Several methods are available for gene transfer.

I. Indirect Method (*Agrobacterium* Mediated Gene Transfer)

The transformation of a plant can be carried out directly by using *Agrobacterium* spp. which is a common bacterium causing crown gall tumour in legumes. This bacterium carries a plasmid with T-DNA which is capable of being integrated into the host chromosome. If a foreign gene is introduced in the plasmid of the bacteria and a plant tissue or cell suspension is grown in culture along with the bacteria, then ultimately the foreign gene can be transferred into the nucleus of the plant or more precisely, in the functional position of the host genome (Figs. 24.2 and 24.3). The two enzymes, restriction enzyme and ligase, play the most significant role in the process of transgenesis.

The Ti plasmids of *A. tumefaciens* has the several features for use as vector:

(i) They contain one or more T-DNA regions.

(ii) The plasmid contains a *vir* region.

(iii) The plasmid has its own origin of replication.

(iv) They contain a *con* region enabling conjugative transfer.

(v) They have the genes for catabolism of opines (a class of amino acids/sugar conjugates).

A. Vector Strategy: Vector strategies used in *Agrobacterium* mediated gene transfer to deliver the transgene into the genome of the plant are cointegrative vector strategy and binary vector strategy.

(a) *Cointegrative vector strategy:* This is a Ti based plasmid vector developed by

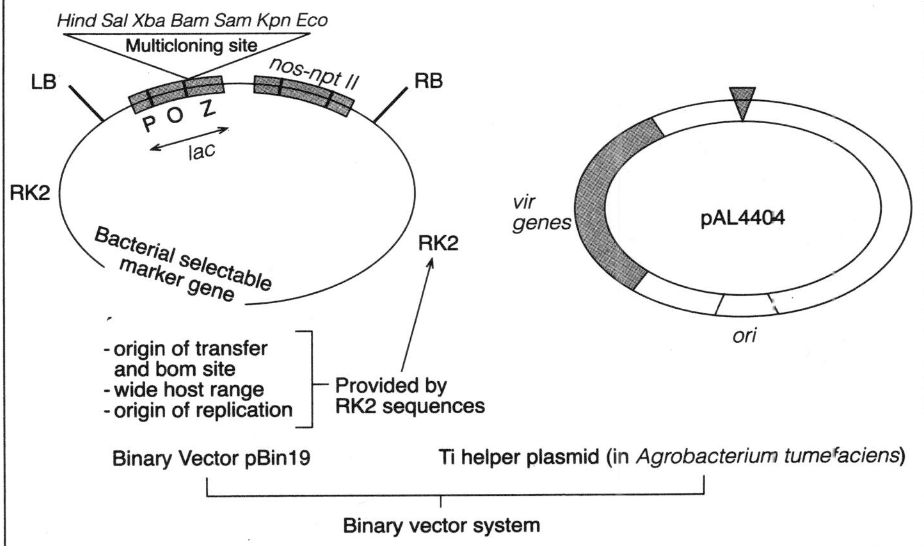

Fig. 24.4: A typical binary Ti vector system (from Chawla)

homologous DNA recombination between engineered plasmid (containing gene to be transferred along with selectable markers) of *E. coli* and resident Ti plasmid (disarmed) of *A. tumefaciens*. This results in integration of the entire engineered plasmid into T-DNA, e.g., pGV 3850::1103.

(b) *Binary vector strategy:* It is a two-plasmid system in *Agrobacterium tumefaciens* in which one plasmid contains the virulence gene (helper plasmid) and another plasmid contains T-DNA borders carrying the selectable marker and the foreign DNA to be transferred (binary vector). Binary vector designed to replicate in both *E. coli* and *Agrobacterium* and thus can shuttle between both, hence also called *shuttle vector,* e.g., Helper plasmid (pAL 4404) + Binary vector (p Bin19) (Fig. 24.4).

B. The Mechanism of T-DNA Transfer and Integration: The transfer of T-DNA and its integration into the plant genome can be divided into the following steps (Fig. 24.5).

(a) *Signal recognition by Agrobacterium: Agrobacterium* perceives signals, such as phenolics (acetosyringone) and sugars released from wounded plant cells which are involved in phytoalexin and lignin biosynthesis, these signals indicate the presence of competent plant cells.

(b) *Attachment to plant cells:* Attachment of *Agrobacterium* to plant cells is a two-step process, involving an initial attachment via a polysaccharide (the product of *attR* locus) which forms a mesh of cellulose fibres. Several chromosomal virulence (*ChrV*) genes are also involved in the attachment of bacterial cell to plant cells.

(c) *Vir gene induction:* Vir A (a membrane linked sensor kinase) senses phenolics like acetosyringone, autophosphorylates and activates Vir G. Then Vir G induces expression of all other *vir* genes. Many sugars like glucose, galactose, xylose — all enhance the *vir* gene induction; it requires also the participation of chromosomal *vir* gene.

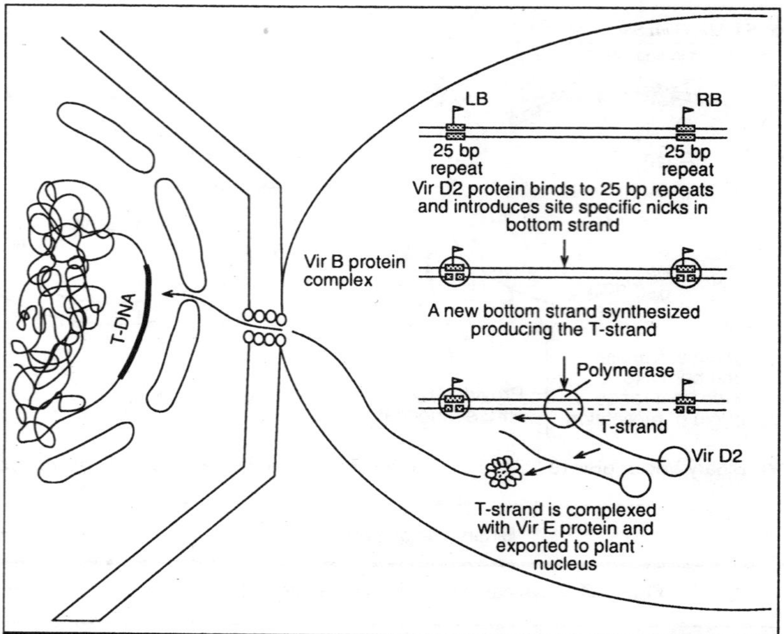

Fig. 24.5: Molecular events in the transfer of the T-DNA to plant cells (Redrawn from D. Grierson. 'Plant Genetic Engineering' 1991)

(d) *T-strand production:* The left and right borders are recognized by $VirD_1/D_2$ complex and $VirD_2$ produces single stranded nicks on the DNA, covalently attaches there and replaces the strand by repair synthesis.

(e) *Transfer of T-DNA out of the bacterial cell:* The T-DNA/Vir D_2 complex is exported from the bacterial cell by a T-pilus composed of proteins encoded by vir B operon and vir D_4.

C. Basic Steps for *Agrobacterium* Mediated Transformation: The steps involved are:

(a) Selection of plant tissue or explant;

(b) Co-cultivation with *Agrobacterium*;

(c) Inhibition of *Agrobacterium* growth;

(d) Selection of transformed tissue;

(e) Regeneration from selected tissue (transgenic plant);

(f) Confirmation of transgenic plant.

(a) **Selection of plant tissue or explant:** Suitable plant tissue, to be used as a source of explants (which has good regeneration ability), is removed from the donor plant and sterilized (if the plant is not grown in sterile condition). The explants may be decapitated seedlings, cells, protoplasts or leaf tissue, callus, etc.

(b) **Co-cultivation with *Agrobacterium*:** The tissue or explant is cut into small pieces and placed into a culture of *Agrobacterium* (which contains the suitable vector containing foreign gene) for about 30 min., a process known as co-cultivation. During this period, the bacteria attach to the plant tissue, and the excess culture is blotted off and placed on medium for co-cultivation.

(c) **Inhibition of *Agrobacterium* growth:** The incubation of the explants with *Agrobacterium* is allowed to continue for 2-3 days to permit the transfer of T-DNA to the plant cells. Then the explants are removed from the medium and washed in an antibiotic solution and further transferred onto antibiotic (bacteriostatic) containing medium to inhibit the growth of *Agrobacterium*.

(d) **Selection of transformed plant cells:** The explants are then transferred onto the selective media containing proper selective agent to encourage the growth of transformed tissue.

(e) **Regeneration from the transformed tissue:** The selected tissue part (putatively transformed), grown on selective media, are then transferred onto the regeneration media for shoot regeneration either by organogenesis or by embryogenesis in presence of proper selective agent. The shoot apices come out, those are then transferred in rooting media to get the whole plant.

(f) **Confirmation of the putatively transformed plant:** The transgene expression is examined either through foreign protein expression or any phenotypic character expression. The presence of foreign DNA can be examined either through PCR or Dot Blot or Southern Blot experiment. The confirmed transgenic plants then are transferred to soil to get the next generation plant. The whole process has been depicted in Fig. 24.6.

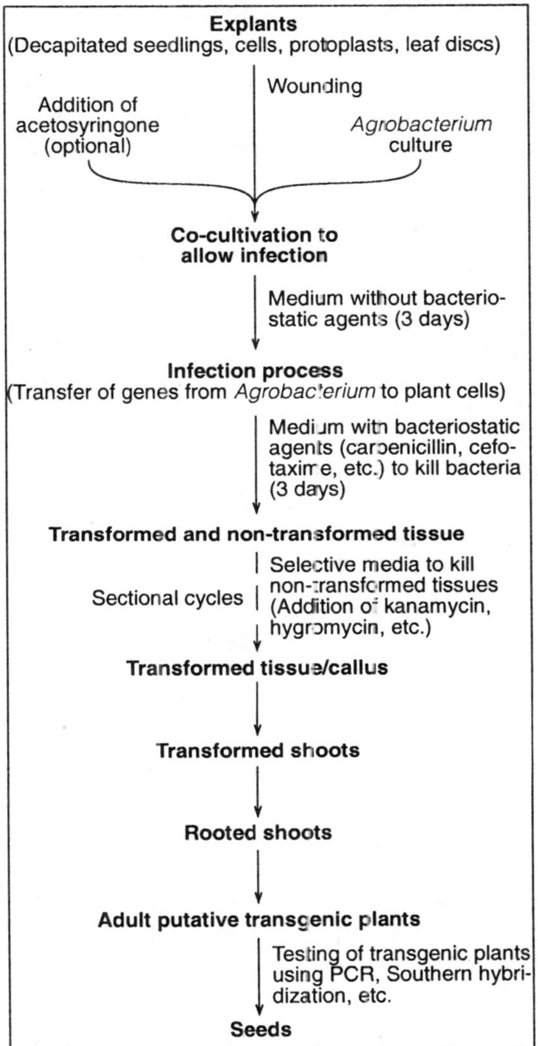

Fig. 24.6: A general scheme of *Agrobacterium*-mediated transformation of explant (from Chawla)

II. Direct Methods of Gene Transfer

1. **Transfer via Electroporation:** Electric field membrane permeabilization is based on the fact that the electric pulses can open the cell membrane and allow penetration of alien DNA. Heat shock, in combination with electroporation has been used resulting in a higher efficiency of the transformation. The method requires protoplasts, because electroporation is only efficient on cell membranes. These protoplasts, harbouring the alien DNA, have to be developed into callus and regenerated into a plant.

2. **Transfer via Polyethylene Glycol (PEG):** The chemical compound polyethylene glycol (PEG) changes the pore size of the cell membranes which enhances the probability of an alien DNA molecule penetrating into the cell. Heat shock may enhance the uptake of DNA. The method is mainly applied to protoplasts. The transfer of DNA via polyethylene glycol has been used for monocots as well as dicots.

3. **Transfer via Biolistics:** The principle of transfer via microprojectile bombardment (Fig. 24.7A) is to shoot particles coated with DNA into selected tissues or cells by particle gun. The gun may be driven by either air-pressure or electric discharge. The particles may consist of either tungsten or gold carrying the DNA. Any growing plant tissue may be used for this method, but the plant material has to be regenerated via callus formation.

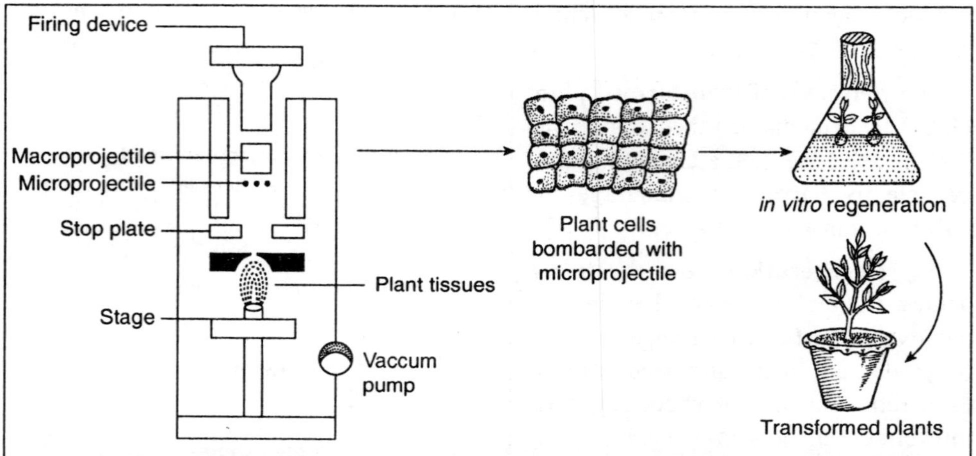

Fig. 24.7A: Overview of microprojectile bombardment by gun powder device

4. **Liposome Mediated Gene Transfer:** Liposomes are small lipid bags, in which a large number of plasmids can be enclosed. Those can be fused with the protoplasts

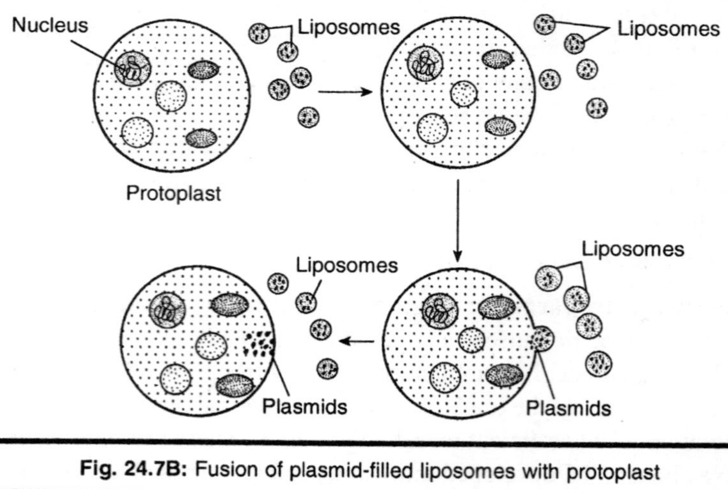

Fig. 24.7B: Fusion of plasmid-filled liposomes with protoplast

using PEG (Fig. 24.7B). This technique offers many advantages: (i) protection of DNA/RNA from nuclease action; (ii) low cell toxicity; (iii) stability and storage of nucleic acids due to encapsulation in liposomes; (iv) high degree of reproducibility; and (v) applicability to a wide range of cell types.

5. **Transfer via Microinjection:** The method involves transfer of a very small solution of DNA into a selected cell by injection with a capillary tube or a micropipette (Fig. 24.7C). The procedure has to be performed under a microscope. Often the specially designed micromanipulator is used for microinjecting the DNA.

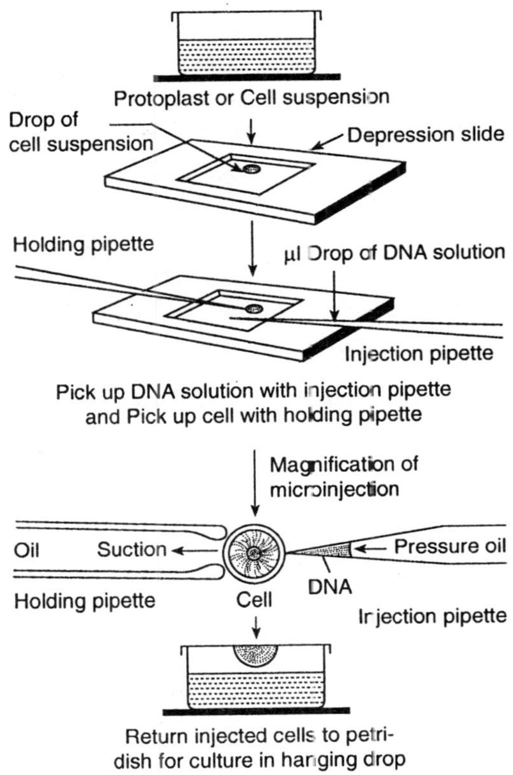

Fig. 24.7C: Microinjection by the holding pipette method (developed by Crossway et al. 1986) followed by hanging drop culture

6. **Transformation using Pollen or Pollen Tube:** There lies a possibility of transfering DNA material with the pollen tube germination method, i.e., the pollen tube with alien DNA can fertilise the egg cell giving rise to transgenic plant.

7. **Silicon Carbide Whiskers:** The use of silicon carbide whiskers in transforming plant cells is very simple and reproducible. These whiskers are very cheap and can penetrate cells making pores for DNA uptake.

8. **Transformation by Ultrasonication:** There are recent reports of transformation of cultured plant tissue when sonicated with alien DNA carrying the marker genes.

9. ***In planta* Transformation:** *In planta* transformation technology is carried out in tissues or cells of growing organs of a plant in field conditions bypassing tissue culture technique. This includes seed and embryo incubation, *in planta* DNA inoculation, floral meristem dipping, *Agrobacterium* spraying, vaccum infiltration, etc.

24.3 CONFIRMATION OF TRANSFERRED GENE

However, even after transformation, it is essential to confirm the integration of genes, for the success of transgenesis. The incorporation can be confirmed through (i) the use of markers, (ii) Southern blotting, (iii) Northern blotting, (iv) Polymerase chain reaction.

(i) **Marker and Reporter Gene:** Transformation in eukaryotic system is a very low frequency event. It is, therefore, important that there should be some means for selecting the transformed plant tissue. There are certain *selectable marker genes* present in vectors that facilitate the selection process. In transformed cells the selectable marker genes are introduced through vector. For each marker gene there is one substrate which is a toxic substance like an antibiotic or any herbicide. The transformed cells are cultured in medium containing the toxic substance or antibiotic which facilitates the selection of transformed tissue to grow as the marker gene confers the resistance property to the transformed cell only.

The most widely used selectable marker gene is the *npt II* (neomycin phosphotransferase) gene that confers resistance to kanamycin, G418, neomycin. Other antibiotic resistance genes against bleomycin (*ble*), spectinomycin and hygromycin (*hpt*) are used quite widely. The resistance genes against some herbicides like chlorosulphuron (*als*) and bialaphos (*bar*) are also in use (Table 24.1).

Table 24.1: Marker genes and Reporter genes used for plant transformation

Marker/Reporter gene	*Abbreviation*	*Source of gene*	*Detection/Assay*
1. Bleomycin resistance	*ble*	*E. coli*	Resistance against bleomycin
2. Hygromycin phosphotransferase	*hpt*	*E. coli*	Resistance against hygromycin
3. Neomycin phosphotransferase II	*npt II*	*E. coli*	Resistance against kanamycin/ geneticin
4. Phosphinothricin acetyl-transferase	*bar/pat*	*Streptomyces hygroscopicus*	Bialaphos, glufosinate resistance
5. β-glucuronidase	*gus*	*E. coli*	Fluorimetric/histochemical/ non-rad technique
6. Luciferase	*luc*	*Photinus pyralis*	Luminescence

In addition there is an alternative procedure where there will be no selection pressure imposed on cells to grow further. The samples from grown tissue are tested for expression of the marker genes which are described as *reporter gene* or *scoreable gene* or *screenable gene*. Reporter genes are used in plant transformation vectors, both as a means of assessing gene expression by promoter analysis and as easily scored indicators of transformation. Ideally, reporter genes should be easy to assay, preferably with a non-destructive assay system and there should be little or no endogenous activity in the plant to be transformed. At present only a small number of reporter genes are in widespread use in plant transformation vectors, these being β-glucuronidase (*gus*)—causing blue colour (Fig. 24.8), green fluorescent protein (*gfp*), luciferase genes (*lux* and *luc*)—causing luminescence and to a lesser degree (although it is widely used in animal systems) the chloramphenicol acetyl transferase gene (*cat*).

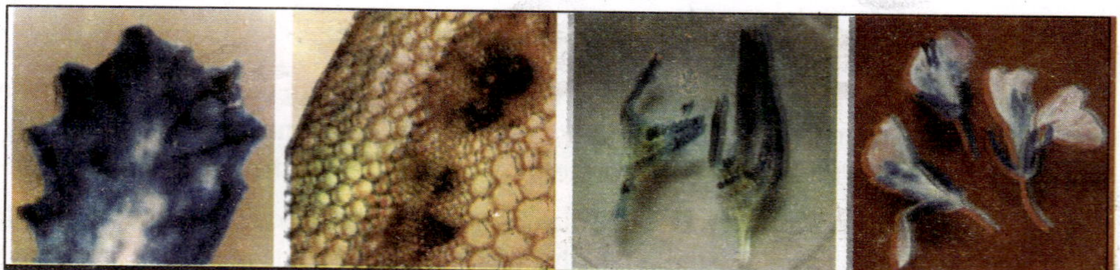

Fig. 24.8: *Gus* gene expression in different plant parts of transgenic plant

(ii) **Southern Blotting or Southern Hybridization:** It is a molecular technique which combines electrophoretic separation of DNA fragments with nucleic acid hybridization. In this method, the desired DNA to be tested is cut by restriction enzymes, gel electrophoresed and denatured gel fragments are transferred to nitrocellulose filter sheets. The denatured DNA probe—the alien DNA, is tagged with radioisotope and added under renaturing condition (lower temperature) to the filter containing target DNA. The hybridized target DNA probe sequence can be detected by radioactive spots or auto-radiography.

(iii) **Northern Hybridization:** It is an analogous method of Southern hybridization where the recipient or target sequence is the mRNA (messenger RNA) and not DNA. Northern hybridization becomes necessary to check that the introduced DNA in the chromosome is able to function or express, and thus undergoes transcription to produce the desired messenger. Northern Blot confirms the presence of desired messenger RNA-complementary to the alien DNA, which is used as probe.

(iv) **Polymerase Chain Reaction (PCR):** The genomic DNA is isolated from the suspected transformed plant tissue; designing the appropriate primers for the transgene, the PCR technique (Refer chapter 22.9) is used for gene amplification. Amplified transgene can be easily detected through gel electrophoresis only. This technique is now easier and quicker method for detection of transgene as it does not require much purified plant DNA and within a very little period of time many samples can be screened for the presence of transgene.

24.4 TRANSGENIC PLANTS

The application of recombinant technique for horizontal transfer of gene in plant system has several limitations. Plant genome is exceedingly complex, involving a huge amount of DNA, which can accommodate sequences equivalent to 6-7 million genes in an average crop plant. Several genes are located outside the nucleus, namely the chloroplastids and mitochondria. Moreover, every gene, in any system, has its promoter, structural sequences and terminator—all of which require to be identified before introduction. For instance, the nitrogen fixing gene of bacteria—*Rhizobium* has 17 genes of 24 kilobases, distributed under eight operational units or operons. All these units require to be identified and isolated, along with their promoter and terminator for introduction. The other limitation in the plant system, is the absence of a large number of vectors, which would carry the desired gene to the recipient. The most commonly used vector is the Ti plasmid of *Agrobacterium*.

However, despite these limitations, the progress achieved so far with the existing technology can be considered as moderate (Table 24.2). In almost all major crops, for crop improvement, some of the agronomic traits could be introduced by transgenic approach where the genetic variability has been utilized from other gene pool. Here, some of the traits improved through transgenic approach are being discussed.

I. Resistance to Biotic Stresses

Resistance to biotic stresses can be categorised under following headings:

1. Insect Resistance
2. Virus Resistance
3. Disease Resistance

Table 24.2: Traits inserted in transgenic plant

Crops	*Trait*	*Gene product*
Cotton, Tobacco, Cowpea	Insect tolerance	Bt toxins Protease inhibitors
Tobacco, Tomato	Virus tolerance	Coat proteins Antisense ribozyme
Rice	Better nutritional quality (Golden rice)	Vitamin A
Potato	Tolerance to fungi	Chitinases Phytoalexins
	Tolerance to bacteria	Lysozymes
Mustard	Male sterility	Ribonuclease Gluconase
	Drought tolerance	Proline Betaine
Wheat, Tobacco	Herbicide tolerance	Superoxide dismutase
	Oxidative stress tolerance	Glutathione reductase
Tobacco	Cold tolerance	Glycerol-1-phosphate acyltransferase
	Salinity tolerance	Mannitol-phosphate dehydrogenase

1. **Insect Resistance:** Transgenes either from microorganism or from plant or other origin can be introduced into plants to increase the level of insect resistance.
   - *Bacillus thuringiensis* (Bt) encodes for the Bt toxin which are active against Lepidopteran insect has been introduced in many of the crop plants like rice, maize, fibre yielding plant cotton, vegetables like tomato and potato against many insects.
   - Cowpea Trypsin Inhibitor (CpTi) from cowpea, α-amylase Inhibitor from *Phaseolus vulgaris,* Lectins from *Galanthes nivalis* (snowdrop)—all these different kinds of proteinase inhibitor genes give resistance against some insect pests. Transgenic tobacco, pea and potato have been obtained carrying these genes show resistance against insects.
2. **Virus Resistance:** A number of coat protein genes from different virus groups have been found to provide resistance by inhibiting the virus replication of the challenge virus during the infection. Many of the transgenic tobacco, potato, tomato, soyabean, rice, maize have been produced using the coat protein gene of TMV (Tobacco Mosaic Virus), PVX (Potato Virus X), TYLCV (Tomato Yellow Leaf Curl Virus), BMV (Bean Mosaic Virus), RSV (Rice Strip Virus), MDMV (Maize Dwarf Mosaic Virus), etc. Some antiviral protein or ribosome inactivating protein (RIPs) from different plant sources have been identified to give the resistance against virus. Transgenic tobacco and potato have been produced with PAP (Pokeweed Antiviral Protein) are found to be resistant to TMV.

3. **Disease Resistance:** A large number of plant defense response genes encoding antimicrobial proteins are now have been cloned. The products of these defense response genes may include (i) Chitinase and PR proteins, (ii) Ribosome inactivating proteins (RIPs), (iii) Antifungal proteins (AFPs), (iv) Enzyme for phytoalexin, (v) Phenolics, Lectins, etc. Here is a list of transgenic plants generated in various crops for resistance against fungal and viral diseases:

Crop	*Gene transferred*	*Controlled pathogen*
	PR-proteins	
Tobacco	Chitinase, PR	*Alternaria longipes* *Peronospora tabacina*
Tomato	Chitinase	*Cercospora nicotianae*
Potato	PR 5	*Phytophthroa infestans*
Rice	Chitinase	*Rhizoctonia solani*
	Antimicrobial proteins	
Potato	Bacteriophage T4-lysozyme	*Erwinia carotovora*
Rice	Cecropin	Bacterial pathogen
Tobacco	Barley RIP	*Rhizoctonia solani*
	Phytoalexins	
Rice	Stilbene synthase	*Pyricularia oryzae*
Tobacco	Stilbene synthase	*Botrytis cineria*

II. Reistance to Abiotic Stresses

Several abiotic stresses such as drought, salinity and extreme temperatures have a common consequences of causing cellular water deficit or osmotic stress. Some osmo-protectants within cells like proline, soluble carbohydrates like polyols and some ammonium compounds like glycine, betaine help to lower the osmotic potential and maintaining turgor within the cell. Here is a list of few transgenic plants derived against resistance to abiotic stresses:

Crop plants	*Compound* (over-expression)	*Origin*	*Type of stress*
Tobacco	Mannitol	Mannitol-1-phosphate dehydrogenase from *E. coli* (mtlD)	Tolerance to salinity
Tobacco	Proline	γ-pyrroline-5-carboxylate synthetase	Tolerance to osmotic stress
Rice	Glycine\ Betaine	Choline Oxidase from *Arthrobacter globiformis*	Tolerance to salinity
Tobacco	Desaturase	o-3 or ω-3 fatty acid desaturase	Chilling stress
Tobacco	Fructan	*Sac* B from *Bacillus*	Resistance to drought

III. Resistance to Herbicide

Transgenic plants against various herbicides such as phosphinothricin (bialaphos), glyphosate, sulfonylurea, imidazolinones, bromoxynil, atrazine, 2, 4-D, etc. have been generated in different food crops, vegetables, horticultural and ornamentals species. Herbicide resistant transgenic plants like cotton, flax, canola, corn and soyabean have already released. The following genes are used from different origin, active against different herbicides.

	Herbicide	*Gene*	*Inhibition*
a.	Bialaphos/Basta Glufosinate	*bar*	Glutamine synthase
b.	Glyphosate	*aro* A *gox*	EPSPS (5 enol pyruvyl shikimate 3-phosphate synthase)
c.	Bromoxynil	*bxn*	Photosystem II

IV. Transgenics for Quality

1. **Transgenic for improved storage:** Transgenic tomato '**Flavr Savr**' with delayed ripening developed by Calgene, UBA using antisense RNA technology of ACC deaminase. Monsanto produced transgenic tomato to express a gene from *Pseudomonas* which produces metabolites other than ethylene from ACC. Another gene SAM hydrolase from bacteriophage T_3 also reduced the ethylene production in tomato giving more shelf life.

2. **Transgenic flowers have longer shelf life:** Transgenic carnations expressing antisence ACC oxidase and producing flowers with little ethylene and marked delay in senescense.

3. **Transgenics for flower color and shape:** Delphinidin is the compound responsible for blue colour in flower, the Flavonoid 3′–5′ hydroxylase (F 3′–5′H) is the enzyme responsible for blue colour formation has been cloned and transgenic violet carnation "Moondust" is available. Modification of intensity of flower color from white to pink has been obtained by antisense RNA technnology of chalcone synthase and transgenics have been obtained in *Petunia* carnation, *Gerbera,* etc.

4. **Transgenics for nutritional quality:** Tailor made starch i.e. *starch with reduced level of amylose* and higher amount of starch have been developed in potato. Trehalose is an additive which improves the taste quality in processed food. Transgenic tobacco has been produced by Calgene producing trehalose. Over-expression SPS (Sucrose Phosphate Synthase) holds promise for achieving the sugar content increase, transgenic tomato has been produced with over-expression of SPS.

Considerable progress has been made in the improvement of nutritional quality of legume *storage protein expression.* Brazil nut 2S albumin protein gene has got expressed in *Arabidopsis,* rapeseed, soyabean, french bean and potato. The soyabean glycine gene and also the lysine have been introduced into rice to increase the protein content and also to make it more digestable.

To save a large population of children from **vitamin A** deficiency, the improved variety of transgenic rice have been developed capable of synthesizing the early intermediate Geranyl-Geranyl diphosphate (GGPP) which can be used to produce the uncolored carotene phytoene by expressing the enzyme phytoene synthase in endosperm. The transgenic '**Golden Rice**' has been obtained by introducing the whole β-carotene synthesis pathway related genes into rice endosperm capable of producing β-carotene, the precursor of vitamin A synthesis.

The **iron content** increase is possible with the use of ferritin transgene from *Aspergillus fumigatus,* transgenic rice plants have been produced with two fold increase in iron content and high activity of phytase.

V. Transgenics for Agricultural Traits

1. **Transgenics for male sterility and production of hybrid seeds** (*using sterility-fertility system*): The character of male sterility have wide use in securing fertile hybrids in interspecific hybridisation. This character has now been widely used for transgenesis to secure male sterile transgenic plant. This method involves induction of **male sterility-fertility restoration.** Bacterial coding sequence for ribonuclease (*barnase*) has been attached with anther specific promoter which will kill the pollen grain by production of cytotoxic element in tapetal cells. Another gene which is a ribonuclease inhibitor (*barstar*) gene has also been isolated from the same bacteria and used to produce transgenic plants that nullify the effect of barnase gene. The transgenic plant with barstar gene was developed which was crossed with the male sterile transgenic plant with (*barnase*) gene. The F_1 progeny is restored due to suppression of cytotoxic ribonuclease activity in the anther. The transgenic male sterile line already have been obtained in *Brassica.*
2. **Transgenics for improvement of uptake capacity of different nutrients** from soil by introducing the transporter gene which reduce the dependence on fertiliser application.
3. **Transgenics for good harvest index** is attempted using the phytochrome genes by regulating the flowering or fruit development at a time.
4. **Transgenics for seedless fruit** i.e., without pollination the ovary may mature into fruit with high pulp quantity.
5. **Transgenics for terminator seed production technology:** A lethal gene (RIP)—Ribosome Inhibiting Protein interferes the synthesis of other proteins in the plant cell, which prevents the germination of seeds when the gene is attached to a particular promoter and expressed only in seed. As a result the desired traits will be expressed in first generation seeds only, not in the successive generations. This mechanism of terminator seed production will help always to maintain the pure line by companies which has marketed it.

VI. Transgenic Plants as Bioreactors

1. **Transgenic plants for PHB synthesis:** Transgenic plants are also being produced to be used as bioreactors for synthesis of various types of chemicals, the technique is called as **Molecular pharming**. Polyhydroxy Butyrate (PHB), an polyester with thermoplastic property, is synthesized by transgenic *Arabidopsis* after introducing the transgene from *Alcaligenes euthophus.*
2. **Transgenics for production of vaccines and other peptides:** The use of transgenic plants as the producer of many peptides of pharmaceutical use and the vaccines are

becoming significant now-a-days. Human Serum Albumin (HSA) is being successfully produced in transgenic tobacco and potato plants.

Transgenic maize plants have been produced capable of producing avidin. Potato tubers have also been used as the bioreactor for high level production of a recombinant single chain FV (SCFV) antibody. The edible plant vaccine against diarrhoea expressed in potato, but as it is cooked before eating, heat may denature the vaccine protein. The vaccine protein gene is now have been introduced into banana which can be eaten raw. Transgenic tobacco plant is capable of producing HBV (Hepatitis B virus) surface protein.

24.5 IMPACT OF TRANSGENIC CROPS ON SOCIETY

The future prospects in this area is very exciting with the indications that novel genes can be introduced to generate plant lines useful for production of materials ranging from pharmaceuticals to biodegradable plastics, vaccines, minerals, vitamins, fatty acids, carbohydrates, etc. The path from inception of the agricultural biotechnology industry to field production or commercial production faces many obstacles ranging from scientific and technological challenges to legal and regulatory hurdles, economic factors and social concerns. Considerable opposition of GM crops is the presence of antibiotic resistant gene within the transgenic whose spreading may result in the development of antibiotic resistant pathogen. Therefore the alternative technique are being introduced apart from the antibiotic resistant marker gene and targeted modification by homologous recombination technique.

Future GM crops will focus on the use of plant genes that control important agronomic traits. However, various concerns with direct relevance to our agricultural system have been raised, particularly the use of terminator seed technology has become an important and valid issue. The new approaches of GM plant development may be directed more towards environmental issues; such as to reduce the contamination of land by residues from explosive (Phytoremediation), detection of toxic chemicals in environment, biofuel production, etc.

There is no report todate of any harmful environmental effect or any other hazard or unforeseen behaviour of transgenic plants in thousands of field trials. There are many instances of commercially accepted GM crops throughout the world.

QUESTIONS

1. Define transgenic plants. Describe different gene transfer methods to develop transgenic plants.
2. What are GM crops? Give suitable examples with desired traits.
3. Distinguish between direct and indirect method of gene transfer. Discuss different direct methods of gene transfer.
4. State vector mediated gene transfer method with special emphasis on *Agrobacterium* mediated transfer.
5. Give an account of transgenic plants developed for human welfare.

25 APPLICATIONS OF PLANT BIOTECHNOLOGY

Biotechnology has been increasingly applied in the different aspects of human welfare like *agriculture, industry, health-care, environment.*

25.1 AGRICULTURE

Biotechnological research is increasingly contributing to the improved production and propagation of new cultivated varieties of plants that have a direct role in agricultural practices in relation to nutritional quality, food output and other useful substances.

1. **Micropropagation:** Mass propagation of crop and forest plants is an improtant application of micropropagation technique. The development of embryos from somatic cells in culture resulted in artificial seed production. This technique involves three stages (a) establishment of culture (b) regeneration of plants (c) transfer of plants from test tube to soil.

   Regeneration of plantlets in cultured plant cell and tissues has been achieved in many trees of high economic value. Many of the studies are aimed at large scale micropropagation of important trees yielding fuel, pulp, timber, oils and fruits. Therefore, clonal forestry and horticulture are gaining an increasing recognition as an alternative for tree improvement. In recent years, the interest has aroused in commercializing the *in vitro* propagation of forest trees. This will bring about refinement in the existing procedures to make micropropagation more cost effective. For betterment and improvement of tree plants of high economic value, genetic transformation and *in vitro* regeneration have been done in many angiospermic and gymnospermic plants.

2. **Induction and Selection of Mutant:** Different physical and chemical mutagens are used in the plant explants of different species to generate mutants. Now the mutants can be used to select out the variant cell lines which are resistant to antibiotics, amino acid analogues, chlorate, nucleic acid base analogue, fungal toxin, environmental

stresses (salinity, chilling, high temperature, aluminium toxicity) and herbicides, etc. Single cell or the protoplast culture systems have proved to be valuable for mutagenesis since the presence of discrete cells in these substances is more effective to cause mutation, and isolation of mutant line is more easier.

3. **Production of Somatic Hybrids:** The protoplasts can undergo fusion under certain favourable conditions and the fused product can give rise to somatic hybrid plant which offers (i) the possibility of hybrid formation of widely unrelated forms, (ii) an asexual means of gene transfer either of whole genome or of partial genome.

Through successful production of somatic hybrid plants at the tetraploid and hexaploid levels, both for inter- and intra-specific fusions, characters from sexually imcompatible wild species are transferred to the cultivar. Other approaches to genetic manipulation include the irradiation of donor protoplasts with useful characters, to fragment their genomes, followed by fusion to tetraploid acceptor protoplasts. Protoplast fusion also provides a means of transfering cytoplasmic traits into another genomic background. Intergeneric somatic hybrids have been produced in many genera like '*Raphanobrassica*', obtained through fusion between *Raphanus sativus* and *Brassica campestris*, '*Solanopersicon*', obtained through fusion between *Solanum tuberosum* and *Lycopersicon esculentum*, etc. The technique of *cybrid* production has been utilized for transfer of cytoplasmic male sterile character, as has been done in case of *Nicotiana, Brassica* and *Petunia*.

4. **Production of Transgenic Plants:** Genetic engineering can be used to introduce genes into a plant, which do not exist in any member of the same plant family. If genetically engineered plants are to be used commercially, then the following criteria are to be satisfied: (a) introduction of the gene(s) of interest to all plant cells; (b) stable maintenance of the new genetic information; (c) transmission of the new gene to subsequent generations; (d) expression of the cloned genes in the correct cells at the correct time.

A number of useful traits, mostly single gene, that have been transferred to get the transgenics for various purposes are:

(a) *Insect-pest resistance plants:* Using gene transfer technique the *Bt* gene (Cry I protein from *Bacillus thuringiensis*) has been transferred to many crop plants like rice, cotton, tomato, potato, etc. and insect resistant plants (*Bt* crops) have been developed.

(b) *Herbicide resistant plants:* Using biotechnological approaches many herbicide resistant crop plants have been obtained as in *Brassica*, tomato, corn, cotton, soyabean, etc. which are resistant against glyphosate (Roundup®), L-phosphinothricin (Basta®), etc.

(c) *Virus resistant plants:* Viral coat protein genes can be introduced to get the virus resistant plants as has been done in tomato, potato, squash, papaya, etc.

(d) *Resistance against bacterial and fungal pathogens:* Several examples are available where the transgenic plants against bacterial and fungal pathogens have been developed. The chitinase gene have been introduced in tobacco to get the resistance against brown spot; acetyl transferase gene has been introduced in tobacco to get the resistance against wild fire disease.

(e) *Improvement in nutritional quality:* Nutritional quality can be improved by introducing the genes for production of cyclodextrins, vitamins, amino acids, etc. Transgenic potato has been obtained to produce cyclodextrin molecule; the transgenic rice named as '*Golden rice*' has been obtained to produce provitamin-A which has opened the way for improving the nutritional standards; Ama-I gene has been introduced in potato. Starch content has been increased in transgenic potato.

(f) *Improvement of quality of seed-protein and seed-oil:* Recombinant DNA technology has been used successfully for improvement of protein quality in seed as has been done in pea plant which is rich in sulphur containing amino acids; lysine rich cereals have also been produced. Oilseed rape has been made transgenic which has the modified seed oil quality, i.e., low erucic acid. Reduced linolenic acid containing flax and high stearic acid containing soyabean and safflower also have been produced.

(g) *Improvement of quality for food-processing:* 'Flavr-Savr' variety of tomato has been raised which shows bruise resistance as well as delayed ripening.

(h) *Male sterility and fertility restoration in transgenic plants:* Male sterile transgenic plants have been produced with '*barnase*' gene which has the cytotoxic product tagged with anther specific TA-29 promoter, and another set of plants have been produced to restore the fertility factor with the help of '*barstar*' gene tagged with the same promoter. F_1 hybrids from these two sets of transgenics should facilitate the hybrid seed production for crop improvement.

(i) *Production of stress tolerant:* Several projects are going on for transgene application to develop the tolerance against different abiotic stresses, e.g., cold (tobacco), drought (mustard), salt (rice).

[For more details refer chapter 24].

25.2 INDUSTRY

Industrial biotechnology involves the commercial exploitation of a variety of processes and techniques related to microorganisms, plants, animals and human beings.

1. **Improvement in Fermentation Products:** This achievement can be done in different ways—by selection of improved strain, by transgene application into the microorganism, by using cheaper raw material, by manipulation of medium constituent as well as by simulation of the reactor (adjustment of different cultural conditions like pH, temp., etc.). Products of microbial fermentation include primary metabolites, secondary metabolites, enzymes, proteins, capsular polysaccharides and cellular biomass (single cell protein). Organic compounds of microbial fermentation includes ethanol, acetone, butanol, gluconic acid, etc.

2. **Microbial Production of Synthetic Fuels:** Important fuels can be produced by using many microbes which include ethanol, methane, hydrogen and hydrocarbons. *Zymomonas mobilis* produces ethanol twice as rapidly as yeasts from carbohydrates. Methane which is used in various industrial purposes can be produced by Clostridia, Bacteriodes, *Sclenomonas, Butyrovibrio*, etc. from the waste.

3. **Microbial Mining or Bioleaching:** The process of bioleaching recovers metals from ores which are not suitable for direct smelting because of their low content. The application of bioleaching process is of particular interest in case of uranium ore. *Thiobacillus ferroxidans* is the commonest organism which is involved in case of copper and uranium ore processing.

4. **Production of Biopesticides and Biofertilizers:** Environment friendly, non-toxic, cost effective pesticides are commercially produced from bacteria (*Bacillus thuringiensis*), fungi (*Beauveria netahizium*) which have been successfully tested for several crops.

   Several N_2 fixing baceria (*Rhizobia, Azotobactor, Azospirillum*) blue green algae (*Anabaena, Nostoc, Aulosira*), mycorrhizae and phosphate solubilising bacteria (*Bacillus, Thiobacillus*) are important commercially used biofertilizers.

5. **Microbial Biomass and Single Cell Protein Production:** Microbial product of commercial significance is the microbial biomass (the microbial cells themselves), e.g., commercially produced yeast cells, bacteria (*Methylophilus methylotrophus*), flavouring cheese from fungal biomass (*Penicillium roquefortii*). Single cell proteins are the dried cells of micoorganisms such as algae, certain bacteria, yeasts, moulds and some higher fungi. The protein percentages for various single cell proteins are high.

6. **Production of Antibiotics:** About 100 antibiotics are produced commercially by microbial fermentation process (Table 25.1).

Table 25.1: Some commercially produced antibiotics

	Antibiotic	*Source organisms*
1.	Penicillin	*Penicillium* spp.
2.	Bacitracin	*Bacillus licheniformis*
3.	Cephalosporin	*Cephalosporium acremonium*
4.	Actinomycin	*Streptomyces caespitosus*
5.	Mitomycin	*S. antibioticus*
6.	Erythromycin	*S. erythreus*
7.	Streptomycin	*S. griseus*

7. **Production of Enzymes:** Bulk of the enzymes are obtained from microbial source by fermentation process. Now several plant enzymes are being used like 'papain' from *Carica papaya*, 'bromelain' from *Ananas comosus*, 'ficin' from *Ficus glabrata*. These enzymes are used in meat tenderizing, as protein hydrolysates in beer industry, in clinical application, etc. Amylase used in textile industry, protease in leather industry, cellulase and xylanase in paper industry, catalase, pectinase, invertase, lactase in food industry are produced through biotechnology.

8. **Production of Secondary Metabolites from Cultured Plant Cell:** In recent years it has been shown that spectrum of compounds can be produced in culture which is beyond the ability of whole plants. By using different precursors several novel compounds of biomedical importance can be obtained. Pharmaceutical compounds like

shikonin is being produced as secondary products with the use of two-stage bioreactor by stimulating the growth phase with the application of different growth regulators. Serpentine can be obtained from *Catharanthus*, pseudoephidrine from *Ephedra*. Plant cells also can be used to accomplish certain changes in the structure and composition of some industrially important chemicals. This conversion by means of a biological system is termed as *biotransformation*, e.g., digoxin, a cardiovascular drug, produced from digitoxin obtained from *Digitalis lanata*.

25.3 HEALTHCARE

A wide range of biological products of human healthcare including polypeptides, proteins, growth factors, hormones, enzymes, vaccines, immunoregulators, antibodies, etc. can be produced through biotechnology.

1. **Human Proteins and Hormones:** There are many human proteins which have long been believed or known to have therapeutic potential and their increased production has been achieved by using recombinant DNA technology (Fig. 25.1).

The gene coding for insulin with two polypeptides has been synthesized. Each synthetic gene was linked to a plasmid near the end of the β-galactosidase gene of *E. coli*. After the gene expression and the translation of mRNA into protein, the two polypeptides were cleaved from the enzyme and linked to form the complete insulin molecule. Different workers also synthesized complementary DNA from RNA of rat pancreas with the help of reverse transcriptase which was inserted into pBR 322 plasmid in the middle of the gene for penicillinase. The plasmid also contained the structural genes for pro-insulin. The hybrid protein synthesized in the bacterial cell was penicillinase + proinsulin from which insulin could be separated by trypsin.

Production of interferon took a vital position when human leucocyte interferon was engineered by yeast cells. A DNA sequence coding for human leucocyte interferon was attached to the yeast alcohol dehydrogenase gene in a plasmid and introduced into cells of *Saccharomyces cerevisae*.

The first human peptide hormone synthesized in a bacterial cell was somatostatin, which is one of a group of hormones secreted by hypothalamus, controls the release of several hormones from the pituitary. The synthetic gene

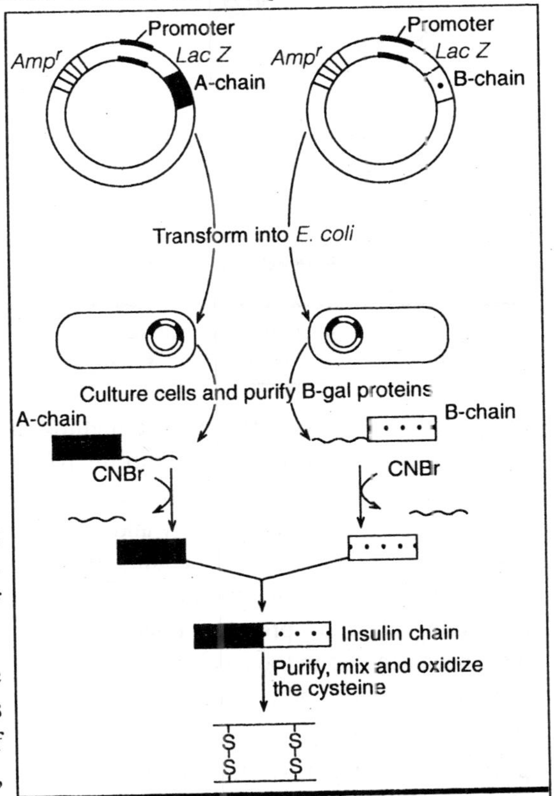

Fig. 25.1: Production of recombinant insulin in *E. coli*

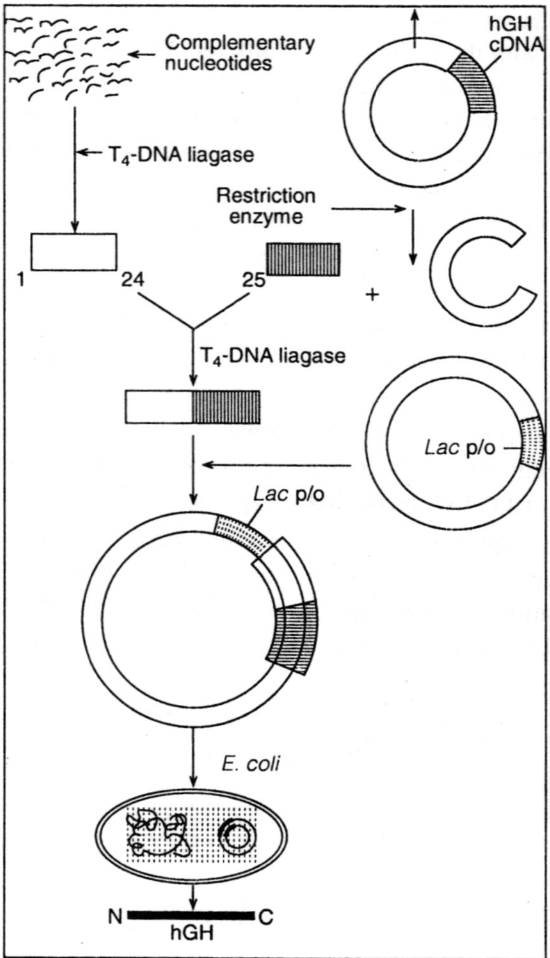

Fig. 25.2: Expression of human growth hormone (hGH) in *E. coli*, *Lac* p/o–lactose promoter/operator

has been inserted into the plasmid, expression vector was constructed from the plasmid pBR 322, to which was added the control region and most of the β-galactosidase gene from the bacterial *lac* operon; and the gene was inserted next to β-galactosidase. After the plasmid was introduced into the cells of the bacterium *E. coli*, the hormone was synthesized as a short peptide tail at the end of the enzyme (Fig. 25.2).

With the advent of techniques of recombinant DNA and gene cloning, several other human hormones are being produced on a commercial scale by isolating specific DNA sequences coding for those proteins/hormones. This is likely to enable clinical application and improve economic provision for their utility in several deficiencies.

2. **Molecular Farming for Healthcare Products** (*Vaccine, Antibody*): Transgenic plants can be used as 'factories' for production of speciality chemicals and pharmaceuticals like sugars, fatty acids, wax materials as well as antibodies, edible vaccines.

The progress is so far reaching that human antibody production through plant seeds has been achieved. The method involves the introduction of heavy and light chains of immunoglobin genes into microbial vectors. In the next step, these are introduced into the leaf cells of two plants and cultured *in vitro* for regeneration of the plants. The plants–one containing heavy and the other with light chain, are then hybridized. The hybrid brings light and heavy chains together, to form the complete immunoglobin (IgA + IgB) in the seeds. This hybrid plant can be utilized for large scale production of seeds containing antibody proteins. Even antibodies presumed to be effective against cancer have been secured in tobacco seeds. The method is thus a synthesis of recombinant DNA, *in vitro* technique and conventional hybridization.

Hepatitis B surface antigen is produced in tobacco, rabies virus glycoprotein is produced in tomato, cholera toxin β-subunit is being produced in potato and tobacco. Transgenic plants are being used as a source of antibodies which provide passive immunization.

One of the recent discoveries in the area of plant biotechnology, is the development of oral vaccine utilizing plant systems. The principle involves, the development of transgenic plants, containing subunits of toxic virus sequences or enterotoxin genes of bacteria like *E. coli* or *Vibrio cholerae*. The oral administration of potato or tobacco transgenic tissues led to the development of immunoglobin G and A antibodies. As such, oral administration of plant tissues for production of antibodies in the system is, in effect, a vaccination–the application of vaccine being oral. Such recombinant vaccines, may prove to be a cheaper substitute for expensive vaccination, both in terms of production and administration.

25.4 ENVIRONMENT

Environmental biotechnology deals with the use of biotechnological tools to protect and conserve the environment. Two important areas are waste treatment and bioenergy production.

1. **Waste Treatment:** The overall production of methane from organic wastes is a complex fermentation involving a number of groups of microorganisms (*Methanogens*, member of Archaebacteria). They utilize the anaerobic environment and organically enriched sediments. The aerobic waste treatment requires a large population of actively metabolizing microorganisms such as *Pseudomonas, Alcaligenes, Achromobacter* and *Brevibacterium,* able to degrade both colloidal and soluble organics with high rate of conversion of CO_2 and water.

Due to anaerobic decomposition of organic waste material, the most important component produced is methane, which can be easily collected and used as biogas fuel. Other than methane, some other products are produced due to the activity of facultative anaerobs.

The nitrogenous waste materials are treated both aerobically and anaerobically alternatively. The degradation of nitrogenous compounds gives ammonia, which is then converted to nitrate by nitrifying bacteria. Then by the action of denitrifying bacteria in anaerobic environment releases N_2 gas in the atmosphere.

2. **Biodegradation:** It is the process by which materials such as oil spill, herbicides, pesticides, etc. are degraded by the action of microbial system. Organic compounds which are naturally occuring (biogenic) are biodegradable while man-made (xenobiotic) compounds may be biodegradable, persistent or recalcitrant.

Genes coding for some enzymes essential for the biodegradation of a number of organic compounds are plasmid borne and organisms have been constructed to degrade difficult waste. A strain of *Pseudomonas putida* is constructed to contain plasmids coding for the breakdown of octane, xylene, metaxylene and camphor. This organism is claimed to be useful for cleansing of oil spills.

3. **Microorganisms in Pollution Control:** Microbial strains can be isolated in order to control various forms of chemical pollution such as biocides, detergents, plastic materials and hydrocarbons. The bacteria belonging to the genus *Pseudomonas* have oxidoreduction or hydroxylation enzymes capable of degrading a large number of hydrocarbon molecules or aromatic compounds that are often highly toxic. Genetic recombination techniques should make it possible to achieve such a result that a

single microbial strain will be capable of decomposing and assimilating numerous compounds, most often non-biodegradable (xenobiotic), produced by the chemical industry.

4. **Biomass Energy Production:** Environmental biotechnology also deals with various sources of energy. Biomass is the living matter or its residues, a perpetual or renewable source of energy. Ethanol is formed from various sources, such as cassava, cereals, potato, sugarcane, pineapple, sugarbeet, etc. It is a solvent and also a substrate for the synthesis of many other components and dyes. *Biogas* formation and methanogenesis are although ancient forms, today they are in great demand. Methane fermentation is widely practised and is one of the easily available sources of energy. Agricultural wastes and farmyard wastes are efficiently used for this purpose. Production of biogas in rural areas has been a priority in the energy policy of India.

5. **Biosensors and Biomonitoring:** Biosensors represent biophysical devices which can detect and measure the quantity of specific substance in a variety of environments and thus may be used in environmental monitoring.

6. **Bioremediation:** Microbes and higher plants which can destroy or reduce the concentration of hazardous wastes from a contaminated site, can be subjected to modify variously through genetic engineering to become efficient and more suitable for bioremediation. Genetically engineered *Pseudomonas* have been used for degradation of PAH (Polyaromatic hydrocarbons) in soil, treating oil spills. Detoxication of soil contaminated with heavy metals can be done by growing certain plant species called **phytoremediation** through *phytoextraction, rhizofiltratioin, phytodegradation* or *Phytostabilization*. Transgenic *Arabidopsis thaliana* plants have been produced for detoxification of soils contaminated with mercury, aluminium, arsenic and transgenic tobacco developed for enhanced TNT (explosive) detoxifying ability.

7. **Restoration of Degraded Lands:** With the advent of biotechnology, unprecedented opportunities for recovery of degraded lands have become available through the manipulation of biological systems. Reforestation though micropropagation, development of stress tolerant plants, use of mycorrhizae, use of microbes for improving soil fertility are some of the measures in this regard.

8. **Biodiversity Conservation:** To avoid the extinction of endangered or threatened plant species, conservation may be carried out by establishing *Gene Banks/Genetic Resource Centres* using biotechnolgical devices like *plant tissue culture.*

Cryopreservation technology is the most popular and effective method for long term storage of plant cell and tissues which involves storage at a very low temperature. The method includes the raising of sterile tissue cultures, addition of cryoprotectant, freezing, storage. The eventual regeneration from the cryopreserved tissue needs thawing, determination of viability and again growth and regeneration ability. The tissues are kept either in Dry ice (–79°C) or Liquid N_2 (–196°C); storage at this temperature halts the metabolic process and biological deterioration. Many of the tropical plants' tissues, embryos have been successfully cryopreserved.

QUESTIONS

1. Discuss the applications of Plant Biotechnology.
2. State the role of biotechnology in Agriculture.
3. Write the application of biotechnology in environmental protection.
4. Mention with examples the use of plant biotechnology in Industry.
5. Discuss the role of biotechnology in human healthcare.

GLOSSARY

Acclimatization : Adaptation of a variety to new environment.

Adaptor : A synthetic single stranded oligonucleotide that produces a molecule containing cohesive end and restriction site.

Addition Rule : The probability that one of several mutually exclusive events will occur is the sum of probabilities of the individual events.

AFLP (Amplified fragment length polymorphism) : A combination of RFLP and RAPD, very sensitive in detecting polymorphism throughout the genome.

Allele Frequency : The proportion of an allele present in a population.

Alternative Hypothesis : Hypothesis complementary to the Null Hypothesis.

Amplification : Formation of many copies of a DNA segment by PCR.

Analysis of Variance : It is directed to study the variations between more than two samples.

Androgenesis Development of haploid plants from the male gametophyte via embryogenesis through anther culture.

Annealing : The process of denaturing and slowly renaturing of ds DNA to form hybrid DNA or DNA-RNA hybrid.

Anther Culture : Culture of anther *in vitro* to cause growth of haploid pollens inside to make haploid plants.

Anthesis : The first opening of a flower.

Antisense RNA : Produced from a gene sequence inserted in the opposite orientation, so that the transcript is complementary to the normal mRNA.

Apogamy : Embryo development from synergids or antipodal cell without fertilisation.

Apomixis : Embryo development without fertilisation.

Apospory : Embryo development from vegetative cell of the ovule.

Arithmatic Mean : Arithmatic average of a set of observations.

Array : The arrangement of values of a set of data in an observation in ascending or descending order of magnitude.

Artificial Seed : Encapsulated somatic embryos capable of producing plantlets.

Aseptic :	Free from all microorganisms.
ASO (Allele specific oligonucleotide) :	Oligonucleotide with a sequence that can match precisely to a particular allele by using stringent hybridisation.
Autoradiograph :	Image produced on X-ray film in response to the emission of radioactive particles.
Auxotroph :	A cell that requires nutritional supplements for growth.
Axenic Culture :	A pure culture of single propagule of an organism not contaminated by any other microorganism.
BAC :	Bacterial Artificial Chromosome.
Back-Cross :	Crossing of hybrid with anyone of its parents.
Back-cross Method :	Breeding method based on repeated crossing of the F_1 and the successive generations to the recurrent parent.
Bar Diagram :	One-dimensional diagram where the length is important, not the width.
Batch Culture :	Starting with single cell propagation by regular transfer of aliquots of suspension to a fresh medium.
BC_1, BC_2, BC_3, etc. :	Progeny from F_1 × Recurrent parent, BC_1 × Recurrent parent, BC_2 × Recurrent parent.
Bimodal Distribution :	A distribution of data characterised by two distinct populations.
Binomial Distribution :	See Bimodal Distribution.
Biodegradation :	The gradual breakdown of compound to its constituents by living organism.
Biodiversity :	The variability among the living organisms from all sources of soil, water, air or associated with other organisms.
Biofertilizer :	Biologically active products used to increase soil fertility.
Biofuel :	Hydrocarbon produced by plants like *Euphorbia*.
Biogas :	A mixture of different gases produced due to microbial action.
Bioinformatics :	Study of information content and information flow in biological system.
Bioleaching/Biomineralisation :	Weathering, leaching and deposition of mineral ores due to bacterial activity.
Biolistic :	Refers to a method of introducing DNA into cells by bombarding them with microprojectile.
Biomass :	The cell mass produced by a population of living organism.
Biometry :	Application of statistical procedures to biology.

Biomonitoring :	Identification and measurement of specific substance with biological devices.
Biopesticides :	Bacteria with the capacity of producing toxin which can kill the crop pests.
Biopolymer :	Any polymer such as protein, nucleic acid, lipid, polysaccharide produced by living organism.
Bioreactor :	Transgenic organisms with capacity of large scale production of valuable proteins.
Bioremediation :	The process of using living organisms to remove contaminants.
Biosensor :	Biophysical devices which can detect the specific substance.
Biotechnology :	Exploitation of biological systems for the production of compounds of human interest.
Biotechnology Breeding :	Breeding methods utilising biotechnology.
Biotic stress :	Stress resulting from living organisms which can harm the other plants.
Biotransformation :	Transformation of cheap compounds into valuable compounds using biological system *in vitro*.
Bivariate Analysis :	The statistical analysis existing between two variables.
Blotting :	Identification of DNA/RNA/Protein using molecular probes.
Blunt end :	DNA termini without overhanging 3′ or 5′ end.
bp (base pair) :	Pairs of nucleotides acids in a double stranded helix (A = T) and (G ≡ C).
Breeder's Seed :	Seeds produced by the breeder.
Broad-host-range plasmid :	A plasmid that can replicate in a number of different bacterial species.
Bt plants :	Plants which carry the toxin producing gene from *Bacillus thuringiensis.*
Bulk Method :	F_2 and subsequent generations are grown in bulk to develop the variety through individual plant selection.
Callus Culture :	Culture of undifferentiated mass of cells.
Caulogenesis :	Shoot induction from callus mass.
c-DNA :	Complementary DNA synthesized from mRNA.
c-DNA Library :	A pool of cDNA clones.
Cell Plating :	Pouring the cell suspension over petriplates.
Cell line :	Cells that acquire the ability to multiply indefinitely *in vitro*.

Cell Suspension Culture : Culture of single cell or small group of cells in suspension.

Central Tendency : Tendency of concentration of the values of variables around some central value of observations.

Centre of Origin : The areas of origin of cultivated plant species.

Certified Seeds : Seeds which are certified by certification agency.

Chemostat : Growth limiting nutrients which keep the growth rate and cell density at a fixed rate.

Chimera : A transgenic plant/animal having different genotypic population of cells.

Chimeric DNA : A recombinant DNA molecule containing unrelated genes.

Chi-square Test : Statistical comparison of observed values and expected values.

Chromosomal Painting : Chromosomal *in situ* hybridisation with multiple sequence specific fluorescent probes.

Chromosome walking : A technique that identifies overlapping cloned sequences which form one continuous segment of a chromosome.

Class Interval : Each group into which the variables are classified from raw data.

Clonal propagation : Vegetative propagation of plants considered to be genetically uniform originated from a single explant.

Clonal Selection : Selection procedure based on clones.

Clone : Collection of genetically identical cells or organisms desired from common and similar genetic ancestor.

Cloning : Integration of foreign gene into the vector and its multiplication along with the DNA of the host.

Coefficient of Variation : Measurement of variability by expressing standard deviation as percentage of the mean.

Cohesive ends : Single stranded complementary nucleotides on the ends of ds DNA molecule.

Colony Hybridisation : Technique of screening DNA libraries using bacteria carrying chimeric vectors which are grown into colonies.

Combining Ability : Ability of a strain to produce superior progeny on hybridization with other strains.

Composite Varieties : Varieties produced by open pollination among a number of outstanding strains.

Contig : A set of overlapping clones that provides a physical map of a portion of chromosome.

Continuous Culture : Steady state maintenance for long period by draining out the used medium and adding fresh medium.

Copy number : The average number of molecules of a plasmid of a gene per genome present in a cell.

Correlation : Relationship between simultaneous variation of two variables.

Correlation Co-efficient : Measurement of the degree of relationship between two variables.

COS site : The 12-base, single strand, complementary extension of lambda phage DNA.

Cosmid : A plasmid vector with COS site of λ-phage and one or more selectable antibiotic marker.

Cryopreservation : Culture or seed or embryo preservation at very low temperature of $-196°C$.

Cryoprotectant : A chemical agent that inhibits the freezing and thawing damage to cells.

Culture Medium : Medium with appropriate constituents for *in vitro* growth of cells, tissues, organs.

Culture Room : Controlled room for growth of excised cells, tissues, organs.

Cumulative Frequency : Summation of frequencies of different values of observations step to step.

Cybrid (Cytoplasmic Hybrid) : Hybrid cell bearing cytoplasm of one cell and nucleus of other cell.

Cytoplasmic Male Sterility : Male sterility showing cytoplasmic inheritance.

Cytoplasmic-genetic Male Sterility : Cytoplasmic male sterility for which a restorer gene is present.

Database : A repository of sequences (DNA or amino acids) which provides a centralised and homogeneous view of its contents.

Degrees of Freedom : Values of a sample which are freely variable without affecting the mean.

Detasseling : Removal of the tassel, i.e., male inflorescence of maize.

Deviation : Value by which the variate differs from the sample mean.

Differentiation : The process of biochemical and structural changes by which cell become specialised in form and function.

Dihaploid : Haploid derived from an amphidiploid.

Diplospory : Embryo sac developing from the megaspore.

Dispersion : It is the measure of variation indicating the spread of the individual values around the central value of a set of data.

Distant Hybridization :	Hybridization between two individuals belonging to two different species or genera.
DNA amplification :	Multiplication of a piece of a DNA by using PCR.
DNA Chip :	A solid support of glass/silicon of a size of slide on which DNA attached in highly organised arrays.
DNA Fingerprint :	Individual specific banding pattern due to DNA polymorphism.
DNA Marker :	Unique DNA sequence used in molecular mapping.
DNA microarray :	Same as DNA chip.
DNA Probes :	Small DNA segments that recognise complementory sequence in DNA/RNA molecules.
DNA Sequencing :	Understanding of ordered base sequence on DNA segments.
Domestication :	The process of bringing a wild species under human management.
Donor Parent :	The parent from which genes are transferred to the recurrent parent by back-crossing.
Dot Diagram :	Graphical method for drawing certain conclusion about the correlation between two variables.
Double Cross :	Cross between two single crosses.
Electrophoresis :	The method of separation of molecules of DNA, RNA or protein based on their relative migration applying electric field.
Electroporation :	A technique where electric impulse is applied to transfer the DNA into protoplast/cell.
Emasculation :	Removal of anthers from a bisexual flower.
Embryo Culture :	Growth of excised embryo *in vitro.*
Embryogenesis :	Process of development of embryo from a fertilized egg cell or any cultured somatic cell.
Embryoids :	Embryo like structure produced as a result of differentiation process such as embryogenesis and androgenesis.
Embryo Rescue :	*Ex situ* growth of otherwise dormant embryo with the help of other's endosperm.
Endonuclease :	An enzyme that catalyses the cleavage of DNA at internal position, cutting at a specific site.
Endosperm Culture :	*In vitro* growth of endosperm to make triploid plants.
EST (Expression Tag Sequence) :	A partial gene sequence unique to the gene which can be used to identify the gene during genome analysis.

Excision :	Cutting of explant or tissue material for culture.
Explant :	The tissue/portion of plant material from a plant/seed and transferred to a culture medium to establish a tissue culture system.
Exponential phase :	The phase of cell growth where they undergo maximum rate of cell division.
Expression vector :	A constructed cloning vector of which coding sequence is properly transcribed and the RNA is efficiently translated.
FISH :	Fluorescence *in situ* hybridisation used for locating the chromosomal segment.
F_1-progeny :	Progeny obtained by crossing two plants with different genotypes.
F_2, F_3, etc. :	Selfed progeny of F_1 and F_2 respectively.
F-statistics :	A test used to determine whether there is homogeneity of variance between two or more sample population.
F-values :	It is the ratio of variances due to treatments and variances due to random differences within the treatment.
Frequency Polygon :	A line graph connecting the mid points of each class in a data set, plotted at a height corresponding to the frequency of the class.
Frequency Distribution :	Grouping of data (observations) in such a manner which can cut down tedious arithmatic calculation.
Friable Callus :	Crubling or fragmenting callus, which can be easily fragmented into single cells or cell clumps.
Functional genomics :	The study of the function of all specific gene/sequences and their expression in time and space in an organism.
Fusion Protein :	A single protein molecule encoded by parts of two or more genes combined together as one unit.
Fusogen :	A fusion inducing agent used for agglutination of protoplast in somatic hybridization.
Gametoclonal variation :	Phenotypic variation apart from normal segregation arising from the culture of gametophytic cells.
Gene :	A hereditary unit; a segment of DNA coding for a specific protein.
Gene Amplification :	Exponential amplification of specific DNA sequence.
Gene Bank :	World collection of germplasm of a species.
Gene Cloning :	See cloning.

Gene Frequency :	The proportion of an allele present in a population.
Gene Transfer :	Technique of horizontal transfer of gene from one organism to another.
Genetic Emasculation :	Use of genetic factors to make the male gamete non-functional.
Genetic Engineering :	Recombinant DNA technology; i.e. isolation of genes, manipulation of them and deliver them into another cell system for a specific product.
Genetic Marker :	See DNA marker.
Genome :	The entire complement of genetic material of an organism or the haploid set of chromosome of any eukaryote.
Genomics :	The study of whole genome sequencing, molecular organisation and information contents and its products.
Genomic Library :	Set of cloned DNA fragments representing entire genome.
Geometric Mean :	A measure of central values computed by taking the Nth root of the product of `N' values representing change.
Germplasm :	Sum total of genes present in a species.
GISH :	Genomic *in situ* hybridization.
GM Crops :	Genetically modified crops.
GMO :	Genetically modified organisms.
Hairpin loop :	A region of single stranded DNA or RNA which has folded itself.
Haploid Plants :	Plants with single genome from anther/pollen culture.
Hardening :	Gradual acclimatization of *in vitro* grown plants to *in vivo* condition to adapt the outdoor conditions.
Hardy-Weinberg Equilibrium :	In the absence of migration, mutation and selection, gene and genotype frequencies remain constant, within certain narrow determinable limits, generation after generation in a large randomly mating population.
Harmonic Mean :	A measure of central tendency computed by taking reciprocal of arithmetic mean of the reciprocal of the value of the variable.
Heterosis :	Superiority of F_1 plants over the parents.
Heterokaryon :	A cell containing two or more nonidentical nuclei as a result of fusion.
Histogram :	A set of vertical bars whose areas are proportional to the frequencies represented.
Homogeneous :	Individuals of the same genotype/phenotype.

Homokaryon : A cell with two identical nuclei as a result of fusion.

Homopolymer : A nucleic acid strand composed of one kind of nucleotide.

Homozygous Diploid : Plant produced through chromosome doubling of haploid cell obtained through anther/pollen culture.

Horizontal Gene Transfer : Foreign gene transfer from one organism to another directly or through vector.

Horizontal Resistance : Resistance against disease governed by polygenes.

Hybrid : Progeny obtained from hybridization of two or more strains.

Hybrid Varieties : F_1 plants obtained from a cross of two strains.

Hybrid-vigour : See Heterosis.

Hybridization : Crossing between two different strains.

Hybridoma : A unique fused cell that produces quantities of a specific antibodies and divide continuously.

Improved Seed : Seeds of improved variety having high genetic and physical purity.

Inbred Lines : Homozygous line developed by continued inbreeding accompanied by selection.

Inbreeding : Mating of closely related individuals with common origin.

Inbreeding Depression : Loss of vigour due to inbreeding.

Independent Events : A set of events are said to be independent if the occurrence of any event does not affect the chance of occurrence of any other event of the set.

Inoculum : A culture containing viable cells of microorganisms.

Insert DNA : A DNA molecule incorporated into a cloning vector.

Inter-specific Hybridization : Hybridization between two species of the same genus.

Interferon : A glycoprotein produced by a virus infected cell which protect cell from another viral attack.

Introduction : Taking a variety into a new area.

Introgression : Transfer of few genes from one species into another species.

Isozyme : Multiple molecular forms of an enzyme that exhibit similar activity.

ISH : *In situ* hybridization, which detects the given sequence of DNA directly on chromosome in cells.

In vitro translation : Protein synthesis directed by purified DNA with bacterial extracts or mRNA, tRNA and ribosomes.

In vitro : Any process carried out in sterile cultures or outside the cell.

In vivo : The natural conditions in which organisms live.

Klenow fragment : DNA polymerase I has both polymerase and 3′ exonuclease activity but not the 5′ exonuclease activity.

Kurtosis : The degree of peakedness of distribution.

Labelling DNA : A radioactive compound attached to DNA/RNA to indicate the presence of a complementary DNA strand in a sample.

Lag phase : Time required by freshly inoculated cells in fresh medium to adapt before start of first division.

Ligation : Joining of two DNA molecules with the formation of phosphodiester bonds by DNA ligase.

Line Diagram : Straight lines whose lengths are proportional to the frequencies.

Linkers : A synthetic double stranded oligonucleotide carrying more than one restriction enzyme sites.

Liposomes : Artificial vesicles of phospholipids.

M_1, M_2, M_3, *etc.* : First, second, third, etc., generation obtained after treatment with a mutagen.

Maintainer Line : Line which is used for maintaining a cytoplasmic male sterile line.

Male Sterility : Absence of functional male gametes.

Marker Gene : Gene used for selecting transformed tissue.

Mass Selection : Seeds of phenotypically selected plants are composited to raise the next generation.

Mean : The average of a set of observations.

Mean Deviation : Measurement of variation which is the arithmetic average of total deviation of the observations.

Median : Middle value when the varieties are arranged in order of their values.

Meristem Culture : *In vitro* culture of excised shoot apical meristem to make virus free plant and for micropropagation.

Metabolomics : Analysis of different metabolite molecules.

Micropropagation : Mass propagation of plants in *in vitro* condition through non-sexual methods.

Mid-point : It is the number falling in the middle of upper and lower level of class interval.

Mode : The class of variate which occurs with the greatest frequency.

Molecular Farming : Manufacture of useful compounds in transgenic organisms.

Molecular markers : The DNA markers revealing variation at DNA level which can be detected by RFLP, RAPD analysis in the progeny.

Molecular Mapping : Genetic map preparation utilizing molecular markers.

Morphogenesis : The development of any morphological structure from an unorganised state to organised state.

Multiline Varieties : Mixtures of several similar purelines having different genes for disease resistance.

Multiplication Rule : The probability of two or more independent events occurring together is the product of the probabilities of individual events.

Mutation Breeding : Breeding method utilising variation caused by mutagen treatment.

Nick translation : A procedure of labelling of DNA *in vitro* using DNA polymerase I.

Normal Distribution : Distribution of a continuous random variable with a bell shaped curve.

Northern Blot : Screening technique in gene cloning involving RNA.

Null Hypothesis : It states that the observed data are in agreement with the expected ratio, i.e., the deviations between the two are due to chance only.

Oligomer : A molecule formed from a small number of monomers.

Ori (origin of replication) : A site of nucleotide sequence from which replication begins.

ORF : Open Reading Frame (a section of sequenced DNA with an initiation and termination codon).

Organ culture : Culture of any plant organs.

Organogenesis : Differentiation of plant organs from callus.

Osmoticum : Sugar or sugar alcohol which maintains the osmotic potential of a nutrient medium equivalent to that of cultured cells.

Outcross : Natural crossing between two different species.

Palindromic sequence : Complementary DNA sequences which are same when read in the opposite direction.

PCR: Polymerase Chain Reaction for gene amplification.

Pedigree Method : Breeding method in which individuals are selected in subsequent generations of F_1 keeping pedigree.

Phagemid :	A cloning vector that contains components derived from both phage DNA and plasmid.
Pie Diagram :	In representing qualitative data, different components are shown by means of sectors of a circle, the angles of the sectors are proportional to the respective measurement of the different components.
Plant Breeding :	Plant science involved in changing the genotype of plants resulting in improved and more useful variety.
Plant Genome Project :	International collaborative research for complete sequencing of whole genome in plant.
Plating efficiency :	The percentage of cells plated on medium which develop colonies.
Plasmid :	Extra-chromosomal self-replicating circular DNA particle.
Pollen Culture:	Culture of individual pollen after its separation from anther to produce haploid plants.
Poly A tailing :	The addition of 'A' residues to 3′ end of eukaryotic mRNAs.
Polylinker :	A segment of DNA which contains several restriction sites.
Polyploid Breeding :	Breeding method using individuals with more than two genomes.
Population :	The sum total of a large number of individuals of one kind, having certain common fundamental characteristics.
Population Genetics :	Genetics of natural population consisting of all the interbreeding individuals that share a common gene pool.
Probability :	Likelihood of the occurrence of any particular form of an event.
Proembryo :	Early embryonic stage developed from zygote or somatic cell.
Progeny Test :	Evaluation of the genotypic value of an individual on the basis of performance of its progeny.
Propagule :	Any plant part which serve for propagation.
Prophage :	Inactive state of bacteriophage genome maintained integrated with bacterial genome.
Proteome :	The entire collection of proteins that an organism can produce.
Proteomics :	Proteome analysis of an organism.
Protoplast Culture :	Culture of isolated, purified protoplast *in vitro.*
Protoplast Fusion :	Union of two genetically dissimilar protoplasts to make somatic hybrid.
Pure Line :	Progeny of a single homozygous self-pollinated plant.

Pure Line Selection : Isolation of purelines from a mixture of population.

QTL (Quantitative Trait Loci) : A locus that affects a quantitative trait.

Random Mating : Mating system in which an individual has equal chances of mating with every other individual of the same population.

Random Sample : A sample in which each individual recorded is independent of all other individuals.

RAPD : Random Amplified Polymorphic DNA.

Raw Data : Data obtained in an experimental observation.

Recipient Parent : The parent to which genes are transferred from donor parent.

Recombinant DNA : Joining of target DNA to vector DNA.

Recombinant DNA Technology : Techniques of gene cloning.

Recurrent Parent : The parent from which genes are transferred to recipient parent.

Recurrent Selection : Selection followed by intermating of the selected plants in cross pollinated plants.

Regeneration : Differentiation of plant organs from dedifferentiated tissue.

Regression : Estimation and prediction of the average value of one variable for a specialized value of the other variable.

Regression Line : A line fitted to a set of data points to estimate the relationship between two variables.

Renaturation : The reunion of two nucleic acid strands after denaturation.

Repetitive DNA : DNA sequences that are present in genome in multiple copies.

Replicon : Synthesis of the genome in which DNA is replicated containing the origin of replication.

Replica plating : Transfer of cells from bacterial colonies growing on one petriplate (master plate) to the other plate.

Reporter Gene : Gene which allows detection of the presence of transgene.

Restorer Gene : Dominant gene that can effectively overcome the effects of male sterile cytoplasm.

Restriction Endonuclease : Endonuclease that cleaves a particular DNA sequence.

Restriction Mapping : Map of a piece of DNA showing recognition sites of specific endonuclease separated by lengths in terms of number of bases.

Reverse Transcriptase : Enzyme synthesizing DNA from RNA.

RFLP : Restriction Fragment Length Polymorphism, DNA fragment of different lengths cut by restriction endonuclease.

Sample : A part of the population selected for study.

Sampling : A set of random observations taken from a population.

SCAR (Sequence characterised amplified region) : A fragment of genomic DNA present at a single locus identified by PCR.

Seed Test : A series of tests on purity and germination of a seed sample to determine its quality.

Self-incompatibility : Lack of seed set on self pollination.

Self-sterility : See self-incompatibility.

Selectable marker : Having a gene product which enables to identify and propagate a particular cell type.

Shuttle vector : A cloning vector that can replicate in two different organisms.

SI-nuclease : An enzyme that degrades single stranded DNA.

Single Cross : A cross between two inbred lines.

Skewness : The event to which a distribution of data points is concentrated at one end or the other, the lack of symmetry.

SNP : Single Nucleotide Polymorphism.

Somaclonal Variation : Existance of altered cells with variant chromosome number and structure *in vitro*.

Somatic Hybridization : See protoplast fusion.

Somatic Embryogenesis : Somatic cell follows the path of embryogenesis.

Southern Blot : Screening technique in gene cloning involving DNA.

Staggered cuts : Symmetrically cleaved phosphodiester bonds that lie on both strands of DNA duplex.

Standard Deviation : Measure of variability of the sample around the mean.

Standard Error : Measure of the variation of the means.

Standard Error of the Difference : The difference between the mean values of two sets of data, the standard deviation of which are known.

Strain : A variety, consisting of individuals with similar phenotype and genotype, released for commercial cultivation.

Structural genomics : The study of structure of all gene sequences.

STS : Sequence Tagged Site.

Suspension culture : A culture system in liquid containing cell aggregates.

Subculture : Subdivision of a culture and its transfer to fresh medium.

Synchronous culture : A culture in which a specified proportion of the cells are at the same phase of the cell cycle.

Synthetic Seed : See artificial seeds.

Synthetic Variety : A variety obtained by mating in all possible combinations of a number of lines.

Systems Biology : Integrated and interacting networks of genes, proteins and biochemical reactions, which put life into the organism.

't'-test : Tests of significance by measuring 't' which is the quantity representing the difference between the sample mean and the true mean expressed in terms of the standard error.

Tailing : Addition of some nucleotides *in vitro* at the 3′ end of duplex DNA.

Taq polymerase : A heat stable DNA polymerase isolate from *Thermus aquaticus* used in PCR.

Test Cross : A cross between a plant and a tester.

Test of Significance : A statistical test used to determine whether observed frequencies between two samples drawn from the same population are actually due to chance or whether they are really significant.

Thawing : A slight melting of pre-cooled specimen.

Three-way Cross : Cross between a single cross and an inbred.

Ti Plasmid : Circular plasmid containing T-DNA.

Tissue Culture : Culture or growth of cell, tissue, organ *in vitro.*

Totipotency : Capacity of the development of the entire plant from any cell.

Transformation : Transfer of a DNA fragment into a cell and DNA acquisition of new genetic marker by the recipient cell.

Transfection : Transfer of DNA into the eukaryotic cell through phage.

Transcriptomics : Analysis of complete set of mRNA transcript.

Transgene : A gene from one source used for incorporation into another source.

Transgenesis : The process through which the genes are directly or through vector transferred from one organism to another resulting in transgenic organism.

Transgenic Plant : Plants (crops) with altered genotype obtained by gene transfer methods.

Transient : Expression of any phenotype for very short duration.

Transposition :	The process of insertion of a transposon into a new site on the same or another DNA molecule.
Variable :	Any quality or quantity which is likely to vary from individual to individual in the same population.
Variance :	Average of the squares of deviations of observations from the mean in a sample drawn from a population.
Variate :	An individual count, measure or value of any variable.
Variation :	Differences among individuals belonging to a single species.
Variety :	A strain released for commercial cultivation.
Unimodal Distribution :	A distribution in which one value occurs more frequently than the rest of the values in the data set.
Ultrasonication :	Disruption of cells or DNA molecules by high frequency.
Vectors :	DNA molecules (plasmid, etc.) used to carry the cloned DNA segment.
VNTR :	Variable Number Tandem Repeat (minisatellites) polygenic markers.
Western Blotting :	Technique for indentifying a particular protein.
Xenobiotic :	A chemically synthesized compound not produced by living organisms.
YAC (Yeast Artificial Chromosome) :	A vector of 100 kb long used for cloning.
YEP (Yeast Episomal Plasmid) :	A cloning vector for yeast that uses plasmid 'ori' gene.

SUGGESTED READINGS

- Allard, R.W., *Principles of Plant Breeding,* John Willey & Sons, Inc. N.Y. 1960.
- Bailey, N.J.J., *Statistical Methods in Biology,* Cambridge University Press.
- Bhojwani, S.S. and Razdan, M.K., *Plant Tissue Culture Theory & Practice*. Elsevier.
- Borer, A. Sentos, F.R. and Bowen, D.B., *Understanding Biotechnology*, Pearson Education, 2003.
- Chaudhuri, H.K., *Elementary Principles of Plant Breeding,* Oxford and IBH Publishing Co., New Delhi, 1971.
- Chawla, H.S., *An Introduction to Plant Biotechnology* (2nd ed.), Oxford & IBH, 2002.
- Collin, H.A. and Edwards, S., *Plant Cell Culture*, Bios Scientific Publishers, 1998.
- De, K.K., *Plant Tissue Culture,* New Central Book Agency, Kolkata, 1992.
- Dixon, R.A. and Gonzales R.A., *Plant Cell Culture: A Practical Approach*, Oxford University Press, 1994.
- Dubey, R.C., *Biotechnology,* S. Chand & Co. Pvt. Ltd., 2006.
- Fisher, R.A. and Yates, F., *Statistical Tables for Biological, Agricultural and Medical Research,* Oliver & Boyd, Edinburgh, 1963.
- Fisher, R.A., *Statistical Methods for Research Workers,* Oliver & Boyd, Edinburgh, 1963.
- Gamborgs, O.L. & Phillips, G.C., *Plant Cell, Tissue & Organ Culture: Fundamental Method,* Narosa Publication House.
- Gupta, P.K., *Biotechnology and Genomes,* Rastogi Publication, Meerut, 2007.
- Ignacimuthu, S., *Plant Biotechnolgy*, Oxford & IBH Publishing Co., 1997.
- Jha, T.B. and Ghosh, B., *Plant Tissue Culture,* University Press, 2003.
- Mahesh, S., *Plant Molecular Biotechnology*, New Age International Limited, 2008.
- Mascarenhas, A.F., *Hand Book of Plant Tissue Culture,* I.C.A.R.
- Misra, B.N. and Misra, M.K., *Introductory Practical Biostatistics*, Naya Prakash, Kolkata, 1983.
- Nicholl, D.S.T., *An Introduction to Genetic Engineering,* Cambridge university Press, 2002.
- Poehlman, J.M. and Borthakur, D.N., *Breeding Asian Field Crops with Special Reference to Crops in India,* Oxford and IBH Publishing Co., New Delhi, 1969.

- Prasad, S., *Elements of Biostatistics,* Rastogi Publications, Meerut, 2004.
- Roy, D., *Plant Breeding : Analysis and Exploitation of Variation,* Narosa Publishing House, 2000.
- Rajdan, M.K., *An Introduction to Plant Tissue Culture (2nd ed.),* Oxford & IBH Publishers, 2003.
- Ramawat, K.G., and Goyal, S., *Comprehensive Biotechnology,* S. Chand and Company Ltd., 2009.
- Sharma, J.R., *Principles and Practice of Plant Breeding,* Tata-McGraw Hill, New Delhi, 1994.
- Simmond, N.W., *Principles of Crop Improvement,* Longman, London, 1979.
- Singh, B.D., *Biotechnology (2nd ed.),* Kalyani Publishers, 2005.
- Singh, B.D., *Plant Breeding,* Kalyani Publishers, New Delhi, 1983.
- Sinha, S.K. and Swaminathan, M.S., *New Parameters and Selection Criteria in Crop Breeding—a contemporary basis,* Pregma Press, London, 1984.
- Sokal, R.R. and Rohef, F.J., *Introduction to Biostatistics,* W.H. Freeman.
- Slatter, A., Scott, N., Flower, N., *Plant Biotechnology,* Oxford University Press, 2008.
- Walker, J.M. & Rapley, R. *Molecular Biology & Biotechnology,* Royal Society of Chemistry, 2000.
- Zar, J.H., *Biostatistical Analysis,* Prentice Hall, 1996.

INDEX